Autonomous Vehicles Technological Trends

Autonomous Vehicles Technological Trends

Editors

Calin Iclodean
Bogdan Ovidiu Varga
Felix Pfister

MDPI • Basel • Beijing • Wuhan • Barcelona • Belgrade • Manchester • Tokyo • Cluj • Tianjin

Editors
Calin Iclodean
Technical University of
Cluj-Napoca Romania
Romania

Bogdan Ovidiu Varga
Technical University of
Cluj-Napoca
Romania

Felix Pfister
IPG Automotive GmbH
Germany

Editorial Office
MDPI
St. Alban-Anlage 66
4052 Basel, Switzerland

This is a reprint of articles from the Special Issue published online in the open access journal *Electronics* (ISSN 2079-9292) (available at: https://www.mdpi.com/journal/electronics/special_issues/AVT_electronics).

For citation purposes, cite each article independently as indicated on the article page online and as indicated below:

LastName, A.A.; LastName, B.B.; LastName, C.C. Article Title. *Journal Name* **Year**, *Volume Number*, Page Range.

ISBN 978-3-0365-7036-5 (Hbk)
ISBN 978-3-0365-7037-2 (PDF)

Contents

About the Editors

Calin Iclodean

Dr. Călin Iclodean is currently an associate professor at Technical University of Cluj-Napoca, Romania, Department of Automotive Engineering and Transports. He obtained a Ph.D. in Mechanical Engineering in 2013, after studying for doctoral studies in Graz, - Austria, at AVL List GmbH. He graduated from the Faculty of Electronics and Telecommunications, Department of Applied Electronics of the Technical University of Cluj-Napoca in 1994, and has more than 25 years of working and business experience in IT&C. He is the author/co-author of 12 published books, including 3 books published by Springer Nature. Dr Iclodean's areas of research are as follows: electric vehicles, autonomous vehicles, fuel cell vehicles, powertrain concept, electronic control units, in-vehicle communication network, energy efficiency, computer modeling, and simulation in the automotive field. His research has received financing from European Investment Bank, national and regional authorities, and economic private entities. He is also active in the field of peer-reviewed publications, where he is a member of a reviewer board for several journals indexed in the Web of Science. He is also a member of the topic editorial board for several journals indexed in the Web of Science and an academic editor for several Special Issues published in journals indexed in the Web of Science.

Bogdan Ovidiu Varga

Prof. Dr. Bogdan Ovidiu Varga is currently a professor at the Automotive Engineering and Transport Department, at Technical University of Cluj-Napoca, Romania, and is the coordinator of the Electric and Hybrid Vehicles course and the director of the Automotive Master Program Advance Techniques in Automotive Engineering. He received his Ph.D. in Mechanical Engineering from the Technical University of Cluj-Napoca after studying in Germany (University of Hohenheim), Italy (Università di Bologna), and Greece (National Technical University of Athens). Since 2011, he has been the technical juridical expert for the Romanian Ministry of Justice in the field of automotive, road, traffic, and equipment for vehicles control, and from 2013, he has worked as an associate member of Federation International des Experts in Automobile. His main research activity covers the areas of smart and green mobility: electric and hybrid vehicles, fuel cell vehicles, energy efficiency of the vehicles, autonomous vehicles, vehicle evaluation, and vehicle diagnosis. He has published several papers in peer-reviewed journals, including the *Energy* journal published by Elsevier and several patterns in the field of electric mobility.

Felix Pfister

Dr.-Ing. Dipl. Felix Pfister currently works for Business Development Powertrain IPG Automotive GmbH, and was previously Business Development Manager at AVL LIST GmbH. His interests include vehicle simulation platforms, ADAS, HIL, ECU control and simulation, vehicle dynamics, and virtual vehicle testing.

Preface to "Autonomous Vehicles Technological Trends"

Twenty years ago, only the most adventurous scientists might have imagined the dramatic modern changes in the automotive industry, where fossil fuels are in the position of being banned and vehicles are driverless. Some scientists still consider that change is occurring too fast, leading to a lack of sustainability. However, the current selection of papers proves the contrary, demonstrating knowledge, direction, and desire for these technologies to be adopted and implemented by the industry. This Special Issue shows a crystalized vision of various fields of industry supporting each other with the clear scope of the development of the most advanced vehicles on the market. Topics presented in this Special Issue include ECUs, live traffic interaction, regulatory bases, pedestrian detection, interior design, trends in the automotive development or software development, safety features, and special maneuvers; all these topics are presented in a mature way with a robust scientific description of the trends and the directions that are necessary to ensure the successful implementation of these vehicles. The selection criteria for the papers was to support the scientific world by offering mature solutions at a minimum level of TLR 4 to validate the provided information in the said articles. The editors hope that this first collection of papers will represent the beginning of a continuous process that, 20 years from now, will provide scientists with an indication of how autonomous vehicles have developed and entered the market, and how they started to represent a safe, reliable, and efficient mean of transport for the future.

Calin Iclodean, Bogdan Ovidiu Varga, and Felix Pfister
Editors

electronics

MDPI

Editorial

Autonomous Vehicles Technological Trends

Calin Iclodean [1,*], Bogdan Ovidiu Varga [1] and Felix Pfister [2]

[1] Department of Automotive Engineering and Transports, Technical University of Cluj-Napoca Romania, 400114 Cluj-Napoca, Romania
[2] IPG Automotive GmbH, Bannwaldallee 60, 76185 Karlsruhe, Germany
* Correspondence: calin.iclodean@auto.utcluj.ro

1. Introduction

Twenty years ago, only the most adventurous scientist might have been in the position of dreaming up such a dramatic change for the automotive industry, where fossil fuels are in a position of being banned and vehicles are driverless. Some of the current scientist still consider that the change is developing too fast and that there is going to be a lack of sustainability. Yet, the current selection of papers proves the contrary, that there is knowledge, there is direction, and there is desire for these technologies to be adopted and implemented by the industry. We have in the current selection of papers a crystalized vision where various fields of industry support each other with the clear scope of setting the tone and tuning the rhythm to the development of the most advanced vehicles on the market. Presented in this Special Issue is the ECU, the live traffic interaction, the regulatory basis, pedestrian detection, interior design, trends in the automotive development or software development, or safety features and special maneuvers; all these topics are presented in a mature way with a robust scientific description of the trends and the directions that are mandatory to be adopted for the successful implementation of these vehicles. The selection criteria for the papers had a goal of implementing a referee that was in the position of supporting the scientific world by offering mature solutions at a minimum level of TLR 4 to validate the provided information in the said articles. The Editors hope that this first collection of papers will represent just the beginning of a continuous process that, 20 years from now, will provide scientists with a red line to detect how autonomous vehicles have developed, entered the market, and how they started to represent a safe, reliable, and efficient mean of transport for the future.

Citation: Iclodean, C.; Varga, B.O.; Pfister, F. Autonomous Vehicles Technological Trends. *Electronics* **2023**, *12*, 1149. https://doi.org/10.3390/electronics12051149

Received: 21 February 2023
Accepted: 22 February 2023
Published: 27 February 2023

2. Short Presentation of the Papers

Ayres et al. [1] presented the vehicle-embedded system, also known as the electronic control unit (ECU), which has transformed the humble motorcar, making it more efficient, environmentally friendly, and safer, but has also led to a system which is highly dependent on software. As new technologies and features are included with each new vehicle model, the increased reliance on software will no doubt continue. It is an undeniable fact that all software contains bugs, errors, and potential vulnerabilities which, when discovered, must be addressed in a timely manner, primarily through patching and updates, to preserve vehicle and occupant safety and integrity. However, current automotive software updating practices are ad hoc at best and often follow the same inefficient fix mechanisms associated with a physical component failure of return or recall. Increasing vehicle connectivity heralds the potential for over the air (OtA) software updates, but rigid ECU hardware design does not often facilitate or enable OtA updating. To address the associated issues regarding automotive ECU-based software updates, a new approach in how automotive software is deployed to the ECU is required. This paper presents how lightweight virtualization technologies known as containers can promote efficient automotive ECU software updates. ECU functional software can be deployed to a container built from an associated image.

Container images promote efficiency in download size and times through layer sharing, similar to ECU difference or delta flashing. Through containers, connectivity and OtA future software updates can be completed without inconveniences to the consumer or incurring an expense to the manufacturer. This paper has been selected as a Feature Paper.

Ji et al. [2] presented the lane-merging strategy for self-driving cars in dense traffic using the Stackelberg game approach. From the perspective of the self-driving car, in order to make sufficient space to merge into the next lane, a self-driving car should interact with the vehicles in the next lane. In heavy traffic, where the possible actions of the vehicle are pretty limited, it is possible to conjecture the driving intentions of the vehicles from their behaviors. For example, by observing the speed changes of the human driver in the next lane, the self-driving car can estimate its driving intention in real time, much in the same way as a human driver. We use the principle of Stackelberg competition to make the optimal decision for the self-driving car based on the predicted reaction of the interacting vehicles in the next lane. In this way, according to the traffic circumstances, a self-driving car can decide whether to merge or not. In addition, by limiting the number of interacting vehicles, the computational burden is manageable enough to be implemented in production vehicles. We verify the efficiency of the proposed method through case studies for different test scenarios, and the test results show that our approach is closer to the human-like decision-making strategy compared to the conventional rule-based method.

Alves et al. [3] presented that environmental concern regularly leads to the study and improvement of manufacturing processes and the development of new industrial products. The purpose of this work is to optimize the amount of injected plastic and reduce the number of parts used in the production of entrance panels to control features inside the car cabin. It focuses on a particular case study, namely the control of opening and closing windows and rotation of the rear-view mirrors of a car, maintaining all of the functionality, and introducing a futuristic and appealing design in line with new autonomous driving vehicles. For this purpose, distinct low-cost touch sensor technologies were evaluated and the performance of several types of sensors that were integrated with plastic polymers of a distinct thickness was analyzed. Discrete sensors coupled to the plastic part were tested and integrated in the injected plastic procedure. In the former, sensitivity tests were performed to find the maximum plastic thickness detectable by the different sensors. For the latter, experiments were carried out on the sensors subject to a very high pressure and temperature inside the molds—the two most relevant characteristics of industrial plastic injection in this context—and functional results were observed later. We conclude that, by changing the way the user interacts with the car cabin, the replacement of conventional mechanical buttons—composed of dozens of parts—by a component consisting of a single plastic part that is associated with conventional low-cost electronics allows for the control of a more diversified set of features, including many that are not yet usual in the interior of automobiles today, but that will eventually be required in the near future of autonomous driving, in which the user will interact less with driving and more with other people or services around her/him, namely of the multimedia type. Additionally, the economic factor was considered, namely regarding the cost of the new technology as well as its manufacturing, replacement, and subsequent recycling processes.

Mounsey et al. [4] presented that pedestrian detection is at the core of autonomous road vehicle navigation systems as they allow a vehicle to understand where potential hazards lie in the surrounding area and enable it to act in such a way which avoids traffic accidents, which may result in individuals being harmed. In this work, a review of convolutional neural networks (CNNs) to tackle pedestrian detection is presented. We further present models based on CNNs and transfer learning. The CNN model with the VGG-16 architecture is further optimized using the transfer learning approach. This paper demonstrates that the use of image augmentation on training data can yield varying results. In addition, a pre-processing system that can be used to prepare 3D spatial data obtained via LiDAR sensors is proposed. This pre-processing system is able to identify candidate regions that can be put forward for classification, whether that be 3D classification or a combination of 2D and 3D classifications via sensor fusion. We proposed a number of models based on transfer learning and convolutional neural networks and achieved over 98% accuracy with the adaptive transfer learning model.

Custódio et al. [5] presented the current design paradigm of car cabin components which assumes that seats are aligned with the driving direction. All passengers are aligned with the driver that, until recently, was the only element in charge of controlling the vehicle. The new paradigm of self-driving cars eliminates several of those requirements, releasing the driver from control duties and creating new opportunities for entertaining the passengers during the trip. This creates the need for controlling functionalities that must be closer to each user, namely on the seat. This work proposes the use of low-cost capacitive touch sensors for controlling car functions, multimedia controls, seat orientation, door windows, and others. In the current work, we have reached a proof of concept that is functional, as shown for several cabin functionalities. The proposed concept can be adopted by current car manufacturers without changing the automobile construction pipeline. It is flexible and can adopt a variety of new functionalities, mostly software-based, added by the manufacturer, or customized by the end-user. Moreover, the newly proposed technology uses a smaller number of plastic parts for producing the component, which implies savings in terms of production costs and energy, while increasing the life cycle of the component.

Marques et al. [6] presented how with the current technological transformation in the automotive industry, autonomous vehicles are getting closer to the Society of Automotive Engineers (SAE) automation level 5. This level corresponds to the full vehicle automation, where the driving system autonomously monitors and navigates the environment. With SAE-level 5, the concept of a shared autonomous vehicle (SAV) will soon become a reality and mainstream. The main purpose of an SAV is to allow unrelated passengers to share an autonomous vehicle without a driver/moderator inside the shared space. However, to ensure their safety and well-being until they reach their final destination, the active monitoring of all passengers is required. In this context, this article presents a microphone-based sensor system that is able to localize sound events inside an SAV. The solution is composed of a micro-electro-mechanical system (MEMS) microphone array with a circular geometry connected to an embedded processing platform that resorts to field-programmable gate array (FPGA) technology to successfully process in the hardware the sound localization algorithms. This paper has been selected as a Feature Paper.

Abdeen et al. [7] presented that the traffic management challenges in peak seasons for popular destinations such as Madinah city have accelerated the need for and introduction of autonomous vehicles and vehicular ad hoc networks (VANETs) to assist in the communication and alleviation of traffic congestions. The primary goal of this study is to evaluate the performance of communication routing protocols in VANETs between autonomous and human-driven vehicles in Madinah city in varying traffic conditions. A simulation of assorted traffic distributions and densities were modeled in an extracted map of Madinah city and then tested in two application scenarios with three ad hoc routing protocols using a combination of traffic and network simulation tools working in tandem. The results measured for the average trip time show that opting for a fully autonomous vehicle scenario reduces the trip time of vehicles by approximately 7.1% in high traffic densities and that the reactive ad hoc routing protocols induce the least delay for network packets to reach neighboring VANET vehicles. From these observations, it can be asserted that autonomous vehicles provide a significant reduction in travel time and that either of the two reactive ad hoc routing protocols could be implemented for the VANET implementation in Madinah city. Furthermore, we perform an ANOVA test to examine the effects of the factors that are considered in our study on the variation in the results.

Ren et al. [8] proposed a novel deep reinforcement learning (DRL) method for optimal path planning for mobile robots using dynamic programming (DP)-based data collection. The proposed method can overcome the slow learning process and improve the quality of the training data inherently in DRL algorithms. The main idea of our approach is as follows. First, we mapped the dynamic programming method to typical optimal path planning problems for mobile robots and created a new efficient DP-based method to find an exact, analytical, optimal solution for the path planning problem. Then, we used high-quality training data gathered using the DP method for DRL, which greatly improves the training data quality and learning efficiency. Next, we established a two-stage reinforcement learning method where, prior to the DRL, we employed extreme learning machines (ELM) to initialize the parameters of actor and critic neural networks to a near-optimal solution in order to significantly improve the learning performance. Finally, we illustrated our method using some typical path planning tasks. The experimental results show that our DRL method can converge much easier and faster than other methods. The resulting action neural network is able to successfully guide robots from any start position in the environment to the goal position while following the optimal path and avoiding collision with obstacles.

Son et al. [9] presented that the concern over safety features in autonomous vehicles is increasing due to the rapid development and increasing use of autonomous driving technology. The safety evaluations performed for an autonomous driving system cannot depend only on existing safety verification methods due to the lack of scenario reproducibility and the dynamic characteristics of the vehicle. Vehicle-in-the-loop simulation (VILS) utilizes both real vehicles and virtual simulations for the driving environment to overcome these drawbacks and is a suitable candidate for ensuring reproducibility. However, there may be differences between the behavior of the vehicle in the VILS and vehicle tests due to the implementation level of the virtual environment. This study proposes a novel VILS system that displays consistency with the vehicle tests. The proposed VILS system comprises virtual road generation, synchronization, virtual traffic manager generation, and perception sensor modeling, and implements a virtual driving environment similar to the vehicle test environment. Additionally, the effectiveness of the proposed VILS system and its consistency with the vehicle test is demonstrated using various verification methods. The proposed VILS system can be applied to various speeds, road types, and surrounding environments.

Ortega et al. [10] presented the overtaking maneuver, that consists of passing another vehicle traveling on the same trajectory but at a slower speed. Overtaking is considered one of the most dangerous, delicate, and complex maneuvers performed by a vehicle, as it requires a quick assessment of the distance and speed of the vehicle to be overtaken, and also the estimation of the available space for the maneuver. In particular, most drivers have difficulty overtaking a vehicle in the presence of oncoming vehicles in other trajectories. To solve these overtaking problems, this article proposes a method of performing safe, autonomous vehicle maneuvers through the PreScan simulation program. In this environment, the overtaking maneuver scenario (OMS) is composed of highway infrastructure, vehicles, and sensors. The proposed OMS is based on the solution of minimizing the risks of collision in the presence of any oncoming vehicle during the overtaking maneuver. It is proven that the overtaking maneuver of an autonomous vehicle is safe to perform through the use of advanced driver-assistance systems (ADAS) such as adaptive cruise control (ACC) and technology-independent sensors (TIS) that detect the driving environment of the maneuver.

Luna-Álvarez et al. [11] presented how in the self-driving vehicles domain, steering control is a process that transforms the information obtained from sensors into commands that steer the vehicle on the road and avoid obstacles. Although a greater number of sensors improves perception and increase control precision, it also increases the computational cost and the number of processes. To reduce the cost and allow data fusion and vehicle control as a single process, this research proposes a data fusion approach by formulating a neurofuzzy aggregation deep learning layer; this approach integrates aggregation using fuzzy measures μ as fuzzy synaptic weights, hidden state using the Choquet fuzzy integral, and a fuzzy backpropagation algorithm, creating data processing from different sources. In addition, implementing a previous approach, a self-driving neural model is proposed based on the aggregation of a steering control model and another for obstacle detection. This was tested in an ROS simulation environment and in a scale prototype. Experimentation showed that the proposed approach generates an average autonomy of 95% and improves driving smoothness by 9% compared to other state-of-the-art methods.

Author Contributions: Writing—original draft preparation, C.I. and B.O.V.; writing—review and editing, C.I., B.O.V. and F.P. All authors have read and agreed to the published version of the manuscript.

Acknowledgments: We would like to thank all the authors for the papers they submitted to this Special Issue. We would also like to acknowledge all the reviewers for their careful and timely reviews to help improve the quality of this Special Issue. Last but not least, we would like to thank the Editorial Team of the *Electronics* journal for all the support provided in the publication of this Special Issue.

Conflicts of Interest: The authors declare no conflict of interest.

References

1. Ayres, N.; Deka, L.; Paluszczyszyn, D. Continuous Automotive Software Updates through Container Image Layers. *Electronics* **2021**, *10*, 739. [CrossRef]
2. Ji, K.; Orsag, M.; Han, K. Lane-Merging Strategy for a Self-Driving Car in Dense Traffic Using the Stackelberg Game Approach. *Electronics* **2021**, *10*, 894. [CrossRef]
3. Alves, C.; Custódio, T.; Silva, P.; Silva, J.; Rodrigues, C.; Lourenço, R.; Pessoa, R.; Moreira, F.; Marques, R.; Tomé, G.; et al. smartPlastic: Innovative Touch-Based Human-Vehicle Interface Sensors for the Automotive Industry. *Electronics* **2021**, *10*, 1233. [CrossRef]
4. Mounsey, A.; Khan, A.; Sharma, S. Deep and Transfer Learning Approaches for Pedestrian Identification and Classification in Autonomous Vehicles. *Electronics* **2021**, *10*, 3159. [CrossRef]
5. Custódio, T.; Alves, C.; Silva, P.; Silva, J.; Rodrigues, C.; Lourenço, R.; Pessoa, R.; Moreira, F.; Marques, R.; Tomé, G.; et al. A Change of Paradigm for the Design and Reliability Testing of Touch-Based Cabin Controls on the Seats of Self-Driving Cars. *Electronics* **2022**, *11*, 21. [CrossRef]
6. Marques, I.; Sousa, J.; Sá, B.; Costa, D.; Sousa, P.; Pereira, S.; Santos, A.; Lima, C.; Hammerschmidt, N.; Pinto, S.; et al. Microphone Array for Speaker Localization and Identification in Shared Autonomous Vehicles. *Electronics* **2022**, *11*, 766. [CrossRef]
7. Abdeen, M.A.R.; Beg, A.; Mostafa, S.M.; AbdulGhaffar, A.; Sheltami, T.R.; Yasar, A. Performance Evaluation of VANET Routing Protocols in Madinah City. *Electronics* **2022**, *11*, 777. [CrossRef]
8. Ren, J.; Huang, X.; Huang, R.N. Efficient Deep Reinforcement Learning for Optimal Path Planning. *Electronics* **2022**, *11*, 3628. [CrossRef]
9. Son, W.; Ha, Y.; Oh, T.; Woo, S.; Cho, S.; Yoo, J. PG-Based Vehicle-In-the-Loop Simulation for System Development and Consistency Validation. *Electronics* **2022**, *11*, 4073. [CrossRef]
10. Ortega, J.; Lengyel, H.; Ortega, J. Design and Analysis of the Trajectory of an Overtaking Maneuver Performed by Autonomous Vehicles Operating with Advanced Driver-Assistance Systems (ADAS) and Driving on a Highway. *Electronics* **2023**, *12*, 51. [CrossRef]
11. Luna-Álvarez, A.; Mújica-Vargas, D.; Rendón-Castro, A.; Matuz-Cruz, M.; Kinani, J.M.V. Neurofuzzy Data Aggregation in a Multisensory System for Self-Driving Car Steering. *Electronics* **2023**, *12*, 314. [CrossRef]

electronics MDPI

Article

Continuous Automotive Software Updates through Container Image Layers

Nicholas Ayres [1], Lipika Deka [2,*] and Daniel Paluszczyszyn [2,3]

[1] Department of Computer Science and Informatics, Cyber Technology Institute, De Montfort University, Leicester LE1 9BH, UK; nick.ayres@dmu.ac.uk
[2] The De Montfort University Interdisciplinary Group in Intelligent Transport Systems (DIGITS), Department of Computer Science and Informatics, De Montfort University, Leicester LE1 9BH, UK; paluszcol@dmu.ac.uk
[3] Royal Academy of Engineering Industrial Fellow, HORIBA MIRA Ltd., Nuneaton CV10 0TU, UK
* Correspondence: lipika.deka@dmu.ac.uk

Abstract: The vehicle-embedded system also known as the electronic control unit (ECU) has transformed the humble motorcar, making it more efficient, environmentally friendly, and safer, but has led to a system which is highly dependent on software. As new technologies and features are included with each new vehicle model, the increased reliance on software will no doubt continue. It is an undeniable fact that all software contains bugs, errors, and potential vulnerabilities, which when discovered must be addressed in a timely manner, primarily through patching and updates, to preserve vehicle and occupant safety and integrity. However, current automotive software updating practices are ad hoc at best and often follow the same inefficient fix mechanisms associated with a physical component failure of return or recall. Increasing vehicle connectivity heralds the potential for over the air (OtA) software updates, but rigid ECU hardware design does not often facilitate or enable OtA updating. To address the associated issues regarding automotive ECU-based software updates, a new approach in how automotive software is deployed to the ECU is required. This paper presents how lightweight virtualisation technologies known as containers can promote efficient automotive ECU software updates. ECU functional software can be deployed to a container built from an associated image. Container images promote efficiency in download size and times through layer sharing, similar to ECU difference or delta flashing. Through containers, connectivity and OtA future software updates can be completed without inconveniences to the consumer or incurring expense to the manufacturer.

Keywords: virtualisation; ECU; automotive E/E architecture; containers; over the air; software updates

Citation: Ayres, N.; Deka, L.; Paluszczyszyn, D. Continuous Automotive Software Updates through Container Image Layers. *Electronics* **2021**, *10*, 739. https://doi.org/10.3390/electronics10060739

Academic Editors: Calin Iclodean, Bogdan Ovidiu Varga and Felix Pfister

Received: 7 March 2021
Accepted: 18 March 2021
Published: 20 March 2021

1. Introduction

In 1886, Karl Benz built what was considered the first modern motor vehicle: the Benz Patent-Motorwagen, the humble car has transformed, not just in looks but function. In 1977, General Motors released the Oldsmobile Toronado, which is regarded as the first car to include an electronic control unit (ECU) [1]; this first implementation managed the electronic spark timing of the combustion process. ECUs benefit the driver with a safer, efficient and more comfortable ride but the benefits can also be seen with regard to the vehicle such as lower CO2 emissions, reduced mechanical wear and higher efficiency in operation. Vehicle systems are no longer mechanically linked together, but rather software driven hardware, connected between driver input and vehicle output. Since ECUs were introduced, software has become an integral part of the motorcar similar to any mechanical component which aids in its function and operation.

According to [2], "over 80% of innovations in the automotive industry are now realised by software-intensive systems". Over 100 million lines of software code across

100 ECUs can be found within the automotive E/E architecture of many modern motor vehicles providing vehicle functions from engine management to passenger comfort [3,4]. These diverse functions make the modern motorcar one of the most software-intensive systems we use in our day-to-day lives [3,5,6]. There are regular and periodic preventative and proactive maintenance procedures of a vehicle's physical components throughout its lifetime [7]. However, the same statement cannot be said concerning automotive software. Despite the requirement for reliable software, bugs and errors are unintentional but appear frequently within software code [8,9]. How and why software code contains errors and flaws are varied [10–12]. Problems are often introduced during the various stages of the software life-cycle. For example, bugs and errors in software code can lead to unexpected and sometimes dangerous results in the output of software-driven devices and functions [13–16]. Current automotive software update practices and procedures are problematic because there is no clearly defined mechanism or standard.

Current software update mechanisms often follow the same return or recall mechanism associated with a physical vehicle component failure. The Original Equipment Manufacturer (OEM) may issue a vehicle recall notice, especially if the fault concerns a safety issue. The "return or recall" process has its own associated problems including cost to the manufacturer and inconvenience to the consumer [13,17,18]. In order to address these limitations, a new approach to automotive software updates is required. Software Over the Air (OtA) update mechanisms can update automotive software without the need to return the vehicle to an authorised garage or dealership [19–21]. Using the increasing deployment of on-board vehicle connectivity, OtA updates can deliver new software as and when required [22]. However, current rigid ECU hardware designs do not facilitate or promote an architecture that can benefit from an OtA software update mechanism.

The focus of this paper is to propose and investigate how specific lightweight virtualisation also known as containers can be deployed within the automotive E/E architecture to promote periodic remote OtA software updates [23]. A container-based ECU can address many of the current software updating issues identified within this paper. It can provide a scalable and updateable solution that is not dependant on many applications of individual ECU hardware systems, which is the standard practice in current automotive E/E architecture design. Automotive functionality hosted within containers is a promising technology which can address many of the current inadequacies in automotive software updating and has the potential to deliver a standardised mechanism to promote continual software updates throughout the vehicle's lifetime as well as a platform that can provide new system functionality to the consumer through aftermarket market sales.

2. Automotive E/E Architecture: Software Associated Issues

Vehicle software is considered to be a significant component of the modern motorcar. As the number of ECUs increases, it inevitably results in more lines of software code to drive those systems. As more lines of code are included it raises specific issues related to an increased dependency on software.

2.1. *Software Bugs and Errors*

Automotive ECU software is often designed, developed, and written by third party suppliers. However, according to [10], "guessing what the designer's intentions were most often results in more bugs". Studies into the quality of software indicate strong correlations between the size of the application and the total number of defects [11]. Reference [12], states that a software system consisting of millions of code lines could have tens of thousands of unknown or undetected bugs. The following chart (Figure 1) highlights the increasing trend of software associated vehicle recalls. In 2018, 8 million vehicles in the U.S. were affected by some form of software defect.

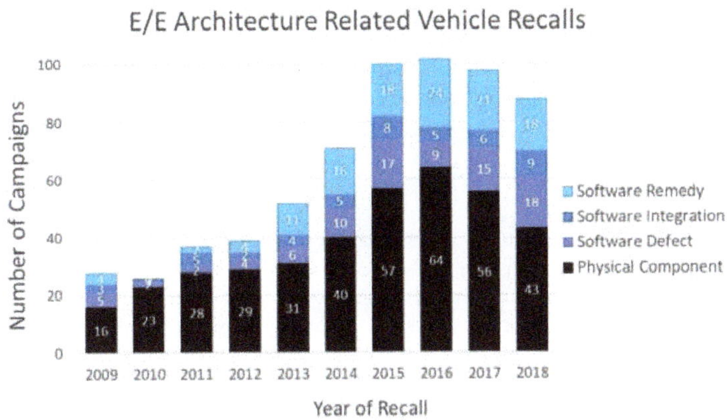

Figure 1. Software related automotive recalls.

As stated in [24], automotive recalls can be classified into four groups, three of which specifically relate to automotive software:

- Integrated electronic components—Failure of a physical, electronic component.
- Software integration—Software interfacing failure between different automotive components or systems.
- Software defect—ECU software failure.
- Software remedy—Fault not solely attributed to software failure but was remedied using a software update/patch.

Depending on the software's application and how critical it is to operational safety bugs and software errors can have disastrous consequences [13]. For example, in 1996, the Ariane 5 Flight 501 rocket disintegrated 40 s after launch due to an undiscovered software error within an arithmetic routine installed in the flight computer. The software bug led to the backup and primary systems crashing, which ultimately led to the rocket's failure [25]. According to [26], the second most common reason for a vehicle-related collision is attributed to automotive software bugs.

Many ECU systems within the modern motorcar are safety-related or considered safety-critical. Any failure in an automotive safety-critical system can potentially endanger vehicle occupant safety. Embedded software bugs and errors can cause control flow errors which result in a flawed execution of the program that can lead to sensor or actuator failure or the system hanging or crashing [27]. To mitigate against these types of errors in dependable and safety-critical systems, expensive hardware-based countermeasures such as triple modular redundancy are often required.

2.2. *Software Associated Security Threats*

Vehicles are no longer closed systems that require direct physical access to gain unauthorised entry to the car. Vehicle connectivity is gaining popularity as it offers vehicle occupants a mechanism to connect to services outside of the vehicle via the Internet. However, the potential to compromise automotive security through vehicle-based connectivity now has the potential to come from anywhere. Nevertheless, even though connectivity systems have been incorporated into vehicles over the last few years, "car hacking" has not been widespread due to the limited potential for cyber-crime and cyber-criminals.

In 2015, two security professionals, Charlie Miller and Chris Valasaek, demonstrated how to compromise a motor vehicle remotely through its connectivity system and vulnerabilities within its software code. They gained access through the HMI unit known as the Uconnect system in the Gran Jeep Cherokee target vehicle. Access to the CANbus vehicle network was possible through the design of this device. This system incorporates an interface for particular vehicle operational and media functions. Due to vulnerabilities in the HMI operating system software, the software update validation mechanism was disabled, which permitted malware injection into the Uconnect software. Once compromised, the system enabled the attackers to remotely inject spoofed CAN frames to ECUs which were responsible for vehicle control. The HMI vulnerability allowed the hackers to interfere with various vehicle subsystems, including interior climate control and vehicle windscreen wipers. They also manipulated safety-critical systems, including shutting down the engine and limited steering control. The Uconnect HMI is a standard product supplied by Fiat Chrysler and is incorporated into numerous vehicle models across several different vehicle makes. This software vulnerability affected 100,000 s of vehicles globally [5,28–30].

As the connected car becomes mainstream, it will ultimately become more of a target for cyber-criminals [31]. Vehicle autonomy and many current ADAS features place the vehicle in level 3 or 4 on the autonomy scale, where level 0 reflects complete driver control and level 5 reflects complete computer control. An intruder's potential to gain remote system access and subsequent unauthorised control of a moving vehicle is an increasing possibility [32]. Vehicle infotainment systems present large attack surfaces that often delivers bi-directional vehicular connectivity. As such, any discovered vulnerabilities within these systems software must be patched promptly to maintain the integrity of the vehicle's subsystems and occupants' safety [33,34].

2.3. *Ageing and Out of Date Code*

Automotive E/E components, including associated ECU software, is often designed and developed years before a particular vehicle model eventually leaves the sales forecourt. The average vehicle has a life expectancy of between 10 to 15 years, and automotive software must mirror this long-time frame. Automotive system longevity significantly differs from many other software-based systems used in our day to day lives. For example, periodic software updates are routinely applied to general-purpose computing and personal smart devices throughout their lifetime. Regular updates address flaws and bugs in software code, provide security and deliver new or additional system functionality [35,36]. According to [37], software can exhibit signs of ageing where old software versions lose market share and customers to new software products. Furthermore, reliability can decrease because of the introduction of bugs and errors during periodic maintenance.

2.4. *Aftermarket Sales and Additional Functionality*

Throughout its life, the modern motorcar requires a robust aftermarket industry to sustain vehicle longevity. Currently, the automotive aftermarket sector is predominately concerned with two main revenue streams; services and parts. The service sector includes the maintenance and repair of vehicles, and accounts for approximately 45% of total European aftermarket revenue. The remaining 55% involves the sale of vehicle parts. The global aftermarket industry in 2015 was worth an approximate $760 bn and accounted for 20% of total automobile revenues [4].

Consumers increasingly demand the features and functions they use on their smart devices to be made available within their vehicles. The automotive industry is looking towards connectivity to provide the consumer with new aftermarket automotive features and functions. Figure 2 highlights the most significant influence over new car purchase decision where 10 means in-car technology has the most significant influence, and 1 refers to the car's performance as the predominant factor. In response to this trend, infotainment systems that offer an "Apple-like" experience are predicted to grow from 18 million units in 2015 to 50 million by 2025 [38].

Three of the six top trends surrounding aftermarket sales refer to new and emerging digital technologies, these include:

- Interface digitisation—by 2035, there will be a predicted shift of between 20–30% from physical component replacement to software upgrades of vehicle components, including new digital services which can be purchased on demand [4].

- Car-generated data—connected vehicles generate considerable amounts of telematics and driver data, approximately 25 GB per hour. Through big data analytics, consumer-generated data can be of substantial value to the manufacturer in determining consumer insights, predictive maintenance and remote diagnostics.
- The increasing influence of digital intermediaries—usage-based companies and technology companies are increasingly using vehicle-generated data. These sectors will require mechanisms to facilitate the retrieval and frequent deployment of automotive software.

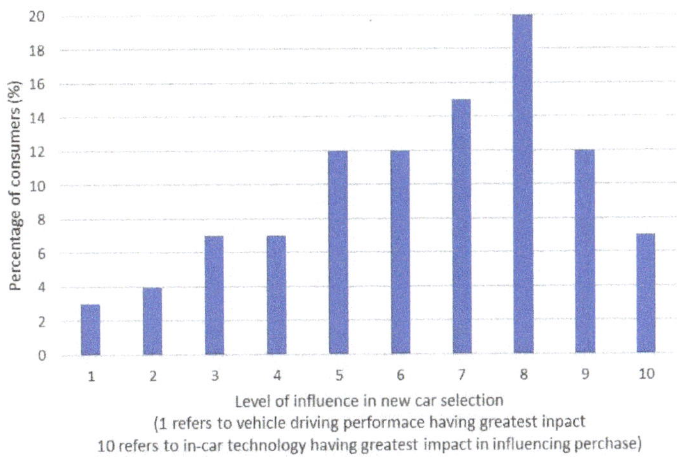

Figure 2. Consumer Preference: New Car Selection.

3. Automotive Software Updating

In the modern motorcar, almost all aspects of vehicle operation require considerable amounts of software code [3,9,39]. However, as with all software, automotive software needs to be periodically updated. In an increasingly software-centric automotive E/E architecture, new software installations may be required several times during a vehicle's lifetime.

The process of online or OtA software updates has been seen in personal computing technology and more recently in our smart devices, which are updated periodically to provide software bug fixes and the latest security patches, and add new software functionality or install newer software and operating system versions. This is enabled through the device's own connectivity hardware. However, this wide-scale software update mechanism is in an infant stage in automobiles.

Any future automotive software update mechanism must present minimal disruption to the customer and be cost-effective to the manufacturer and supplier. There are several principle reasons why it is vital to periodically update automotive software, these include:

- Addressing system failure through software errors and bugs.
- Patching or enhancing the system and software security.
- Adding value post-sale through aftermarket content.

Current automotive software update practices and procedures are problematic because there is no clearly defined mechanism or standard. Historically, when a common fault was discovered within a particular installed physical component of a vehicle, the OEM could issue a vehicle recall notice [40], especially if the fault reflects a severe safety issue.

The current mechanisms for automotive software updates are ad hoc at best. This paper has identified three common mechanisms, including:

- Manufacturer-initiated vehicle recall process.
- Guided user intervention.
- Over the air update.

3.1. Software Update Mechanism: Manufacturer-Initiated Recall Process

Vehicle recalls are relatively common. For example, since 1966 in the U.S., over 390 million vehicles have been recalled due to safety issues [41]. Like a physical component, a software-related problem, depending upon the severity, needs to be addressed and resolved. The recall mechanism, for both physical and software-related issues, requires the vehicle's return to a qualified engineer to rectify the problem [13]. Vehicle recalls are an expensive exercise for the manufacturer [13,17,18,42]. They are also a disruptive and time-consuming procedure for the customer [42,43].

The process of a physical component fix may differ from a software fix. Physical components are replaced with new ones, often because of mechanical wear or a fault in the original component design or construction. In contrast, a software fault may require specialist equipment and a new software version installed on the existing hardware. However, this is not always possible with older embedded systems. Legacy ECU systems have their code pre-set at component manufacture. According to [44], high hardware optimisation often results in ECUs with minimal resources where limited storage, memory and processing capacity cannot accommodate additional lines of new software code. Limitations in ECU hardware resources require a similar exchange of hardware to repair a software-related fault. As such, like a physical component, ECU hardware exchange may be the only option to repair a software-related defect. This has led to a state where more than 50% of error-free hardware is replaced with entirely new hardware to resolve a software-related issue [44].

Incurred manufacturer maintenance costs can be high if a previously undetected software error or design flaw requires a vehicle recall [45]. A much higher cost multiplier to repair a software fault post-production is applied when compared with identifying the same fault much earlier in the software development life cycle. Cost is not the only factor in this update process. Customer confidence and brand loyalty can also be affected by software bugs and errors [6]. In recent years this has been an issue with the highly publicised Grand Jeep Cherokee cyber-attack [5,28,30].

3.2. Software Update Mechanism: Guided User Intervention

This mechanism uses a physical input port installed inside the vehicle. Many modern cars provide a physical connection port for their owners' portable electronic devices, such as external Global Positioning System (GPS) and personal mobile devices, including MP3 players, mobile phones and tablets. These devices are often connected to the vehicle via a universal serial bus (USB) port. Using this port, vehicle owners are able to undertake their own software update either by inserting a supplied preloaded removable storage device or downloading a specific update from the manufacturer onto a USB device. Notably, Fiat Chrysler employed this type of update following the 2015 Grand Jeep Cherokee remote cyber-attack. Using the postal system, Fiat Chrysler distributed preloaded USB memory sticks with updated software to 1.4 million affected customers [30]. However, there are problems associated with this type of update mechanism. These include the following:

- Limited port functionality.
- Inaccessible code.
- Basic understanding of computing technology.
- Willingness to undertake the task.

If any of the above prerequisites cannot be achieved, the software update will not be completed and it will be left unresolved. This software update method relies heavily on the customer having a particular level of technical knowledge and a willingness to perform the update process themselves. For example, there may be a reluctance to complete a necessary software update task due to a fear that their actions could "break the car", rendering it unserviceable and them responsible for any additional repair costs. Furthermore, there are inherent security risks. This method is open to potential exploitation from malicious threat actors that could enable unauthorised vehicle system access or the introduction of malware into the vehicle through compromised storage devices or software download files [46].

3.3. Software Update Mechanism: Over the Air (OtA) Update

OtA mechanisms can update automotive software without the need to return the vehicle to an authorised garage or dealership, or relying on the customer to update themselves. Using on-board vehicle connectivity, OtA updates can deliver new software as and when required. There are several options which can provide OtA software updates:

3.3.1. Dedicated Short-Range Communication (DSRC)

DSRC is an 803.11p-based wireless communication technology used for vehicle to infrastructure (V2I) and vehicle to vehicle (V2V) communication to aid and support ADAS and autonomous driving technologies. This communication technology can be used to transfer software updates between fixed infrastructure or vehicles [47–49]. However, the primary issue with DSRC and automotive software updates is the relatively short time frames involved in V2I and V2V, especially when vehicles are travelling in the opposite direction.

3.3.2. Cellular Networks

In contrast to DSRC, cellular network technology (3G, 4G, and 5G) can provide a stable high bandwidth communication mechanism. Software updates are downloaded by connecting to a particular cell tower within range, regardless of vehicle speed and travel direction. However, coverage may be restricted due to geographical limitations. Nevertheless, by using the extensive scope of cellular networks, future automotive software updates can be transmitted and downloaded to the target vehicle regardless of that vehicle's location. New software, when released, can also be downloaded.

3.3.3. Fixed Location Wireless Local Area Network (WLAN)

This is another potential option for receiving software downloads. Updates can be sent to the target vehicle while parked, for example, at home or at work. Tesla has been using this OtA update mechanism from 2017 by using P2P wireless connections to download software from Tesla servers to target vehicles [50].

Whichever form of OtA update mechanism is chosen, requires a vehicle connectivity solution. There are three modes of connectivity operation, depending upon the connection hardware type employed in the vehicle:

- Mirrored—applications stored on a paired portable smart device are replicated onto the vehicle's HMI unit. The application processing is usually performed on the smart device with screen updates sent to the HMI via a physical or wireless connection [5].
- Tethered—this type of connection uses the paired device's communication technology. Applications are installed to the vehicle's HMI unit and application data processing is performed within the car.
- Embedded—a vehicle with this type of connectivity does not rely on a paired smart device but uses its own connectivity hardware and installed applications.

There has been a widespread introduction of Long-Term Evolution (LTE) technology within the motor vehicle using one of the three aforementioned connectivity types in recent years.

4. The Significance of OtA Software Updates

There are several benefits associated with OtA software updates, making it a promising technology for the automotive industry. Through using lightweight virtualisation technology and OtA software updates can provide several specific benefits:

- Reduction in vehicle recalls and associated costs.
- Vehicles can be updated in locations other than a dealership or maintenance garage.
- Centralised software—software updates can be distributed directly to the target vehicles without distributing to dealers and maintenance garages.

- Time to market—new software can be distributed as and when required rather than waiting for the customer to return the vehicle or waiting for periodic maintenance schedule.
- Convenience—updates can be performed at the customer's desired location and discretion, reducing vehicle downtime.
- Mandatory updates—new and updated software, especially where safety is concerned, can be pushed to the target vehicle without waiting for customer participation.
- Increase in safety—OtA software updates can reduce the time a vehicle is operated under faulty conditions.
- Proven technology—OtA updates are widespread in the telecommunication industry, which has provided users with new updated software via OtA mechanisms.

Reference [51] has predicted that vehicle connectivity could be available in all new motor vehicles by 2025. In 2015, [21] suggested OtA software updates were an attractive technology for the OEM and the customer, with cost savings expected to reach $35 billion by 2022.

5. Current Automotive Software Re-Flashing Techniques

The current practice of updating automotive ECUs involves software flashing or re-flashing techniques [20,52,53]. The operating system and functional software of an ECU are generally held within embedded FLASH memory. Depending upon the model, modern motor vehicles can have hundreds of megabytes of FLASH memory spread across their ECUs. Under the return or recall mechanism, flashing or re-flashing software is often completed by authorised personnel requiring the vehicle to be offline. New software is delivered to the target ECU in one of two formats—full binary and diff/Delta file [32].

5.1. *Full Binary Re-Flashing*

ECU firmware is updated in its entirety through a process known as re-flashing, which conforms to ISO 14229-3/UDS and ISO 15765-2/DoCAN. As part of this process, the entire ECU software image is replaced with a newer version and the time taken to update the software can often take hours to complete. This in part depends on the size of the software update, the destination memory, protocol and whether encryption is used. The previously installed software has no relevance on the new update, which can be beneficial if the previous version requires replacing in its entirety rather than upgrading specific parts. The size of the image binary impacts the time taken to transmit and download the file. The new updated software image must also be stored within the target ECU, which requires redundant storage, potentially of an undetermined fixed amount in order to accommodate any future software update.

5.2. *Difference/Delta File*

Diff/delta file flashing is a concept that compares the base file with the new version file and creates a delta or difference file, thus reducing the size of the update [50]. Compared with a full binary software update, a diff/delta software update is approximately below 10% of the full binary file size. Diff/delta files are much quicker to transmit, decreasing overall transmission time by up to 90%. This method requires considerably less redundant storage but it is reliant on the previous ECU software version. A patching algorithm block erases the old data and writes new data in its place.

6. Container-Based Software Updating

Current software upgrades and bug fixes require the car to be shut down while being updated and subsequently brought back online when complete. Looking towards the automotive industry's future, container-based ECU software can provide a platform to facilitate software updates [23].

Containers, as depicted in Figure 3, represents a newer virtualisation technology which differs from conventional hosted (Type 2) and bare-metal (Type 1) virtualisation technologies. Individual functionality can be provided by small programs hosted within multiple containers that do not require the heterogeneity provided by full system virtualisation which comes at a cost when considering small scale embedded computing devices.

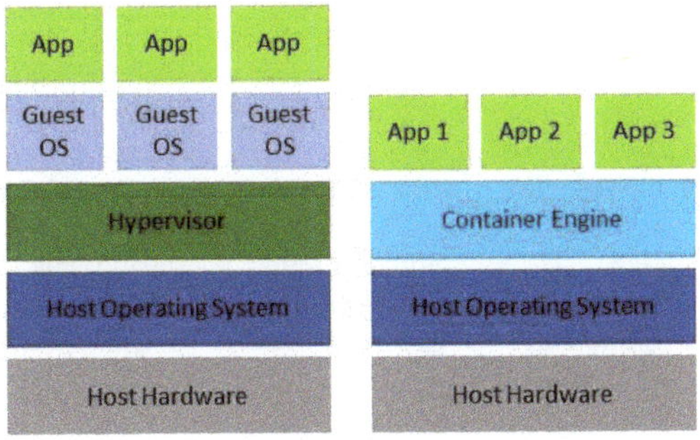

Figure 3. Full System Virtualisation and Container Architectures.

Using ECU container virtualisation in conjunction with OtA updates can address the problems associated with customer disruption and the intrinsic delay between the availability of a new software update and the deployment of that update to the target vehicle. Through vehicle connectivity, new automotive software updates can be pushed or pulled to the target vehicle at any time. For example, a software update pull request could be initiated by the consumer as part of an aftermarket software upgrade or additional automotive functionality. A push update could be applied by the Original Equipment Manufacturer (OEM) or vehicle manufacturer, to resolve an identified software bug or vulnerability thus circumventing the return/recall process and the inherent delays this entails. However, the current primary focus of OtA is on applying a new update when the vehicle is solely offline. The modern motorcar is a system which operates using many subsystems of mixed-criticality. When in operation, there are numerous safety-critical and continual service systems that require a real-time response, for example, engine management and occupant safety systems. The criticality of the software-related issue often determines the required type of software update response. This research has identified three distinct container-based automotive software update modes: offline, online and dynamic.

6.1. *Offline Update*

Offline updates are initialised when the vehicle is powered down. Once the software update verification and initial container creation are complete, any updates applied are available when the vehicle is next started, similar to a system proposed by [54]. This process mirrors the current return or recall procedure but does not incur any associated disruption to the manufacturer or consumer, or recall costs [19,20,55]. Furthermore, this type of update mechanism can be used for multiple system updates, which may affect numerous subsystems across different automotive domains or involve safety-critical systems that cannot be updated safely with the vehicle in operation.

6.2. *Online Update*

New software updates can be pushed or pulled to the vehicle using onboard connectivity and applied while the vehicle is powered up but not in operation. The update process is initiated and a new container is created from the new updated image. The affected subsystem is then temporarily shut down before the new container's initialisation with the updated software. This update method could be applied to any automotive system but only where a system's required initialisation does not incur long time delays. For example, small and frequent periodic updates and software security patches would be ideal candidate systems and functions.

6.3. *Dynamic Update*

DSU do not require the system to be taken offline [56]. As such, they provide an essential service where systems must offer a 100% uptime [57,58]. Taking a system offline to fix bugs, improve system performance, or extend functionality causes delay and disruption. Driverless vehicular technology promises non-stop long-haul trucks and round-the-clock lift-hailing rides and therefore the window to administer software updates become shorter and downtime is a significant disruption [59–61]. For the purposes of this research, a DSU refers to a vehicle sub-system that can be updated and made available once completed, without the vehicle requiring shut down and while it is still in a mode of operation. This type of automotive update is suited ideally to any automotive function which is not involved in vehicle operation or safety. Potential systems could include security software updates and patches, any software relating to autonomous driving functions which are not in operation and passenger-related systems relating to comfort, heating and occupant-vehicle interaction.

7. Implementation and Evaluation of Container-Based Software Updating

Containers offer many benefits to current and future automotive E/E architectures [23]. For example, they provide a standardised environment that can facilitate automotive embedded software updates and their hardware is not fixed to a particular version or type of software. Consolidation is a crucial benefit of container ECUs where multiple containers operate on larger, more resource capable embedded hardware platforms. Containers are constructed from images based on a layered architecture, image layers represent specific data, software, hardware and network configuration parameters.

A container image incorporates one or more layers (as can be seen in Figure 4), which define all required software, libraries and binaries, and configuration settings for any subsequent containers created from that image. Therefore, a container-based ECU must also conform to the three principles of safety, security and transparency:

- Safety—new software containers can be rolled back to the 'last known good' image and known safe containers can be reinstated.
- Security—new container images can be either pulled from an authorised repository to the target vehicle or pushed by the manufacturer. All image layers use, for example, SHA256 encryption and the checksum's validation before the image goes 'live'.
- Transparency—new container images, once validated, can be checked within a sandbox area of the vehicle's automotive E/E architecture before deploying live containers, ensuring the updated system's safe and continued service.

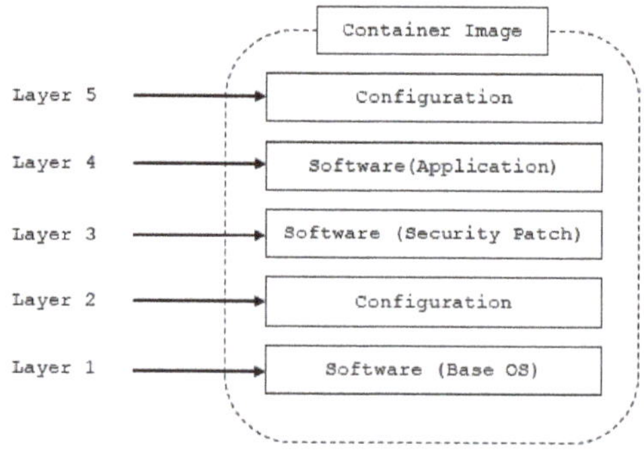

Figure 4. Container Image Layers.

Multiple containers can be created from the same image, which consists of several read-only layers. Any change to the image is specific to a particular layer. If a change is made within the image this alteration is contained within the appropriate layer the change refers to. Small image configuration changes or an update to a specific piece of software within the image will prompt the system to download only the layers which pertain to those particular changes. A further benefit of this layered approach is the ability to share image layers between separate images. Multiple images that share common layers promote efficiency. A layered design boosts image download speed and minimises the overall image footprint and storage requirements.

To evaluate the benefits of the proposed approach, a test system which closely resembles a typical ECU hardware architecture is required. Previous research into embedded systems and engine management has successfully used the ARM processor-based Raspberry Pi to simulate an ECU [62,63] . The Raspberry Pi version 3B hardware specification used is suitably equipped to host the container software [64–66]. The ECU testbed operating system was Raspbian Lite which used Docker, a container virtualization technology. The high level-programming language chosen is Python as it provides flexibility in accessing the GPIO pins of the Raspberry Pi.

The following test case illustrates how container-based ECUs can promote software update efficiency through image layer duplication. Layer duplication in this context refers to any container image which has the same software version or set of configurations. In this test case two separate image downloads are presented with and without layer duplication. The following results when layer duplication is used show a reduction in time to download and required storage footprints in a potential future automotive software update procedure.

7.1. Individual Container Image Downloads

The example in Figures 5 and 6 illustrates two separate alpine-python images that consist of three individual layers which define the configuration and required software for any container created from that image. The two images alpine-python2 and alpine-python3, both of which share a common Linux based OS (alpine), can be seen in the unique identifier layers cbdbe7a5bc2a and 136e07eea1d6. However, each image has a different version of application software (python2 and python3) with the unique identifier layers f890c68la889 and la5281d56ld0 respectively.

Figure 7 displays the time taken to download and extract both of the alpine images including the total size on disk the two images require in MB. This individual image download is a standard procedure in full binary software updating. If both images are downloaded independently, each image is downloaded in its entirety and bears no relationship with the other image, even though they both share the same underlying OS (alpine).

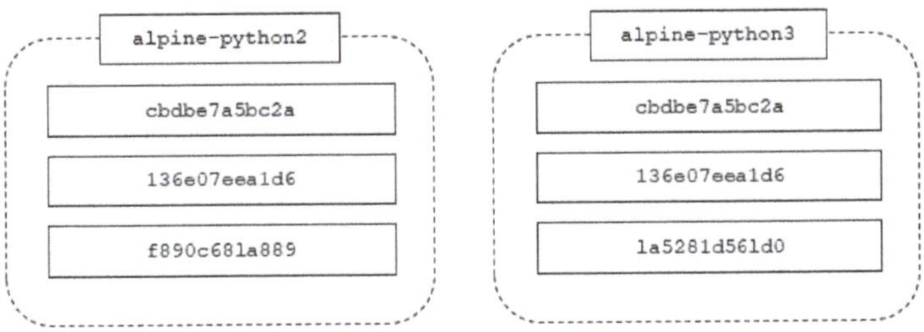

Figure 5. Independent Image Download Layers.

```
master01:~/images docker pull digitalgenius/alpine-python2-pg
Using default tag: latest
latest: Pulling from digitalgenius/alpine-python2-pg
cbdbe7a5bc2a: Pull complete
136e07eea1d6: Pull complete
f890c681a889: Pull complete
Digest: sha256:c4cff01d2c59c13fb729c158bb309194fb469bc28117e70b535bf48d4aac82de
Status: Downloaded newer image for digitalgenius/alpine-python2-pg:latest
docker.io/digitalgenius/alpine-python2-pg:latest
```

```
master01:~/images# docker pull digitalgenius/alpine-python3-pg
Using default tag: latest
latest: Pulling from digitalgenius/alpine-python3-pg
cbdbe7a5bc2a: Pull complete
136e07eea1d6: Pull complete
1a5281d561d0: Pull complete
Digest: sha256:0b1512a41129f127bd512185873e351e89cd647a6b32856c8f4ba4aef1b9921b
Status: Downloaded newer image for digitalgenius/alpine-python3-pg:latest
docker.io/digitalgenius/alpine-python3-pg:latest
```

Figure 6. Independent Image Download.

Image Name	Version	Compressed Image size	Total size on disk after download and extraction	Time taken to download and extract image	Additional download time required
alpine-python	2	115.2MB	883MB	23.4335s	-
alpine-python	3	153.53MB		31.3837s	7.9502s

Figure 7. Image Download, Extraction Times and Storage Requirements.

7.2. Container Image Sharing Downloads

The following test case uses the same alpine-python images. However, before a new image is downloaded, the container host will examine all locally stored images and check each image layer unique ID key which was generated during image creation. Any duplicate layers that share the same image layer ID are ignored. Only those unique layers relating to the new image are downloaded. In this test, an existing image contains two identical layers in common with the new image. Therefore, only one layer of the new image is downloaded, which is observed in Figure 8 (layers indicated by the red outlines in the figure) and Figure 9.

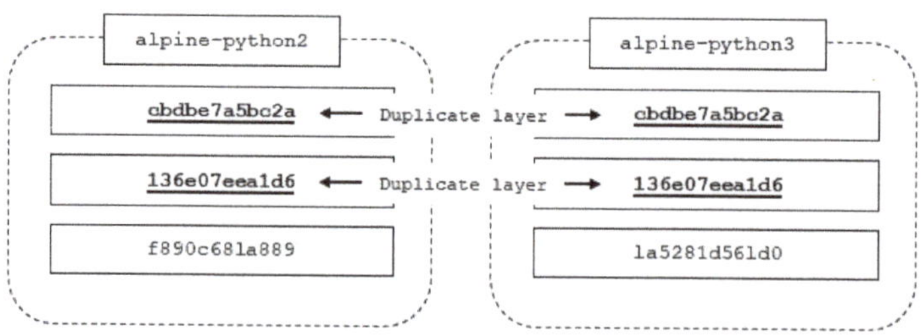

Figure 8. Image Repository Comparison Download.

```
master01:~/images/alpine-python2-pg
default tag: latest
latest: Pulled from digitalgenius/alpine-python2-pg
cbdbe7a5bc2a:
136e07eea1d6:
f890c681a889:
Digest: sha256:c4cff01d2c59c13fb729c158bb309194fb469bc28117e70b535bf48d4aac82de
Status: Current stored image from digitalgenius/alpine-python2-pg:latest
docker.io/digitalgenius/alpine-python2-pg:latest
```

```
master01:~/images# docker pull digitalgenius/alpine-python3-pg
Using default tag: latest
latest: Pulling from digitalgenius/alpine-python3-pg
cbdbe7a5bc2a: Already exists
136e07eea1d6: Already exists
1a5281d561d0: Pull complete
Digest: sha256:0b1512a41129f127bd512185873e351e89cd647a6b32856c8f4ba4aef1b9921b
Status: Downloaded newer image for digitalgenius/alpine-python3-pg:latest
docker.io/digitalgenius/alpine-python3-pg:latest
```

Figure 9. Image Repository Comparison Download.

During the alpine-python3 image download, the two duplicate layers (cbdbe7a5bc2a and 136e07eea1d6) are not downloaded. These two layers represent the alpine OS which both python images share. Only the updated python version layers f890c68la889 and 1a5281d561d0 are pulled from a repository. A repository in this context could be the manufacturer, third party supplier, or one which is stored locally within the vehicle similar to the container image distribution acceleration mechanism proposed by [67–69]. The benefits of layer sharing between container images include reducing the download time for any software update and minimising overall image footprint. Using the performance monitoring tools within the container software the results in Figure 10 highlight the reduced size on disk of both images, which was observed at 44.96%. A reduction in download times between the alpine-python3 images was 5.6346 s across the two tests. Reducing storage requirements for individual software images benefits automotive systems by minimising associated storage hardware costs. Furthermore, layer sharing promotes quicker download speeds which reduces the impact on OtA bandwidths.

Image Name	Version	Layer	Layer Size	Size on Disk after download and extraction	Time Taken to Download and Extract Image	Additional Download Time Required
alpine-python	2	cbdbe7a5bc2a	2.81MB	486MB	23.4335s	-
		136e07eea1d6	38.92MB			
		f890c681a889	73.47MB			
alpine-python	3	1a5281d561d0	111.80MB		25.7491s	5.6346s

Figure 10. Shared Image Layers Size on Disk and Download Times.

7.3. Image Update: Result and Discussion

The test cases in Section 7 examines the benefits surrounding software update size and speed of download. The two alpine-python images share a common OS (alpine); however, the application software within each image comprises of different versions. The test case first examined the specific download and extract times for each image when downloaded independently. The total download and extract time for the two images was 54.8172 s. The overall size on disk for the two images was recorded at 883 MB. These results provided a baseline for a standard software version upgrade, which is similar to full binary re-flashing techniques. The second part of the test used the same two images but used image layer sharing provided by the container software. There were similarities between the two images as they were both built using the same OS (alpine). As such, because both images share two of the three image layers. Due to layer duplication the OS layers of the new image were not downloaded. This reduced the overall download time by 5.6346 s, which was a 10.28% reduction in total overall download and extraction times. Furthermore, when using container layer sharing, overall storage requirements were reduced considerably by 397 MB or 44.96% when compared with independent image downloads

With each new vehicle model, more and more software are included, adding to the burden of addressing software-related problems. Automotive software update practices have followed the same vehicle recall procedure as when a physical component fails. However, the recall process incurs consumer inconvenience, system downtime, high monetary costs for the manufacturer and potentially reduced brand reputation and customer loyalty. The motor industry is attempting to address the limited available options regarding automotive software updates by using new automotive enabled connectivity.

To illustrate the benefits of container-based software updating, the image download test outlined in Sections 7.1 and 7.2 was conducted. As new software is made available it can be pulled from a remote central software repository. It is envisaged that any future software download would be conducted through automotive connectivity using a static or mobile-based communication network and the Internet, or a central repository which is either hosted with the vehicle manufacturer or third-party supplier.

Consideration has been given to develop the experimental setup to represent an actual ECU hardware environment. The necessary container image layers used within this test case were downloaded over the existing University Internet connection media. Hence, the reported time to download is as it would be in a real scenario using ground/location-based WiFi. However, it is envisioned that vehicular on-board connectivity solutions could be used to connect to other communication technologies including DSRC or the LTE network (4G and 5G). These could be used to provide a mobile connection and download method. Furthermore, within this test-case scenario, bandwidth is not a primary consideration as any future software download can take place over a time-period, unless in certain, very rare, safety critical scenarios, which is not within the scope of the proposed approach. Container-based software image layers have minimum storage overhead compared to the traditional ECU architecture-based updates, as only layers pertaining to the update is downloaded as opposed to all the layers within the new software image. The efficiencies provided by container image layers and layer duplication can minimise the size requirements of redundant storage associated with future software updates, given that component costs can escalate sharply due to high vehicle production numbers and the lifespan of a vehicle model, both of which are an important consideration in ECU design.

8. Conclusions

Container-based ECUs, as proposed and evaluated within this paper, can promote automotive software updates, particularly OtA software updates in conjunction with vehicle connectivity. This can reduce significantly the need to recall vehicles when encountering a software-related problem because the new software can be deployed to a target vehicle remotely, as and when required. Notably, by using containers in this way, overall vehicle security can be maintained and any potential software vulnerabilities can be addressed. This has significant implications for the automotive industry.

The number of vehicles recalls in the U.S. associated with a software fault has risen dramatically by 1400% since 2010. Vehicle recalls are highly disruptive to the consumer, expensive for the manufacturer and can, in some cases, reduce brand reputation. It is estimated that resolving a software-related error post-sale is 30 times more expensive than compared with fixing the same issue during the early stages of the SDLC. Compounding this, the current automotive software update practices and procedures are not keeping pace with the rapid increase in the number of lines of software code. The research findings presented in this paper demonstrate that these problems may be overcome by using container-based ECUs, whereby errors, bugs and vulnerabilities cannot only be addressed promptly and effectively, but also throughout the vehicle's lifetime. Additionally, container-based OtA software updates can significantly reduce disruption to consumers. Consumers are also able to incorporate additional or new functionality into a container-based ECU, which can generate additional revenue for the manufacturer in terms of their aftermarket sales. The proposal holds promise of a paradigm shift in the automotive E/E architecture and the way software update is performed.

Author Contributions: Conceptualization, N.A. and L.D.; methodology, N.A., L.D. and D.P.; software, N.A.; validation, N.A., L.D. and D.P.; investigation, N.A.; data curation, N.A.; writing—original draft preparation, N.A.; writing—review and editing, N.A., L.D. and D.P.; visualization, N.A.; supervision, L.D. and D.P.; project administration, L.D.; funding acquisition, L.D. All authors have read and agreed to the published version of the manuscript.

Funding: This research received no external funding. The submitted research is part of N.A.'s PhD thesis and the PhD Bursary has been funded internally by De Montfort University, UK.

Data Availability Statement: Data available on request due to restrictions.

Conflicts of Interest: The authors declare no conflict of interest.

References

1. Bereisa, J. Applications of Microcomputers in Automotive Electronics. *IEEE Trans. Ind. Electron.* **1983**, *30*, 87–96. [CrossRef]
2. Haghighatkhah, A.; Banijamali, A.; Pakanen, O.P.; Oivo, M.; Kuvaja, P. Automotive software engineering: A systematic mapping study. *J. Syst. Softw.* **2017**, *128*, 25–55. [CrossRef]
3. Petri, R.; Springer, M.; Zelle, D.; McDonald, I.; Fuchs, A.; Krauß, C. Evaluation of lightweight TPMs or automotive software updates over the air. In Proceedings of the World's Leading Automotive Cyber Security Conference, Detroit, MI, USA, 1–2 June 2016.
4. Breitschwerdt, D.; Cornet, A.; Kempf, S.; Michor, L.; Schmidt, M. *The Changing Aftermarket Game and How Automotive Suppliers can Benefit from Arising Opportunities*; McKinsey & Company: New York, NY, USA, 2017.
5. Coppola, R.; Morisio, M. Connected car: Technologies, issues, future trends. *ACM Comput. Surv.* **2016**, *49*, 1–36. [CrossRef]
6. Riggs, C.; Rigaud, C.E.; Beard, R.; Douglas, T.; Elish, K. A Survey on Connected Vehicles Vulnerabilities and Countermeasures. *J. Traffic Logist. Eng.* **2018**, *6*, 11–16. [CrossRef]
7. Levitt, J. *Complete Guide to Preventative and Predictive Maintenance*, 1st ed.; Industrial Press: New York, NY, USA, 2003.
8. Hangal, S.; Lam, M.S. Tracking down software bugs using automatic anomaly detection. In Proceedings of the 24th International Conference on Software Engineering. ICSE 2002, Orlando, FL, USA, 25 May 2002; pp. 291–301.
9. Onuma, Y.; Terashima, Y.; Nozawa, M. Improved Software Updating for Automotive ECUs. *Atlanta* **2016**, *2*, 319–324.
10. Noergaard, T. *Embedded Systems Architecture A Comprehensive Guide for Engineers and Programmers*; Elsevier: Oxford, UK, 2005.
11. Ebert, C.; Jones, C. Embedded software: Facts, figures, and future. *Computer* **2009**, *42*, 42–52. [CrossRef]
12. Heiser, G. Hypervisors for consumer electronics. In Proceedings of the 2009 6th IEEE Consumer Communications and Networking Conference, Las Vegas, NV, USA, 10–13 January 2009; pp. 1–5.
13. Sax, E.; Reussner, R.; Guissouma, H.; Klare, H. *A Survey on the State and Future of Automotive Software Release and Configuration Management*; KIT: Amsterdam, The Netherlands, 2017; pp. 1–19.
14. Martyn, A. Automatic Braking Systems in Some Nissan Rogues are Going Rogue. 2019. Available online: https://www.consumeraffairs.com/news/automatic-braking-systems-in-some-nissan-rogues-is-going-rogue-safety-group-says-032719.html (accessed on 12 March 2021).
15. Buckland, K. Toyota Issues Second Prius Recall in a Month on Crash Risk. 2018. Available online: https://www.bloomberg.com/news/articles/2018-10-05/toyota-issues-second-prius-recall-in-a-month-on-crash-risk (accessed on 23 January 2021).
16. Shepardson, D. Fiat Chrysler Recalls 5.3 Million Vehicles for Cruise Control Defect. 2018. Available online: https://www.reuters.com/article/us-fiat-chrysler-recall/fiat-chrysler-recalls-4-8-million-u-s-vehicles-for-cruise-control-defect-idUSKCN1IQ1QY (accessed on 14 January 2021).
17. Drolia, U.; Wang, Z.; Pant, Y.; Mangharam, R. AutoPlug: An automotive test-bed for electronic controller unit testing and verification. In Proceedings of the 2011 14th International IEEE Conference on Intelligent Transportation Systems (ITSC), Washington, DC, USA, 5–7 October 2011; pp. 1187–1192.

18. Lönn, H.; Freund, U. Automotive architecture description languages. In *Automotive Embedded Systems Handbook*; CRC Press: Boca Raton, FL, USA, 2009.
19. Furst, S.; Bechter, M. AUTOSAR for connected and autonomous vehicles: The AUTOSAR adaptive platform. In Proceedings of the 2016 46th Annual IEEE/IFIP International Conference on Dependable Systems and Networks Workshop (DSN-W), Toulouse, France, 28 June–1 July 2016; pp. 215–217.
20. Onuma, Y.; Nozawa, M.; Terashima, Y.; Kiyohara, R. Improved software updating for automotive ECUs: Code compression. In Proceedings of the 2016 IEEE 40th Annual Computer Software and Applications Conference (COMPSAC), Atlanta, GA, USA, 10–14 June 2016; Volume 2, pp. 319–324.
21. Braun, L.; Armbruster, M.; Gauterin, F. Trends in vehicle electric system design: State-of-the Art Summary. In Proceedings of the 2015 IEEE Vehicle Power and Propulsion Conference (VPPC), Montreal, QC, Canada, 19–22 October 2015; pp. 1–6.
22. Rouse, M. OTA Update (Over-the-Air Update). 2018. Available online: https://searchmobilecomputing.techtarget.com/definition/OTA-update-over-the-air-update (accessed on 4 November 2019).
23. Ayres, N.; Deka, L.; Passow, B. Virtualisation as a Means for Dynamic Software Update within the Automotive E/E Architecture. In Proceedings of the 2019 IEEE SmartWorld, Ubiquitous Intelligence & Computing, Advanced & Trusted Computing, Scalable Computing & Communications, Cloud & Big Data Computing, Internet of People and Smart City Innovation (SmartWorld/SCALCOM/UIC/ATC/CBDCom/IOP/SCI), Leicester, UK, 19–23 August 2019; pp. 154–157.
24. Steinkamp, N.; Levine, R.; Roth, R. Automotive Defect and Recall Report, Stout Risius Ross. 2019. Available online: https://www.stout.com/zh-cn/insights/report/2019-automotive-defect-and-recall-report (accessed on 19 February 2021).
25. Lions, J.L.; Luebeck, L.; Fauquembergue, J.L.; Kahn, G.; Kubbat, W.; Levedag, S.; Mazzini, L.; Merle, D.; O'Halloran, C. Ariane 5 Flight 501 Failure Report by the Inquiry Board. 1996. Available online: http://sunnyday.mit.edu/nasa-class/Ariane5-report.html (accessed on 15 January 2021).
26. Lin, P.-S.; Wang, Z.; Guo, R. *Impact of Connected Vehicles and Autonomous Vehicles on Future Transportation*; ASCE Library: Hsinchu, Taiwan, 2016.
27. Thati, V.B.; Vankeirsbilck, J.; Pissoort, D.; Boydens, J. Hybrid Technique for Soft Error Detection in Dependable Embedded Software. In Proceedings of the 2019 IEEE XXVIII International Scientific Conference Electronics (ET), Sozopol, Bulgaria, 12–14 September 2019.
28. Miller, C.; Valasek, C. *Adventures in Automotive Networks and Control Units*; DEF CON: Las Vegas, NV, USA, 2013.
29. Woo, S.; Jo, H.J.; Kim, I.S.; Lee, D.H. A Practical Security Architecture for in-vehicle CAN-FD. *IEEE Trans. Intell. Transp. Syst.* **2016**, *17*, 2248–2261. [CrossRef]
30. Automotive IQ. Automotive Software Development Reliability and Safety. 2017. Available online: https://www.automotive-iq.com/electrics-electronics/reports/automotive-software-development-reliability-safety-1 (accessed on 21 February 2021).
31. Boucherat, X. Make it Safe, Make it Profitable: The Writing's on the Wall for the Connected Car. 2016. Available online: https://www.automotiveworld.com/articles/make-safe-make-profitable-writings-wall-connected-car/ (accessed on 10 January 2021).
32. Howden, J.; Maglaras, L.; Ferrag, M.A. The Security Aspects of Automotive Over-the-Air Updates. *Int. J. Cyber Warf. Terror.* **2020**, *10*, 64–81. [CrossRef]
33. Happel, A.; Ebert, C. *Security in Vehicle Networks of Connected Cars*; Springer: Wiesbaden, Germany, 2015.
34. Alam, M. The Software Defined Car: Convergence of Automotive and Internet of Things. In *Wireless World in 2050 and Beyond: A Window into the Future!* Springer Series in Wireless Technology; Springer: Berlin, Germany, 2016; pp. 83–92.
35. Chowdhury, T.; Lesiuta, E.; Rikley, K.; Lin, C.W.; Kang, E.; Kim, B. Safe and Secure Automotive Over-The-Air Updates. In *Computer Safety, Reliability and Security*; Springer: Cham, Switzerland, 2018; pp. 172–187.
36. Quain, J.R. With Benefits and Risks Sofwtare Updates are Coming to the Car. 2018. Available online: https://digitaltrends.com/cars/over-the-air-software-updates-cars-pros-cons/ (accessed on 16 February 2021).
37. Parnas, D.L. Software aging. In Proceedings of the 16th International Conference on Software Engineering, orrento, Italy, 16–21 May 1994; pp. 279–287.
38. Breitschwerdt, D.; Cornet, A.; Michor, L.; Müller, N.; Salmon, L. Performance and disruption—A perspective on the automotive supplier landscape and major technology trends. *Hg. v. McKinsey and Company, zuletzt geprüft am* **2016**, *7*, 2018.
39. Holmes, F. Over-the-Air Updates Moving from 'Nice to Have' to 'Vital'. 2018. Available online: https://www.automotiveworld.com/articles/over-the-air-updates-moving-from-nice-to-have-to-vital/ (accessed on 11 January 2021).
40. Halder, S.; Ghosal, A.; Conti, M. Secure over-the-air software updates in connected vehicles: A survey. *Comput. Netw.* **2020**, *178*, 107343. [CrossRef]
41. NHTSA. National Highway Traffic Safety Administration. 2020. Available online: Https://nhtsa.org (accessed on 10 January 2021).
42. Mckenna, D.; Automotive, B.U.; Semiconductors, N.X.P. Making Full Vehicle OTA Updates a Reality. NXP. 2016. Available online: http://www.nxp.com/automotivesecurity (accessed on 4 March 2021).
43. Odat, H.A.; Ganesan, S. Firmware over the air for automotive, fotamotive. In Proceedings of the IEEE International Conference on Electro/Information Technology, Milwaukee, WI, USA, 5–7 June 2014; pp. 130–139.
44. Broy, M. *Challenges in Automotive Software Engineering*; ACM: Shanghai, China, 2006; pp. 33–42.
45. Kopetz, H. *Design Principles for Distributed Embedded Applications*; Springer: New York, NY, USA, 2011.

46. Checkoway, S.; McCoy, D.; Kantor, B.; Anderson, D.; Shacham, H.; Savage, S. *Comprehensive Experimental Analyses of Automotive Attack Surfaces*; USENIX: San Francisco, CA, USA, 2011.
47. Guo, J.; Balon, N. *Vehicular Ad Hoc Networks and Dedicated Short-Range Communication*; University of Michigan: Ann Arbor, MI, USA, 2006.
48. Vegni, A.M.; Biagi, M.; Cusani, R. Smart Vehicles, Technologies and Main Applications in Vehicular Ad hoc Networks. 2013. Available online: https://www.intechopen.com/books/vehicular-technologies-deployment-and-applications/smart-vehicles-technologies-and-main-applications-in-vehicular-ad-hoc-networks (accessed on 2 March 2021).
49. Patterson, A. The Evolution of Embedded Architectures for the Next Generation of Vehicles. *ATZelektronik Worldwide* **2017**, *12*, 26–33. [CrossRef]
50. Stegar, M.; Boano, C.A.; Niedermayr, T.; Karner, M.; Hillebr, J.; Roemer, K.; Rom, W. An Efficient and Secure Automotive Wireless Software Update Framework. *IEEE Trans. Ind. Informatics* **2017**, *14*, 2181–2193. [CrossRef]
51. Gissler, A. Automotive World Ltd, The Auto Industry must Get Connected to Fend Off Marginalisation. 2016. Available online: https://www.automotiveworld.com/articles/auto-industry-must-get-connected-fend-marginalisation/ (accessed on 15 December 2020).
52. Habermas, C.S. General Motors Corporation. Method and System for Remote Reflash. U.S. Patent 7,366,589 B2, 29 April 2008.
53. Link, M.C.; Hughes Telematics, Inc. Methods and Systems for Software Upgrades. U.S. Patent US 2009/0119657 A1, 7 May 2009.
54. Tobolski, T.; Esselink, C.E.; Westra, M.R.; Ellis, J.T. Silent in-Vehicle Software Updates. U.S. Patent No. US10140109B2, 27 November 2018.
55. Herberth, R.; Körper, S.; Stiesch, T.; Gauterin, F.; Bringmann, O. Automated Scheduling for Optimal Parallelization to Reduce the Duration of Vehicle Software Updates. *IEEE Trans. Veh. Technol.* **2019**, *68*, 2921–2933. [CrossRef]
56. Seifzadeh, H.; Abolhassani, H.; Moshkenani, M.S. A survey of dynamic software updating. *J. Softw. Evol. Process.* **2013**, *25*, 535–568. [CrossRef]
57. Neamtiu, I.; Hicks, M.; Stoyle, G.; Oriol, M. Practical dynamic software updating for C. *ACM Sigplan Not.* **2006**, *41*, 72–83. [CrossRef]
58. Hayden, C.M.; Smith, E.K.; Hardisty, E.A.; Hicks, M.; Foster, J.S. Evaluating Dynamic Software Update Safety Using Systematic Testing. *IEEE Trans. Softw. Eng.* **2012**, *28*, 1340–1354. [CrossRef]
59. Hörl, S. Agent-based simulation of autonomous taxi services with dynamic demand responses. *Procedia Comput. Sci.* **2017**, *109*, 899–904. [CrossRef]
60. Shankwitz, C. *Long-haul Truck Freight Transport and the Role of Automation: Collaborative Human—Automated Platooned Trucks Alliance (CHAPTA)*; Western Transport Institute: Bozeman, MT, USA, 2017.
61. Simpson, J.R.; Mishra, S.; Talebain, A.; Gollias, M. An Estimation of the Future Adoption Rate of Autonomous Trucks by Freight Organizations. *Res. Transp. Econ.* **2019**, *76*, 100737. [CrossRef]
62. Walter, J.; Fakih, M.; Grüttner, K. Hardware-based real-time simulation on the raspberry pi. In Proceedings of the 2nd Workshop on High Performance and Real-time Embedded Systems, Vienna, Austria, 20 January 2014.
63. Vaughan, A.; Bohac, S.V. An extreme learning machine approach to predicting near chaotic HCCI combustion phasing in real-time. *arXiv* **2013**, arXiv:1310.3567.
64. Krylovskiy, A. Internet of things gateways meet Linux containers: Performance evaluation and discussion. In Proceedings of the 2015 IEEE 2nd World Forum on Internet of Things (WF-IoT), Milan, Italy, 14–16 December 2015; pp. 222–227.
65. Hurst, W.; Shone, N.; El Rhalibi, A.; Happe, A.; Kotze, B.; Duncan, B. Advancing the micro-CI testbed for IoT cyber-security research and education. *Cloud Comput.* **2017**, *2017*, 139.
66. Johnston, S.J.; Cox, S.J. The Raspberry Pi: A Technology Disrupter, and the Enabler of Dreams. *Electronics* **2017**, *3*, 51. [CrossRef]
67. Suarez, A.J.; Windsor, S.K.; Hayrapetyan, N.; Gerdesmeier, D.R.; Prakash, P.K.; Amazon Technologies Inc. Software Container Registry Container Image Deployment. U.S. Patent 10,002,247, 19 June 2018.
68. Suarez, A.J.; Windsor, S.K.; Hayrapetyan, N.; Gerdesmeier, D.R.; Prakash, P.K.; Amazon Technologies Inc. Software Container Registry Service. U.S. Patent 10,261,782, 16 April 2019.
69. Zhao, A.; Cao, Y.; Peng, L.; Junping, Z.H.A.O.; Durazzo, K.; EMC IP Holding Co LLC. Container Image Distribution Acceleration. U.S. Patent 10,291,706, 14 May 2019.

electronics

MDPI

Article

Lane-Merging Strategy for a Self-Driving Car in Dense Traffic Using the Stackelberg Game Approach

Kyoungtae Ji [1], Matko Orsag [2] and Kyoungseok Han [1,*]

[1] School of Mechanical Engineering, Kyungpook National University, Daegu 41566, Korea; wlrudxo644@knu.ac.kr

[2] Department of Control and Computer Engineering, University of Zagreb, Zagreb 10000, Croatia; matko.orsag@fer.hr

* Correspondence: kyoungsh@knu.ac.kr

Abstract: This paper presents the lane-merging strategy for self-driving cars in dense traffic using the Stackelberg game approach. From the perspective of the self-driving car, in order to make sufficient space to merge into the next lane, a self-driving car should interact with the vehicles in the next lane. In heavy traffic, where the possible actions of the vehicle are pretty limited, it is possible to conjecture the driving intentions of the vehicles from their behaviors. For example, by observing the speed changes of the human-driver in the next lane, the self-driving car can estimate its driving intention in real time, much in the same way of a human driver. We use the principle of Stackelberg competition to make the optimal decision for the self-driving car based on the predicted reaction of the interacting vehicles in the next lane. In this way, according to the traffic circumstances, a self-driving car can decide whether to merge or not. In addition, by limiting the number of interacting vehicles, the computational burden is manageable enough to be implemented in production vehicles. We verify the efficiency of the proposed method through the case studies for different test scenarios, and the test results show that our approach is closer to the human-like decision-making strategy, as compared to the conventional rule-based method.

Keywords: self-driving car; game theory; decision-making; stackelberg game; lane-merging; intention estimation

Citation: Ji, K.; Orsag, M.; Han, K. Lane-Merging Strategy for a Self-Driving Car in Dense Traffic Using the Stackelberg Game Approach. *Electronics* **2021**, *10*, 894. https://doi.org/10.3390/electronics10080894

Academic Editors: Calin Iclodean, Bogdan Ovidiu Varga and Felix Pfister

Received: 6 March 2021
Accepted: 5 April 2021
Published: 8 April 2021

1. Introduction

The recent development of self-driving cars has shifted the concept of partially autonomous driving from purely imaginary to the real. However, in order to achieve fully autonomous driving (i.e., Level-5 [1]), developers should still overcome many technical difficulties. One of the most challenging tasks is to describe the interaction between a self-driving car and human-driven vehicles [2,3]. In city driving, the vehicle often faces complex traffic situations that should be appropriately addressed. For instance, in congested traffic, human drivers constantly interact with other vehicles to create flexibility [4] by guessing the driving intentions of other drivers. Thus, autonomous vehicles should also act similarly to human drivers when facing complex traffic situations instead of conservative motions. Otherwise, to ensure safety, very conservative decisions such as waiting for traffic to ease is most likely to be made, which are not efficient [5]. Therefore, it is important to reflect the interactions between autonomous vehicles (AV) and interacting vehicles in the decision-making logic. In this way, human-like decision-making can be realized, which is essential when human-driven vehicles and self-driving cars share roads in the near future.

To resolve the technical problems mentioned above, we propose the game theoretic decision-making strategy that enables the self-driving car to consider the interactions with the surrounding vehicles. In particular, we model human thinking processes using game theory as a good candidate to handle heavy traffic conditions in which vehicles affect each other [6]. In this approach, the game participants are assumed to be rational players that make decisions maximizing their own utility [7]. The latter includes setting the player's particular objective. Thus, the appropriate decisions for self-driving cars can be made provided so that they play the game with surrounding vehicles. These vehicles are considered rational decision makers maximizing the utility [8,9].

The efficiency of game theory in modeling vehicle's decision-making process has been verified in previous studies. The most common approaches include the level-K reasoning framework and Stackelberg game approach.

The level-K framework (also referred to as the hierarchical reasoning game theory) is a method to model interaction between players using the hierarchical depth of thought [10,11]. The key idea of the level-K framework is that each player has a depth of strategic thought from level-0 to a specific number K, and the K-level player makes the decision, assuming that the other players choose the particular actions, which are based on the (K-1)-level depth of thought [12]. More specifically, the players assume themselves as the most advanced ones who can think one level ahead of others. In [13,14], the level-K framework decision making is proposed at the unsignalized intersection where many interactions between the vehicles occur. Other researchers also considered the lane-changing problem in highways using the level-K framework [15]. Although, a level-K framework is a promising technique to describe the interactions between the multiple agents, it should model the depth of thought for all agents in the strategic game. Therefore, if a self-driving car faces multiple vehicles, a heavy computational burden is required to model the depth of all vehicle's thoughts [16].

In contrast, the Stackelberg game (also referred to as the leader-follower game) is a hierarchical game where each player is assigned the roles of either leader or follower. By modeling the utility function that needs to be maximized by the each game participant, the interactions between each vehicle can be modeled effectively. Compared to the level-K framework, the Stackelberg game does not need to model the depth of thought for all game players hierarchically, so the computation is less complex. The lane-changing scenario is modeled using Stackelberg game theory in [17] and the surrounding drivers' intentions such as "aggressive driver" and "cautious driver" are estimated in real time [18,19]. In addition, the modeling of multi-vehicle interactions at uncontrolled intersections is considered in [20]. However, these approaches do not consider active interaction which is essential for lane-merging in dense traffic condition.

In this paper, we develop the decision-making strategy for a vehicle merging into another lane in dense traffic where all vehicles interact with each other. In such dense traffic, lane-merging is not possible unless there is a concession between interacting vehicles. So we consider the active interaction that changes the behavior of the interacting vehicle. For the manageable computation, we exploit the Stackelberg game approach. In real-world driving, the human-drivers consider only interacting vehicles, not all vehicles on the road. Similarly, in this paper, the self-driving cars consider interactions only with a single interacting vehicle in the next lane. In this way, the computational load of the proposed method is manageable enough to be implemented in the hardware.

It is also worth noting that the estimation of surrounding vehicle intentions is essential to make the appropriate decision in real time [21]. In our problem (i.e., dense traffic condition), we assume that the intentions of the drivers in the next lane are limited to "yield" or "ignore" [22]. By imposing a certain quantity (i.e., politeness) to the surrounding vehicles, the AV that needs to merge estimates the politeness of the interacting vehicle in real time and decides whether to merge [23]. However, human drivers utilize a strategy that is not exactly known to self-driving cars. To reflect this aspect, in our verification environments, the interacting vehicles in the next lane behave based on the car-following model that the self-driving car does not know. To verify the effectiveness of our game theoretic decision-making strategy, the performances for the various test scenarios are compared to those of the rule-based approach [24].

The characteristic features of the proposed strategy are summarized as follows.

1. The complex vehicle interactions in dense traffic are effectively modeled using the Stackelberg game approach with manageable computation;
2. To establish reliable and verifiable test environments, we propose the modified car-following model that can change the target leading car depending on varying circumstances;
3. The intentions of the surrounding vehicles are effectively estimated in real time by monitoring the speed variations of the interacting vehicles.

The remainder of the paper is organized as follows. In Section 2, we provide a problem definition for the lane-merging scenario in heavy traffic. In Section 3, we introduce the vehicle model, action space, and the driving strategy of the surrounding vehicles. The key result of this paper, game theoretic decision-making strategy is presented in Section 4. The efficiency of our approach is verified through the case studies in Section 5. Finally, we make conclusions and provide a future outlook in Section 6.

2. Problem Statement

Here we describe the interactions between two agents in a strategic game. In Figure 1, AV in the side lane is called an "ego-car", which is a controlled host vehicle and the surrounding vehicles are the human-drivers modeled by a car-following model described in Section 3.3. As illustrated in Figure 1, the interactions between the vehicles in dense traffic are modeled using game theory. More specifically, we propose a lane-merging decision-making strategy for an autonomous vehicle in dense traffic where AV essentially does not have enough space to merge into the next lane from the side lane. In such an environment, lane-merging is not encouraged due to the risk of collision. However, for extreme cases, when traffic is not relaxed for a long time, the driver have to wait indefinitely unless aggressive behavior is considered. In addition, when in emergency situations such as hospital transport, aggressive behavior is required to a certain degree. Therefore, the lane-merging method for traffic congestion can be an option for self-driving cars, especially where traffic congestion happens frequently.

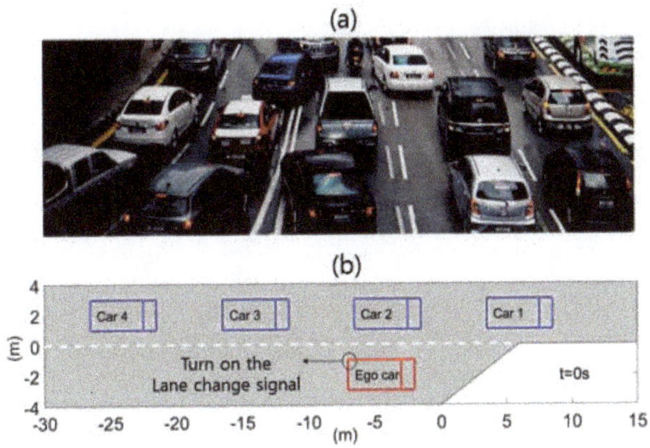

Figure 1. (**a**) Lane-merging scenario in the heavy traffic (photo is from Unsplash) and (**b**) reproducing the real-world driving scenario.

Under such heavy traffic conditions, the AV should interact with vehicles in the next lane to efficiently change the lane [25]. For example, if the AV in the side lane waits until enough space becomes available to merge into the next lane, the decision and control algorithm that is generally not willing to take a risk (i.e., lane-changing with an insufficient distance) may not merge into the next lane unless safety is ensured. However, in reality, human drivers in the side lane often attempt to influence interacting vehicles to get an opportunity to merge [26]. The notable waysi in which drivers interact with others are hand gestures (motion indicating that they about to change the lane) and others. Obviously, in general traffic, not only surrounding vehicles but also the AV influences the behaviors of interacting vehicles. Therefore, such interaction modeling is essential to adequately describe real-world traffic.

From the perspective of an AV, it can secure sufficient distance to merge rather than wait forheavy traffic to be relaxed by affecting the behaviors of surrounding vehicles. To achieve this goal, the AV should predict the response of other vehicles to its actions, which is very uncertain in reality. From the perspective of the surrounding vehicles in the next lane, they should decide whether to yield to AV while obeying traffic regulations. For instance, even if the vehicle in the next lane is willing to make the safety gap by decelerating, the traffic condition does not allow to decelerate considering the relative distance or velocity with the behind vehicle. The latter is a very common situation for heavy traffic. For instance, the perspective of Car 3 is shown in Figure 1. Car 3 makes the decision based on its relationship to the AV, its relative distance, and velocity to Car 4.

In general, the decision-making of the drivers is determined by their dispositions e.g., aggressive, cautious dispositions. However, for the limited traffic condition considered (lane-merging in dense traffic), a decision of the vehicle is mainly governed by the traffic conditions rather than its driving disposition. Generally, the resulting decisions of the vehicle appears in the form of driving intentions. In the merging scenario, these intentions include "yield" or "ignore" (from the perspective of the vehicle in the next lane). For instance, when AV expresses a lane-merging intention by turning on the lane-changing signal, the reaction of Car 3 can be the deceleration to express "yield" or maintain the speed to express "ignore" (Figure 1). To build reliable and verifiable scenarios where the human drivers usually consider only the adjacent vehicles, the AV considers only one interacting vehicle.

3. Vehicle Model and Action Space

In this section, we introduce the model that represents the vehicle dynamics and the decision-making process for all interacting vehicles.

3.1. Vehicle Dynamics

For simplicity, we consider a point-mass vehicle model with continuous time:

$$\begin{aligned} \dot{x} &= v_x, \\ \dot{v}_x &= a_x, \\ \dot{y} &= v_y, \end{aligned} \tag{1}$$

and discretize it using the Euler forward method [27]:

$$\begin{aligned} x(t+1) &= x(t) + v_x(t)\triangle t, \\ v_x(t+1) &= v_x(t) + a_x(t)\triangle t, \\ y(t+1) &= y(t) + v_y\triangle t, \end{aligned} \tag{2}$$

where state $\mathbf{x} = [x, v_x, y]'$ is defined by the longitudinal position at the center of gravity, velocity, and lateral position at the center of gravity. Moreover, control $\mathbf{u} = [a_x, v_y]'$ is defined by the acceleration, and lateral velocity. Finally, $\triangle t$ and t stand for the time step size and is the discrete time instance, respectively.

3.2. Action Set

According to the Stackelberg game approach [28], the game players are assumed to choose a discrete action and execute it for the entire control cycle. To focus on vehicle interactions rather than dynamics itself, the finite discrete actions are assumed for all game participants (interacting vehicles) as follows:

1. "Maintain" : Maintain the speed.

2. "Accelerate" : Constant acceleration with a_x $(\mathrm{m/s^2})$ until vehicle speed reaches to the maximum speed $v_{\max}$.
3. "Decelerate" : Constant deceleration with $-a_x$ $(\mathrm{m/s^2})$ until vehicle speed drops to zero.
4. "Lane-merge" : Changing lane with v_y $(\mathrm{m/s})$, i.e., movement in a lateral direction.

Here, we set the constant acceleration, maximum velocity, and the lane-merge velocity to: $a_x = 0.97\ \mathrm{m/s^2}$, $v_{\max} = 2.5\ \mathrm{m/s}$, and $v_y = 2\ \mathrm{m/s}$, respectively. The latter is only available to the ego-car.

Based on the action space, the strategy space for the leader and follower, i.e., the AV and interacting vehicle, is defined as,

$$S = \Gamma_l \times \Gamma_f \tag{3}$$

$$\Gamma_l = \begin{cases} \{\mathrm{L}, \mathrm{M}\} & \text{, Lateral Motions} \\ \{\mathrm{A}, \mathrm{M}, \mathrm{D}\} & \text{, Longitudinal Motions} \end{cases}$$

$$\Gamma_f = \{\mathrm{A}, \mathrm{M}, \mathrm{D}\},\ \text{Longitudinal Motions}$$

where S is a strategy space, Γ_l and Γ_f are action sets of the leader and follower, L and M denote the "Lane-merge" and "Maintain", and A, M, and D are the "Accelerate", "Maintain", and "Decelerate" in the longitudinal direction.

Obviously, the lateral motions are only available to the ego-car, and the surrounding vehicles are assumed to move forward in a longitudinal direction. In other words, for each time step, the leader decides whether to change lanes or not, while the follower decides its longitudinal motion based on the circumstances, such as traffic conditions.

3.3. *Intelligent Driver Model*

As mentioned earlier, in real-world driving, the ego-car reacts to the behaviors of the surrounding vehicles and vice versa. Therefore, to establish reliable and verifiable lane-merging scenarios for the formulated problem, the modeling of these interactions between the vehicles is important, which distinguishes our test environment itself from others where the vehicle motion is not interactive [29].

In game theoretic interaction modeling, it is assumed that all vehicles choose their actions based on the game theory so that all actions are limited by defined strategy space S. However, the surrounding vehicles, i.e., vehicles in the next lane, choose their actions based on their own strategies that the ego-car does not know exactly. The latter is reasonable because, generally, the drivers do not know the future behaviors and/or trajectories of others. Instead, they can predict the behavior (velocity and acceleration) of other vehicles from their observations.

To reflect this human-like decision-making process for interacting vehicles, a widely-used longitudinal car-following model referred to as the intelligent driver model (IDM) is introduced [30] :

$$\dot{v} = \frac{dv}{dt} = a_{\max}\left\{1 - \left(\frac{v}{v_0}\right)^{\delta} - \left(\frac{s^*(v, \triangle v)}{s}\right)^2\right\}, \tag{4}$$

where v_0, $\triangle v$, $a_{\max}$, δ, s, and s^* stand for target speed, velocity difference (approach rate), maximum acceleration, constant acceleration component, gap, and desired gap with the front vehicle, respectively.

The desired gap s^* is a function of v and $\triangle v$ and given by:

$$s^*(v, \triangle v) = s_0 + vT + \frac{v \triangle v}{2\sqrt{a_{\max} b}}, \tag{5}$$

where s_0 is the minimum gap between ego and front vehicles, T is a safe time headway, and b is the desired deceleration that makes a driver feel comfortable.

If there is no car ahead, s^* is ignored, i.e., $s^* = 0$, thus IDM become a function of the v and v_0. All parameter values for the IDM are summarized in Table 1. The preferred time headway in dense traffic is defined based on [31] and the established models aim to describe the car-following in heavy traffic.

Table 1. Parameter values for the intelligent driver model (IDM).

Parameter	Given Value
Desired velocity v_0	2.5 m/s
Safe time headway T	1.2 s
Maximum acceleration a_{max}	0.97 m/s^2
Desired deceleration b	1.67 m/s^2
Acceleration exponent δ	4
Jam distance s_0	1 m

The conventional IDM is a mathematical model that is based on psychical properties such as the relative distance and speed between vehicles. Thus, the intention of the drivers cannot be described in (4) and (5). To tackle this problem, we impose politeness in the conventional model, and IDM is modified to adequately react to the surrounding circumstances. In particular, when the ego-car sends a lane-changing signal to an interacting vehicle, the latter chooses its action depending on its specified politeness $p_i \in [0,1]$. If the p_i of the ith interacting vehicle is close to 1, then the interacting vehicle is likely to allow the ego-car to change lanes by reducing the speed. Once the ego-car merges into the next lane successfully, the interacting vehicle now follows the ego-car based on (4) and (5). Otherwise (p_i is close to 0), the interacting driver ignores the signal from the ego-car and follows the car ahead.

The speed control procedure of the modified IDM is described in Algorithm 1. Here $S_t = [\{x^1(t), v_x^1(t), y^1(t)\}, \cdots, \{x^n(t), v_x^n(t), y^n(t)\}]'$ is the state tuple of the interacting vehicles at time step t, and $S_t' = \{S_t, s^e\}'$ is the state tuple including the state of the ego-car $s^e = (x^e(t), v_x^e(t), y^e(t))$. Moreover, $M = \{i | i \in \{1, \cdots, n\}\}$, where i is the index of the interacting vehicle and n is the number of the interacting vehicle, P is the set of the specified politeness to the interacting vehicle ($P = \{p_i | i \in M\}$). Additionally, $S_{\text{flag}} \in \{0,1\}$ is a flag of the lane-changing signal that the ego-car sends to the interacting vehicle. For example, when the ego-car turns on the lane-changing signal, $S_{\text{flag}} = 1$, otherwise $S_{\text{flag}} = 0$. Finally, f_{IDM} represents a conventional IDM in (4) and (5).

It is assumed that only one interacting vehicle can see the lane-changing signal from the ego-car. Thus, if *Observe*(S_{flag}) is true (i.e., ego-car turns on the lane-changing signal and only one interacting vehicle observes it), the behavior of the interacting vehicle is determined by the assigned politeness. The process of decision making for the interacting vehicle behavior is as follows.

To include the stochastic component of the driver's behavior, we first generate the random number $p_{\text{rand}} \in [0,1]$ and compare it with the assigned politeness of the ith vehicle $p_i \in [0,1]$. If the p_i is larger than p_{rand}, it is assumed that the interacting vehicle now considers the ego-car as its leader car (Line 6) and takes an action based on f_{IDM} (Line 7). More specifically, the interacting vehicle is willing to allow the lane-merging of the ego-car. For example, if the interacting vehicle recognizes that the ego-car is too close, it decelerates to keep a desired distance from the ego-car. In the case that p_i is smaller than p_{rand}, the interacting vehicle ignores the ego-car's lane-merging intention and follows the original front vehicle in the same lane (Line 9). In addition, when *Observe*(S_{flag}) $= 0$ (non-interacting vehicle that cannot see the lane-merging signal from the ego-car), the vehicle speed is controlled based on the conventional IDM (Line 12). This procedure is repeated for every time step to create a reasonable test environment. Following Algorithm 1, the politeness is imposed to the conventional car-following model.

Algorithm 1: IDM with Politeness

Input S_t, M, P and s^e
for $i \in M$ **do**
  if *Observe*(S_{flag}) **then**
    $p_{\text{rand}} = rand[0,1]$;
    if $p_i > p_{\text{rand}}$ **then**
      $S'_t \leftarrow S_t \cup s^e$;
      $s^i_{t+1} = f_{\text{IDM}}(s^i_t \mid s \in S'_t)$;
    else
      $s^i_{t+1} = f_{\text{IDM}}(s^i_t \mid s \in S_t)$;
    end
  else
    $s^i_{t+1} = f_{\text{IDM}}(s^i_t \mid s \in S_t)$;
  end
end
$S_{t+1} = \{\forall s^i_{t+1} \in S_{t+1} \mid i \in M\}$;
Output S_{t+1}

4. Game Theoretic Lane-Merging Strategy

4.1. Utility Function

In game theory, the participants are considered as rational decision-makers whose goal is to maximize their utility function (achieve a certain numerical design value). Here we define an appropriate utility function, and the ego-car assumes that the interacting vehicle's behavior aims to maximize the utility.

The objective of the ego-car is to merge into the next lane while maintaining safety. At the same time, the interacting vehicles also try to adjust their speeds to avoid a collision. These objectives for all game participants can be described by the utility function $U_{\leq 0}$ [32]:

$$U = w_1 C + w_2 V + w_3 H. \tag{6}$$

where w_1, w_2, and w_3 are the non-negative weights for each term depending on its importance, C, V and H denote "Collision," "Velocity," and "Headway" functions defined below.

The collision detection function $C \in \{-1, 0\}$ is equal to -1 when the collision occurs. Otherwise, it is set to 0. Additionally, we set the follower's weight, w_1, as a varying parameter depending on the politeness:

$$w_1 = \begin{cases} w_c \times p_i & \text{, for follower} \\ w_c & \text{, for leader} \end{cases} \tag{7}$$

where w_c is a constant collision penalty.

Once we introduce p_i, there is room for the follower to choose the less conservative action, even if the collision is expected due to the action of the ego-car. For example, when the p_i is close to zero, the follower's behavior is dominated by the functions V and H. It means that an impolite driver usually prevents the merging of ego-car by choosing the aggressive actions, such as "Accelerate" and "Maintain."

The function V is the normalized difference between the current and the target speeds:

$$V = -\left\| \frac{v - v_0}{v_0} \right\|_2. \tag{8}$$

If v is different from v_0, V is always negative.

The function H is defined based on the headway:

$$H = \begin{cases} -1 & \text{, if headway} \in \text{"close"} \\ 0 & \text{, if headway} \in \text{"sufficient."} \end{cases} \tag{9}$$

The comparison of the current distance headway, s, and desired distance headway, s_0, allows for the vehicle to determine whether the headway is closed or not. When the headway is "close," the vehicle is likely to decelerate and keeps a sufficient distance to the front car.

The defined utility (6) can be utilized for all game participants because the common goal of all vehicles is to drive safely by taking the appropriate actions moment. Once the follower recognizes the leader's lane-merging intention, it tries to maximize its utility based on the specified politeness. For instance, although the leader car takes action (L) first, the follower with low politeness takes an action that prevents the leader's merging by choosing the non-conservative action. For the leader car, by estimating the follower's driving intention in real time, the prediction for the follower's behavior is available, which will be described in the next subsection. Other parameters such as "comfort" are not considered due to the limitations of the scenario: Dense traffic, slow driving [33].

4.2. Stackelberg Game

In the Stackelberg game approach, the game participants are assigned as the leader and the follower. A leader takes action first, and a follower makes a decision after observing the leader's behavior ("first-mover advantage" [34]). We assign the leader role to the ego-car and the follower role to the vehicle in the next lane.

To successfully merge into the next lane, the leader vehicle should predict the reaction of the follower according to the leader's action. We assume that the behavior of the interacting vehicle is based on the basic principle of game theory: *"All game participants make their decisions in such a way as to maximize their own utility U"*. Thus, the future behavior of the follower can be predicted to a certain degree.

For example, from the perspective of the follower, the optimal action $\gamma^{f,*}$ should maximize its utility $U^f(\gamma^l, \gamma^f)$. Note that the utility is influenced by the leader's action γ^l as well as the follower's action γ^f, which is true in reality. From the perspective of the leader, he can predict the follower's behavior based on $U^f(\gamma^l, \gamma^f)$ while maximizing its utility $U^l(\gamma^l, \gamma^f)$. Since the follower's optimal action $\gamma^{f,*}$ may be not unique in some cases, the leader assumes the worst-case scenario (the follower can take the action that is the worst in terms of $U^l(\gamma^l, \gamma^f)$ maximization).

The optimal action for the leader $\gamma^{l,*}$ referred to as the Stackelberg equilibrium [6] is given by:

$$\gamma^{l,*} \in \text{argmax}\left(\min_{\gamma^f \in S^f(\gamma^l)} U^l(\gamma^l, \gamma^f) \right), \tag{10}$$

$$S^f(\gamma^l) \stackrel{\text{def}}{=} \left\{ \zeta \in \Gamma_f : U^f(\gamma^l, \zeta) \geq U^f(\gamma^l, \gamma^f), \forall \gamma^f \in \Gamma_f \right\}, \tag{11}$$

where $\zeta \in \Gamma_f$ is an optimal action of the follower that maximizes U^f after observing the leader's action γ^l, $S^f(\gamma^l)$ is the strategy space for the given leader's action and follower's optimal action ζ, and $\gamma^{l,*}$ is optimal action of the leader based on the maximin strategy.

For a better understanding of the above-described algorithm, we give a simple example of the Stackelberg game in Figure 2. In the overtaking case, the leader car in the left line wants to go back to the right line and the follower car in the right line wants to drive faster. The possible decisions of the leader and follower are limited to discrete actions in Section 3.2 and denotes the initial letters in the heading as (4). To simplify, Γ_l and Γ_f are given as $\{A, L, D\}$ and $\{A, M, D\}$ respectively, so that the tree diagram has nine leaves. The numbers at the leaves are the payoff of the leader and follower for each action combination in S. The leader can predict the decision of the follower based on the game theory approach.

If the leader chooses action L, then $S^f(\gamma^l)$ is $\{M\}$ to maximize its own utility. Therefore, the leader can predict that the U^l should be 0.7 when choosing action L. Likewise, the utility of action A and D can be predicted if the leader can predict the follower's action based on $S^f(\gamma^l)$. However, when the leader chooses action D, the decision to maximize the follower's utility is not unique, i.e., $S^f(\gamma^l) = \{A, M\}$. To avoid the risk of the worst case, the leader assumes that the follower chooses the action A, which is the worst for the leader's utility. The expected utility of the leader's decision is 0.6 with action A, 0.7 with action L, and 0.8 with action D. In this manner, we can find that the optimal action $\gamma^{l,*}$ that maximizes the utility is action D. It means the optimal action of the leader is deceleration so as to allow the follower to go ahead.

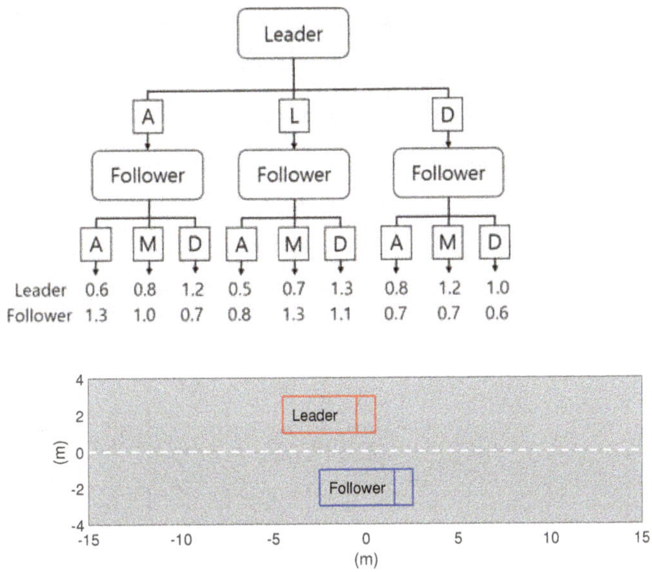

Figure 2. Simple example of the Stackelberg game in an overtaking scenario.

4.3. Real Time Politeness Estimation

As defined earlier, politeness is the numerical value representing the intention of the interacting vehicle to yield the ego-car. When the politeness is low, the interacting vehicle is likely to ignore the lane-changing signal from the leader car and follows the front car based on the IDM. If the follower's intention is to yield to ego-car with high politeness, the follower considers the ego-car as its target vehicle to follow. Therefore, for safe lane-changing, the leader car should estimate the interacting vehicle's driving intention (i.e., politeness).

The politeness can be estimated by observing the current interacting vehicle's acceleration and is given by:

$$P(t+1) \leftarrow \frac{P(t) + \alpha}{1 + \beta} \tag{12}$$

where $P(t) \in [0, 1]$ is the estimated politeness at time step t, $\alpha \in [0, \beta]$ and $\beta \in (0, \infty]$ are the tunable weighting parameters that determine the update rate, and the initial politeness $P(0)$ is set to a relatively low value.

For example, when the acceleration of the interacting vehicle is observed (i.e., $a_f(t) \geq 0$ and $v_f(t) \neq 0$), the politeness should be decreased. That is, if the interacting vehicle accelerates in dense traffic, where there is no sufficient gap between the vehicles, we assume that its driver is aggressive. In this case, we set $\alpha = 0$. In contrast, if the deceleration of the interacting vehicle (i.e., $a_f(t) < 0$ or $v_f(t) = 0$) is observed, the politeness increases until it reaches 1 by setting α equal to β. The ego-car estimates the follower's intention in real time by comparing the $P(t)$ with the threshold $P_{\text{th}} = [P^l_{th}, P^u_{th}]$, where $P^l_{th} = 0.2$ and $P^u_{th} = 0.8$. For instance, if $P(t)$ is less than P^l_{th}, then the ego-car determines the intention of follower as "ignore" and vice versa.

4.4. Game Process

The process of the merging strategy is shown in Figure 3. As the game starts, the leader and the follower are assigned based on their states. Specifically, the ego-car is always the leader, and the vehicle behind the ego-car in the next lane is considered as the follower (the Stackelberg game settings). We assume that ego-car has already reached the end of the side lane so that lane merging is needed as soon as possible. After the target vehicle (follower) that the ego-car should interact with is selected, the ego-car estimates the politeness of the target vehicle by observing its acceleration and follows the optimal action based on Stackelberg equilibrium for every time step.

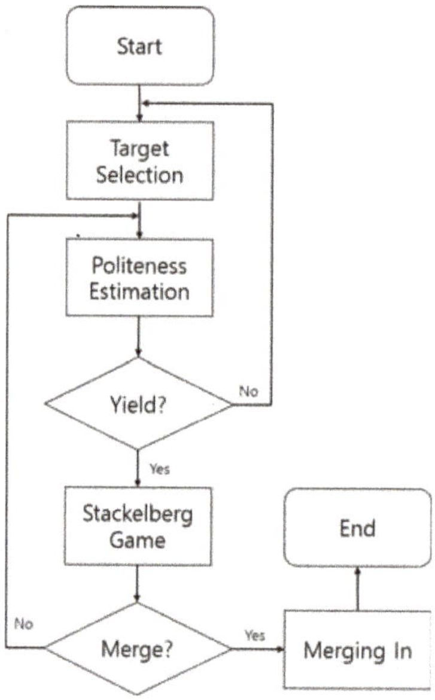

Figure 3. Game process.

If the estimated politeness of the target vehicle is high enough ($P(t) > P^u_{th}$), it is considered as "Yield" intention and the optimal action of the ego-car ($\gamma^{l,*}$) is "Lane Change." In this case, based on the Stackelberg game approach, the ego-car tries to change the lane. By contrast, even though $P(t) < P^u_{th}$ is given, if $\gamma^{l,*}$ is determined as "Maintain," the ego-car waits for the next time step and repeats the same procedure of the "Politeness Estimation" (Figure 3). In addition, when the target vehicle ignores the signal from the ego-car for some reason, the latter finishes the interaction with the target vehicle and starts the game again with the other target vehicle. If the ego-car fails lane merging by interacting with all the vehicles in the next lane, lane merging is possible when the traffic condition in the next lane is relaxed rather than based on the game theoretic approach.

5. Case Studies

In order to verify the effectiveness of the proposed approach, we conduct case studies for various test scenarios and compare our approach with the conventional rule-based decision making. The simulations are performed on Matlab R2020a platform under desktop specification (Intel i5-9500 CPU, Ram 16GB, Windows 10) and no computational difficulty is found. Figure 4 and Table 2 visualize and give the initial vehicle state conditions. There are four vehicles in the next lane. Car 3 and Car 4 are the interacting vehicles and all the vehicles move slowly due to the dense traffic. Therefore, space for the ego-car to merge into the next lane is not sufficient. In addition, we assume that all vehicles can observe the state of other vehicles, i.e., state tuple $S_t^{'}$ is available for all vehicles. Although the perception part is one of the major components in the autonomous driving technology, we exclude it from the scope of our study and focus on the decision and control parts.

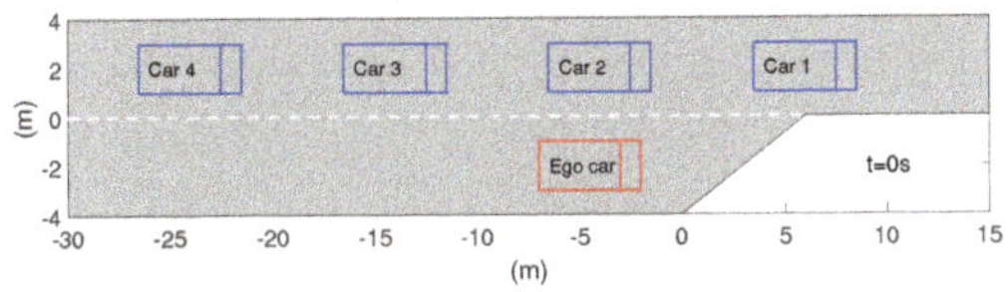

Figure 4. Initial condition for the case studies.

Table 2. Initial conditions for simulations.

	x_0 (m)	y_0 (m)	v_0 (m^2)
Car 1	6	2	2.5
Car 2	−4	2	2.5
Car 3	−14	2	2.5
Car 4	−24	2	2.5
Ego car	−4.5	−2	0

5.1. Test Environment Setup

We consider three scenarios by assigning the different politeness values to the four vehicles in the next lane. For a fair comparison, the same initial conditions are used for all scenarios. The differently assigned politeness values mean that the interacting vehicles can interpret the same traffic condition differently. For example, even for the same condition, the reactions of the interacting vehicles to the action of the ego-car are different, which is true in reality.

It is worth noting that instead of setting the extreme values (0 and 1), we consider the politeness of 0.1 or 0.9 for all vehicles. It gives room to act against unexpected situations like jaywalking. For example, even though the politeness of the interacting vehicle is 0.1, the vehicle has a 10% chance to behave cautiously in an emergency situation.

5.2. Rule-Based Lane Merging

In this subsection, the rule-based lane-merging approach is introduced, and the decision-making performance is compared to that of our approach in the next subsection. Rule-based lane-merging is a quite conservative decision-making strategy since it prefers to obey the traffic rules rather than interacting with the vehicles. For example, only the physical properties such as the relative distance and velocity between the vehicles are considered when making the decision [16]. From the perspective of the ego-car using the rule-based lane-merging approach, the surrounding vehicles in the next lane are considered as the moving obstacles, and the gap between the obstacles seems too tight to attempt a cut-in.

5.3. Case Studies

The results of the case studies for different scenarios are shown in Figures 5–7, where the snapshots are visualized for every 5 s during the entire simulation time (15 s). In Scenario 1, since the politeness of Car 3 is relatively high, we reckon that Car 3 is likely to allow the lane-changing of the ego-car as shown in Figure 5. As expected, starting from the initial traffic state in Figure 4, the ego-car successfully merges into the next lane around

$t = 5$ s. In contrast, Car 3 in Scenario 2 (Figure 6) ignores the ego-car's lane-merging signal due to its low politeness. Instead, the ego-car interacts with Car 4 whose politeness is high, and the ego-car can change lanes around $t = 10\ s$ as illustrated in Figure 6.

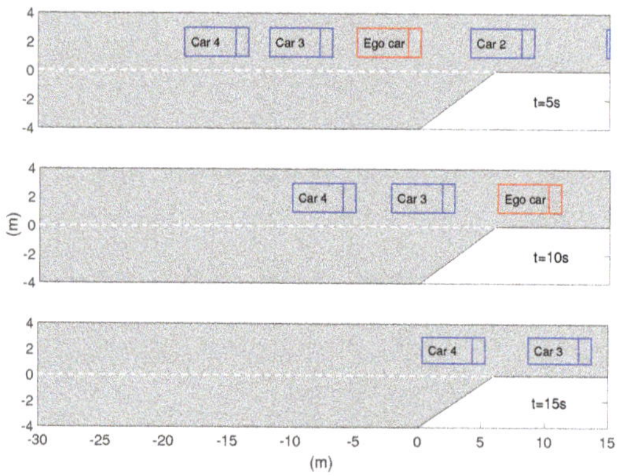

Figure 5. Scenario 1: $p_i = \{0.9, 0.1, 0.9, 0.9\}$.

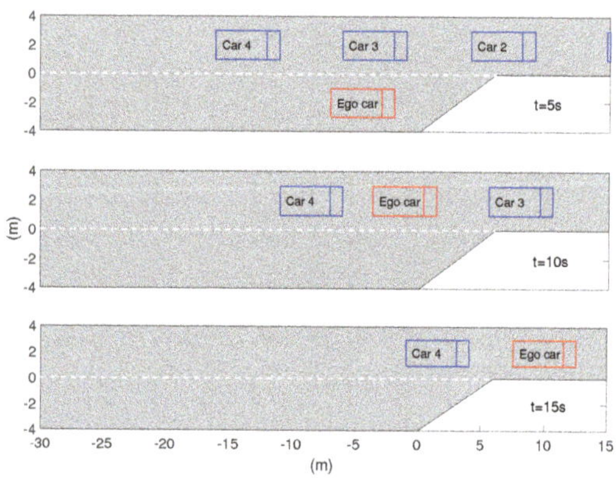

Figure 6. Scenario 2: $p_i = \{0.1, 0.9, 0.1, 0.9\}$.

In Scenario 3 (Figure 7), the ego-car fails to merge into the next lane when all interacting vehicles have (Car 3 and Car 4) aggressive drivers. In this case, lane merging is only possible once the traffic condition is relaxed ($t = 15$ s). The production vehicle may not want to take a risk when they face these conflicts. Thus, it is most likely that the production AV in the side lane may behave as the ego-car in Figure 7, which is not very efficient.

Next, we estimate politeness values for all interacting vehicles using (12). The corresponding plots are shown in Figure 8. We make a neutral guess by assigning the initial politeness $P_0 = 0.5$, which is tunable. As can be seen in Figure 8a, the ego-car changes lanes successfully when the estimated politeness is larger than the upper threshold, i.e., $P(t) > P_{th}^{u}$, and the optimal action $\gamma^{l,*}$ is calculated as "Lane change." The estimation of the Car 4's politeness is not performed since the ego-car does not interact with it after completing the mission (i.e., Lane-merging).

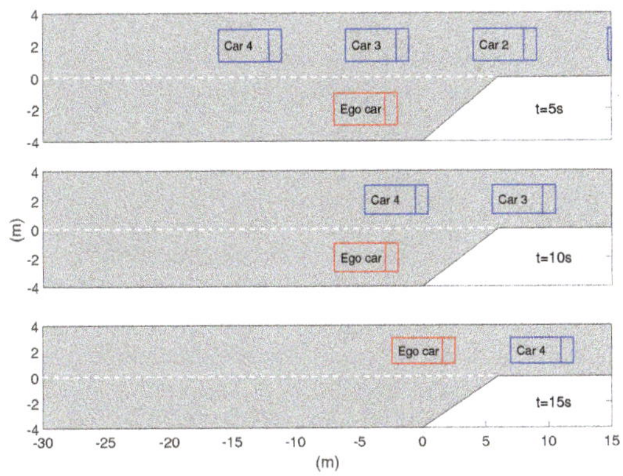

Figure 7. Scenario 3: $p_i = \{0.9, 0.1, 0.1, 0.1\}$.

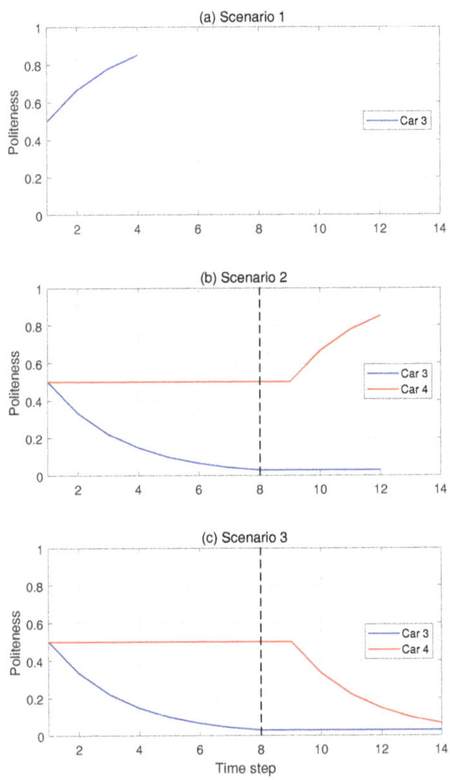

Figure 8. Estimated politeness of target vehicles for each scenario.

Figure 8b shows the estimated politeness for Car 3 and Car 4. For the first few steps, the ego-car interacts with Car 3, and the estimated politeness of Car 3 decreases as the steps progress. Once the $P(t)$ of Car 3 reaches the lower bound P^l_{th}, the ego-car gives up lane-merging attempts and changes the target vehicle. Around 8 s, the target vehicle is changed from Car 3 to Car 4, and the ego-car repeats the same procedure. Similar to Figure 8a, the estimated politeness of Car 4 increases, and the ego-car attempts to change lanes when the predicted action of the follower is "Yield," i.e., $P(t) > P_{th}^{u}$. The vertical dotted line indicates the moment of time when the target vehicle is changed.

The estimated politeness for Scenario 3 is illustrated in Figure 8c. Similar to the interaction with Car 3 in Figure 8b, both estimated politeness values decrease as the ego-car interacts with Car 3 and Car 4. Therefore, the ego-car fails to change lanes (Figure 7). Thus, we confirm that the estimated politeness values correspond to those from Table 3, even though the ego-car does not know the decision-making strategy of the interacting vehicle exactly, which is in fact based on the modified IDM.

For a fair comparison, the rule-based lane-merging strategy that is only based on the relative distance is implemented in our scenarios with the same initial condition. In this case, the vehicles in the next lane always ignore the lane-changing signals from the ego-car and move based on the conventional IDM. The snapshot of this case is omitted since it does not differ from that of Scenario 3 in Figure 7 regardless of the interacting vehicle's politeness. Instead, as shown in Figure 9, we illustrate the relative distance between the ego-car and vehicle in the next lane.

Table 3. Assigned politeness.

	Scenario 1	Scenario 2	Scenario 3
Car 1	0.9	0.1	0.9
Car 2	0.1	0.9	0.1
Car 3	0.9	0.1	0.1
Car 4	0.9	0.9	0.1

Using the rule-based approach, the ego-car calculates the distance to each interacting car and compares it with the threshold (dashed line). In Figure 9, $x_t^{ego} - x_{t+1}^{i+1}$ is the distance between the ego-car and the vehicle behind the ego-car in the next lane (i.e., $(i+1)^{th}$ vehicle). As the $(i+1)^{th}$ vehicle approaches the ego-car, the relative distance decreases until the $(i+1)^{th}$ vehicle passes the ego-car. Thus, the ego-car calculates the predicted relative distance one step ahead. In contrast, the relative distance between the ego-car and the vehicle in front of the ego-car in the next lane ($(i)^{th}$ vehicle) described by $x_t^i - x_t^{ego}$ increases since the vehicle ahead is moving away.

Figure 9. Calculated distance for rule-based approach.

$x_t^{ego} - x_{t+1}^{i+1}$ and $x_t^i - x_t^{ego}$ changes drastically at the 8-th time step. It corresponds to the $(i+1)^{th}$ vehicle approaching and going beyond the ego-car at this moment. After this step, the $(i+1)^{th}$ vehicle becomes the ith vehicle, since the ego-car fails to change the lane.

At the same time, the ego-car changes the $(i+2)^{th}$ vehicle to a new interacting vehicle $((i+1)^{th})$.

We assume the safety threshold of 7 m because the vehicle length is 5 m. Therefore, the safe gap between i^{th} vehicle and $(i+1)^{th}$ vehicle is set to 2 m. The ego-car initiates the lane-merging when both relative distances exceed the safety threshold. In this dense traffic, unlike the results of the game theoretic decision-making strategy, there is no chance to merge into the next lane. That is, the ego-car is tuned to behave cautiously so that risky decision making is avoided, which is general in production vehicles.

Based on the results of the case studies, we confirm that the Stackelberg game approach in Figures 5 and 6 is much closer to human decision making in dense traffic compared to the rule-based approach (see Figure 7). Since an AV being too cautious in its decision making may not be preferred by the driver in the AV, human-like decision making should be considered. The AV may share the road with human-driven vehicles until the penetration rate of the AV in the road reaches 100%. Therefore, including vehicle interactions in decision-making algorithm is quite promising. Moreover, it can also be extended to other driving situations with little modifications.

6. Conclusions

This paper presented the lane-merging strategy for a self-driving car in dense traffic using the Stackelberg game approach, which included the driving intention of the surrounding vehicles. By monitoring the speed variations of the interacting vehicle, the self-driving car could estimate its politeness, representing driving intention. Based on the Stackelberg game theory, the decision of the self-driving car is made in such a way as to maximize utility function that is affected by the self-driving car as well as the interacting vehicle. Furthermore, to describe the reasonable behavior of the human driver, we present the modified car-following model that responds to the self-driving car's action. The proposed method is verified through case studies in various driving conditions. Compared to the rule-based lane-merging strategy, the decision made by our approach is much closer to that of the human driver in real-world driving. To extend the proposed method for different driving scenarios, future work will include the generalization of the proposed logic in other situations where the self-driving car frequently interacts with vehicles (e.g., intersection, take over, cut-in, etc). In addition, rather than discrete actions, continuous action will be considered to capture accurate vehicle dynamics.

Author Contributions: K.J., K.H., M.O.; writing—original draft, K.J., K.H. and M.O; writing—Review and editing, K.J., K.H. and M.O. All authors have read and agreed to the published version of the manuscript.

Funding: This work was supported by the National Research Foundation of Korea (NRF) grant funded by the Korean government (MSIT) (No. NRF-2021R1C1C1003464 and NRF-2020K1A3A1A39112277).

Conflicts of Interest: The authors declare no conflict of interest.

References

1. An, H.; Jung, J.I. Decision-making system for lane change using deep reinforcement learning in connected and automated driving. *Electronics* **2019**, *8*, 543. [CrossRef]
2. Dingus, T.A.; Klauer, S.G.; Neale, V.L.; Petersen, A.; Lee, S.E.; Sudweeks, J.; Perez, M.A.; Hankey, J.; Ramsey, D.; Gupta, S.; et al. *The 100-Car Naturalistic Driving Study, Phase II-Results of the 100-Car Field Experiment*; Technical Report; Department of Transportation, National Highway Traffic Safety Administration: Washington, DC, USA, 2006.
3. Yurtsever, E.; Lambert, J.; Carballo, A.; Takeda, K. A survey of autonomous driving: Common practices and emerging technologies. *IEEE Access* **2020**, *8*, 58443–58469. [CrossRef]
4. Li, G.; Cheng, J. Exploring the effects of traffic density on merging behavior. *IEEE Access* **2019**, *7*, 51608–51619. [CrossRef]
5. Kalra, N.; Paddock, S.M. Driving to safety: How many miles of driving would it take to demonstrate autonomous vehicle reliability? *Transp. Res. Part A Policy Pract.* **2016**, *94*, 182–193. [CrossRef]
6. Basar, T.; Olsder, G.J. *Dynamic Noncooperative Game Theory*; SIAM: Philadelphia, PA, USA, 2013; ISBN 978-0-898-714-29-6.
7. Myerson, R.B. *Game Theory*; Harvard University Press: Cambridge, MA, USA, 2013.
8. Yoo, J.; Langari, R. A Game-Theoretic Model of Human Driving and Application to Discretionary Lane-Changes. *arXiv* **2020**, arXiv:2003.09783.

9. Sadigh, D.; Sastry, S.; Seshia, S.A.; Dragan, A.D. *Planning for Autonomous Cars that Leverage Effects on Human Actions*; Robotics: Science and Systems: Ann Arbor, MI, USA, 2016; Volume 2.
10. Backhaus, S.; Bent, R.; Bono, J.; Lee, R.; Tracey, B.; Wolpert, D.; Xie, D.; Yildiz, Y. Cyber-physical security: A game theory model of humans interacting over control systems. *IEEE Trans. Smart Grid* **2013**, *4*, 2320–2327. [CrossRef]
11. Lee, R.; Wolpert, D. Game theoretic modeling of pilot behavior during mid-air encounters. In *Decision Making with Imperfect Decision Makers*; Springer: Berlin/Heidelberg, Germany, 2012; pp. 75–111.
12. Camerer, C.F.; Ho, T.H.; Chong, J.K. A cognitive hierarchy model of games. *Q. J. Econ.* **2004**, *119*, 861–898. [CrossRef]
13. Li, N.; Kolmanovsky, I.; Girard, A.; Yildiz, Y. Game theoretic modeling of vehicle interactions at unsignalized intersections and application to autonomous vehicle control. In Proceedings of the 2018 Annual American Control Conference (ACC), Milwaukee, WI, USA, 27–29 June 2018; pp. 3215–3220.
14. Tian, R.; Li, N.; Kolmanovsky, I.; Yildiz, Y.; Girard, A.R. Game-theoretic Modeling of Traffic in Unsignalized Intersection Network for Autonomous Vehicle Control Verification and Validation. *IEEE Trans. Intell. Transp. Syst.* **2020**, 1–16. [CrossRef]
15. Sankar, G.S.; Han, K. Adaptive Robust Game-Theoretic Decision Making Strategy for Autonomous Vehicles in Highway. *IEEE Trans. Veh. Technol.* **2020**, *69*, 14484–14493. [CrossRef]
16. Garzón, M.; Spalanzani, A. Game theoretic decision making for autonomous vehicles' merge manoeuvre in high traffic scenarios. In Proceedings of the 2019 IEEE Intelligent Transportation Systems Conference (ITSC), Auckland, New Zealand, 27–30 October 2019; pp. 3448–3453.
17. Yoo, J.H.; Langari, R. A Stackelberg game theoretic driver model for merging. In Proceedings of the Dynamic Systems and Control Conference. American Society of Mechanical Engineers, Palo Alto, CA, USA, 21–23 October 2013; Volume 56130, p. V002T30A003.
18. Zhang, Q.; Langari, R.; Tseng, H.E.; Filev, D.; Szwabowski, S.; Coskun, S. A Game Theoretic Model Predictive Controller With Aggressiveness Estimation for Mandatory Lane Change. *IEEE Trans. Intell. Veh.* **2019**, *5*, 75–89. [CrossRef]
19. Yu, H.; Tseng, H.E.; Langari, R. A human-like game theory-based controller for automatic lane changing. *Transp. Res. Part C Emerg. Technol.* **2018**, *88*, 140–158. [CrossRef]
20. Li, N.; Yao, Y.; Kolmanovsky, I.; Atkins, E.; Girard, A.R. Game-theoretic modeling of multi-vehicle interactions at uncontrolled intersections. *IEEE Trans. Intell. Transp. Syst.* **2020**, 1–15. [CrossRef]
21. Dong, C.; Dolan, J.M.; Litkouhi, B. Intention estimation for ramp merging control in autonomous driving. In Proceedings of the 2017 IEEE Intelligent Vehicles Symposium (IV), Los Angeles, CA, USA, 11–14 June 2017; pp. 1584–1589.
22. Bae, S.; Saxena, D.; Nakhaei, A.; Choi, C.; Fujimura, K.; Moura, S. Cooperation-aware lane change maneuver in dense traffic based on model predictive control with recurrent neural network. In Proceedings of the 2020 American Control Conference (ACC), Denver, CO, USA, 1–3 July 2020; pp. 1209–1216.
23. Kesting, A.; Treiber, M.; Helbing, D. General lane-changing model MOBIL for car-following models. *Transp. Res. Rec.* **2007**, *1999*, 86–94. [CrossRef]
24. Chandra, R.; Selvaraj, Y.; Brännström, M.; Kianfar, R.; Murgovski, N. Safe autonomous lane changes in dense traffic. In Proceedings of the 2017 IEEE 20th International Conference on Intelligent Transportation Systems (ITSC), Yokohama, Japan, 16–19 October 2017; pp. 1–6.
25. Evestedt, N.; Ward, E.; Folkesson, J.; Axehill, D. Interaction aware trajectory planning for merge scenarios in congested traffic situations. In Proceedings of the 2016 IEEE 19th International Conference on Intelligent Transportation Systems (ITSC), Rio de Janeiro, Brazil, 1–4 November 2016; pp. 465–472.
26. Saxena, D.M.; Bae, S.; Nakhaei, A.; Fujimura, K.; Likhachev, M. Driving in dense traffic with model-free reinforcement learning. In Proceedings of the 2020 IEEE International Conference on Robotics and Automation (ICRA), Paris, France, 31 May–31 August 2020; pp. 5385–5392.
27. Han, K.; Nguyen, T.W.; Nam, K. Battery Energy Management of Autonomous Electric Vehicles Using Computationally Inexpensive Model Predictive Control. *Electronics* **2020**, *9*, 1277. [CrossRef]
28. Zhang, J.; Zhang, Q. Stackelberg game for utility-based cooperative cognitiveradio networks. In Proceedings of the Tenth ACM International Symposium on Mobile ad hoc Networking and Computing, New Orleans, LA, USA, 18–21 May 2009; pp. 23–32.
29. Jin, I.G.; Schürmann, B.; Murray, R.M.; Althoff, M. Risk-aware motion planning for automated vehicle among human-driven cars. In Proceedings of the 2019 American Control Conference (ACC), Philadelphia, PA, USA, 10–12 July 2019; pp. 3987–3993.
30. Treiber, M.; Hennecke, A.; Helbing, D. Congested traffic states in empirical observations and microscopic simulations. *Phys. Rev. E* **2000**, *62*, 1805. [CrossRef] [PubMed]
31. Ayres, T.; Li, L.; Schleuning, D.; Young, D. Preferred time-headway of highway drivers. ITSC 2001. In Proceedings of the 2001 IEEE Intelligent Transportation Systems. (Cat. No. 01TH8585), Oakland, CA, USA, 25–29 August 2001; pp. 826–829.
32. Li, N.; Oyler, D.W.; Zhang, M.; Yildiz, Y.; Kolmanovsky, I.; Girard, A.R. Game theoretic modeling of driver and vehicle interactions for verification and validation of autonomous vehicle control systems. *IEEE Trans. Control. Syst. Technol.* **2017**, *26*, 1782–1797. [CrossRef]
33. Arbis, D.; Dixit, V.V. Game theoretic model for lane changing: Incorporating conflict risks. *Accid. Anal. Prev.* **2019**, *125*, 158–164. [CrossRef] [PubMed]
34. Kerin, R.A.; Varadarajan, P.R.; Peterson, R.A. First-mover advantage: A synthesis, conceptual framework, and research propositions. *J. Mark.* **1992**, *56*, 33–52. [CrossRef]

electronics

Article

smartPlastic: Innovative Touch-Based Human-Vehicle Interface Sensors for the Automotive Industry

Cristiano Alves [1,2,†], Tiago Custódio [1,2,†], Pedro Silva [3,*,†,‡], Jorge Silva [3,†,‡], Carlos Rodrigues [3,‡], Rui Lourenço [3,‡], Rui Pessoa [3,‡], Fernando Moreira [3,‡], Ricardo Marques [3,‡], Gonçalo Tomé [3,‡] and Gabriel Falcao [1,2,†]

1 Instituto de Telecomunicações, 3030-290 Coimbra, Portugal; cristianoalves1717@gmail.com (C.A.); tiagocustodio1@gmail.com (T.C.); gff@co.it.pt (G.F.)
2 Departamento de Engenharia Electrotécnica e de Computadores, Faculdade de Ciências e Tecnologia da Universidade de Coimbra, University of Coimbra, R. Silvio Lima, 3030-290 Coimbra, Portugal
3 CIE Plasfil, 3090-380 Figueira da Foz, Portugal; jorge.silva@plasfil.pt (J.S.); carlos.rodrigues@plasfil.pt (C.R.); rui.lourenco@plasfil.pt (R.L.); rui.pessoa@plasfil.pt (R.P.); fernando.moreira@plasfil.pt (F.M.); ricardo.marques@plasfil.pt (R.M.); goncalo.tome@plasfil.pt (G.T.)
* Correspondence: pedro.silva@plasfil.pt; Tel.: +351-233-401-218
† These authors contributed equally to this work.
‡ Current address: Zona Industrial Da Gala, Lote 6, 3090-380 Figueira da Foz, Portugal.

Citation: Alves, C.; Custódio, T.; Silva, P.; Silva, J.; Rodrigues, C.; Lourenço, R.; Pessoa, R.; Moreira, F.; Marques, R.; Tomé, G.; Falcao, G. smartPlastic: Innovative Touch-Based Human-Vehicle Interface Sensors for the Automotive Industry. *Electronics* **2021**, *10*, 1233. https://doi.org/10.3390/electronics10111233

Academic Editors: Calin Iclodean, Bogdan Ovidiu Varga and Felix Pfister

Received: 29 April 2021
Accepted: 14 May 2021
Published: 22 May 2021

Abstract: Environmental concern regularly leads to the study and improvement of manufacturing processes and the development of new industrial products. The purpose of this work is to optimize the amount of injected plastic and reduce the number of parts used in the production of entrance panels to control features inside the car cabin. It focuses on a particular case study, namely the control of opening and closing windows and rotation of the rear-view mirrors of a car, maintaining all of the functionality and introducing a futuristic and appealing design inline with new autonomous driving vehicles. For this purpose, distinct low-cost touch sensor technologies were evaluated and the performance of several types of sensors that were integrated with plastic polymers of distinct thickness was analyzed. Discrete sensors coupled to the plastic part were tested and integrated in the injected plastic procedure. In the former, sensitivity tests were performed for finding the maximum plastic thickness detectable by the different sensors. For the latter, experiments were carried out on the sensors subject to very high pressure and temperature inside the molds—the two most relevant characteristics of industrial plastic injection in this context—and functional results were observed later. We conclude that, by changing the way the user interacts with the car cabin, the replacement of conventional mechanical buttons—composed of dozens of parts—by a component consisting of a single plastic part that is associated with conventional low-cost electronics allows the control of a more diversified set of features, including many that are not yet usual in the interior of automobiles today, but that will eventually be required in the near future of autonomous driving, in which the user will interact less with driving and more with other people or services around her/him, namely of the multimedia type. Additionally, the economic factor was considered, namely regarding the cost of the new technology as well as its manufacturing, replacement, and subsequent recycling processes.

Keywords: sensors; touch sensor; touch button; capacitive sensor; polymer; injected plastic; plastic button; automotive industry; self-driving car

1. Introduction

The interfaces currently in automotive vehicles mostly depend on mechanical buttons to act on the control of several functionalities inside the car cabin, which include opening and closing door windows, adjusting the position of rear-view mirrors, handling multimedia systems, air conditioning, navigation, etc. [1].

Because they are composed of several moving parts, mechanical buttons support a limited number of uses—or cycles—due to the wear and tear that results from their continuous utilization [2]. Other technological areas, such as multimedia, information technology, telecommunication systems, or even medical devices, have overcome this obstacle over the past two decades, changing the paradigm of the interfaces of their products by adopting technologies that are based on touch sensors [3,4].

A sensor is a device that reacts to stimulus [5]. It is used to measure physical quantities, such as temperature, humidity, radiation, or light, among many other possibilities. In the context of this study, we will always be referring to electrical sensors, whcih is, devices that, in the face of a variation of a given physical quantity, manifest themselves in a variation of an electrical quantity (e.g., voltage, current, impedance). This variation is always the result of a stimulus.

However, a simple sensor is rarely able to effectively measure a quantity, due to limitations that typically show their influence in the form of non-linear behavior [6], the presence of noise, or degradation of performance throughout use over the years. In order to overcome these limitations, it is common to use a complementary electronic signal conditioning system in association with the sensor, such as amplifiers, filters, or even microcontrollers. By using microcontrollers, we gain the ability to process, store, and communicate data between different devices. One particular disadvantage lies in the need to implement additional electronics and the corresponding increase in production costs.

There are several touch sensor technologies available today, which provide different properties and costs:

- inductive sensors;
- infrared sensors;
- ultrasound sensors;
- resistive sensors; and,
- capacitive sensors.

Different sensors show distinct advantages and disadvantages. The analysis of each sensor technologies is performed in the context of this article. The tests and analysis conducted concluded on the appropriateness of coupling capacitive touch sensors [7,8] with a piece of injected plastic several millimeters thick, to serve as an interface [9] for controlling the opening and closing of door windows, as well as for positioning rear-view mirrors from inside a vehicle's cabin. We used a microcontroller-based system to acquire the touch signals and execute the control algorithm. We have found that this approach adds value to the product, as it is programmable and, thus, presents the necessary flexibility to adapt to different models of the same car range, as well as other vehicles in new contexts and applications. Because it is programmable, it can also be updated by reprogramming the car's control software (without hardware intervention) in order to integrate new context-dependent interactions, such as controlling the multimedia system, the temperature of the car, air conditioning, navigation, the driving inherent functionalities of the car, or even to accommodate future requirements of the vehicle [10] that are unforeseen at this stage of development. The integration of these systems with self-driving cars where disruptive technologies and functionalities may have to be integrated soon us ine aspect that deserves particular attention. Implementing these new functions as simply as possible can make more sense with touch-based sensors that are placed in unusual positions inside the cabin, e.g., on the steering wheel, roof, bank seat, etc. [11]. A secondary aspect that assumes increasing importance and, thus, requires even more attention from the automotive manufacturers regards the continuous integration of the smartphone with the car cabin [12]. Additionally, in this context, the integration of touch-based sensors can favor this approach.

Moreover, the association of this technology with injected plastic molding allows for rethinking the design of components for the car cabin control panels as both decorative and functional elements—now designed together and not as separate parts—that are geometrically aligned and combined to produce a more compact desired functionality and visual effect [13].

This article presents the following contributions

- the analysis and comparative study of touch sensors for the automotive industry;

- study on the incorporation and functionality of capacitive sensors in injected plastic;
- the incorporation of microelectronics to control capacitive sensors;
- development and validation of an experimental case-study to control the opening and closing of car door windows, and the position of electric rear-view mirrors; and,
- production, implementation, and testing of the prototype developed in an industrial context, on a real car manufactured by an international group.

This work is also part of the 'Collective Efficiency Strategies', being totally aligned with the 'Mobinov—Automobile Cluster Association', which identified "to contribute to making Portugal a reference in research, innovation, design, development, manufacture and testing of products and services of the automotive industry" and "strengthen the competitiveness of a fundamental sector of the economy, promoting an increase in exportation" as some of its main goals [14].

This article is structured, as follows. Section 2 describes the materials and methods used, analyzing, in depth, the available touch sensor technologies that are available and the criteria used to select, among these technologies, the most suitable for the current context in the automotive industry. Section 3 analyzes the approaches and methods used to integrate injectable plastic with touch sensors and which polymers are the most suitable for this purpose. The experimental results are discussed in Sections 4 and 5 closes the article.

2. Materials and Research Methods

A sensor device responds to a change in the surrounding physical environment, which translates to a variation of an electrical charge. The present work addresses a specific family of sensors in view of their use when integrated with injected plastic parts to control the functionalities in the car cabin—touch sensors.

2.1. Touch Sensors

Touch sensors allow the detection of human touch or the proximity of an object for human interaction. Its adoption is expanding in an increasing range of devices, products, and technologies, as it replaces buttons and mechanical switches, as well as potentiometers and encoders. Such levels of replacement imply several benefits, such as increased equipment durability and reliability when compared to devices made up of a set of mechanical parts. The appearance of touch-based devices can also be more appealing and attractive to the end user.

Its operation is based on the principle of a normal mechanical button. A flow of electrical current is generated in the sensor whenever it is touched (no pressure or movement of mechanical parts is required). The current flow is interrupted when the sensor is no longer touched. The automatic detection of this current variation allows the system to be controlled to act accordingly (e.g., opening or closing door windows).

There are several types of touch sensors, namely inductive, infrared, ultrasound, resistive, or capacitive ones. They all present advantages and disadvantages, which are analyzed below.

Inductive sensors need to be excited at both the sensor and touched object (for example, see the case of RFID cards [15]). Ultrasonic sensors [16] are large in size, and they require additional electronics and processing on the receiver end. In the same way, infrared sensors have limitations between the emitter and receiver that are essentially related to distance and the need for additional electronics. Resistive sensors are generally cheaper and allow activation with any object that precedes the sensor, however they are not very accurate. A capacitive sensor consists of only one conductive electrode [17], which, in the presence of a contiguous object, alters its electrical characteristics, namely the capacitance. This change can be accurately measured. When considering that the touch is to be performed by a human finger, and while taking the analysis of important criteria imposed by industry, such as cost, precision, sensitivity, and dimension into account, the sensors that assume greater prominence and relevance in view of the intended use are of the resistive or capacitive type.

The next subsections provide detailed analysis of each one of these sensors, with a view to the selection of the most appropriate device for the present automotive context.

2.1.1. Inductive Sensors

Inductive sensors are widely used to validate touch devices. Good examples include the metro ticket, house keys, car keys, and many other beacons that we use on a daily basis. These types of sensors are not ideal for applications that are activated by human touch, as their operating principle resides in the specific resonance frequency that is created through an electronic circuit by operating a coil with an oscillator. Faraday's law explains the sensor's working principle:

$$e = -N\frac{d\Phi}{dt}, \tag{1}$$

where e represents the electromotive force, N is the number of turns of wire, Φ is the magnetic flux, and t is the time constant. A sensor is composed by several components:

By observing the diagram presented in Figure 1, we conclude that the use of inductive sensors for the present application is not suitable for several reasons, namely: cost, need to incorporate additional electronics, complexity, and the presence of magnetic fields is not viable with the interaction of the human body.

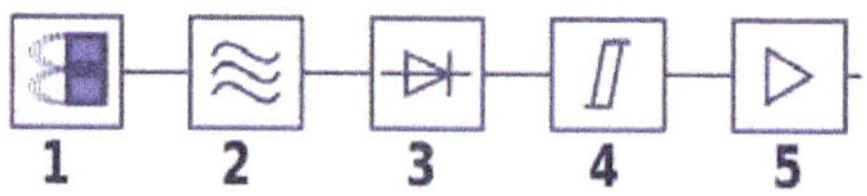

Figure 1. Block diagram of an inductive sensor consisting of 1–Sensor Field; 2–electronic oscillator; 3–Demodulator; 4–Flip-Flop; 5–Output [18].

2.1.2. Infrared Sensors

Infrared screens have light-emitting diodes, as well as receivers parallel to these diodes, along the frame of the entire panel. This way, the beams are interrupted whenever a human finger touches the screen (see Figure 2). We can perceive the coordinates of the touch by knowing which specific beams are interrupted.

There are numerous advantages associated with this type of panel. The smooth recognition of touch and gestures, as well as short response times, are two crucial properties in this type of technology. It is usually applied in large screens, allowing, for example, outdoor usage. There are also disadvantages with the use of this technology, such as, for example, the greater probability of false activations. Whenever a foreign body comes into contact with the interface, the infrared beam is blocked and an activation occurs [19]. Another disadvantage lies in cost, given that it is significantly more expensive than other devices offering a similar purpose.

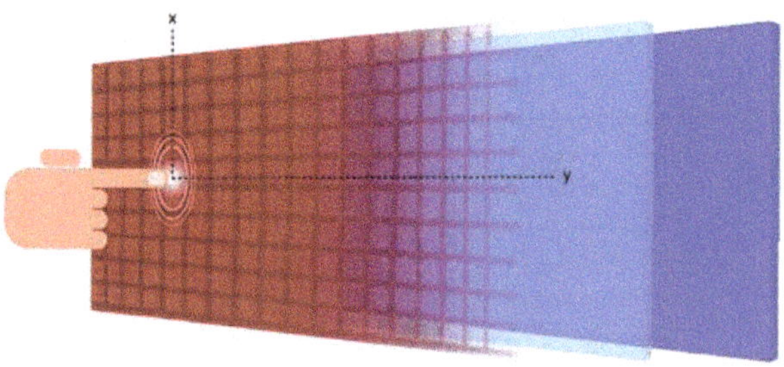

Figure 2. Panel with infrared technology.

2.1.3. Ultrasound Sensors

Ultrasound sensors detect objects by measuring the travel time of an acoustic signal that is emitted by it in order to discern whether there is an object in front of the sensor. The most well known application is the parking sensor array in the back of vehicles.

Ultrasound touch panels are currently under development, with the intention that it will be possible to accurately determine the touch as well as the magnitude of the touch. This technology can also be used to detect a fingerprint on a panel, as it allows for a scan of the object that performs the touch. This technology has already been applied to LCDs that analyze users' fingerprints through the screen. It is often referred to as a technology that allows the 3D detection of a touch, because, in addition to the touch coordinate, it also allows obtaining depth information.

Through intense miniaturization, it is possible to embed an ultrasound transducer in a silicon chip, making it possible to produce sensors as small as 3 mm^2 [20], which, if arranged in a grid pattern, become suitable for detecting touch and discern gestures. In the realm of innovation, this technology enables the use of uncommon materials for the touch interface, such as glass and metal. Because it remains a new technology, it is still expensive and inaccessible for experimentation.

2.1.4. Resistive Sensors

Resistive sensors detect the pressure on the surface and, thus, obtain better performance when it comes to the use of pens, gloves, or devices to perform the touch.

These sensors are made with two layers of a conductive material (or non-conductive, covered with a conductive film), which are separated by small spaces. The instant the body presses the surface of the sensor, the upper layer flexes and comes into electrical contact with the bottom layer, creating a potential difference. The contact point of the electrodes present in the layers is automatically obtained based on X-Y coordinates due to the particular voltage drop generated.

There are different types of resistive touch sensors:

- 4 wires;
- 5 wires; and,
- 8 wires.

The 4 and 5 wire resistive touch panels are similar, with the electrodes on the bottom layer, the former being less expensive, and the latter more resistant. It should be noted that the 4-wire panel also allows for measuring the pressure of the touch against its surface. The lower the resistance measured at the touch, the greater the pressure exerted [21]. The 8-wire touch panels represent higher complexity, as they provide greater measurement meridians along the panel.

Figure 3 illustrates the functioning of a resistive panel:

Figure 4 illustrates a resistive panel, where two electrodes are visible on each surface arranged perpendicularly, forming a matrix. The electrodes are connected by four wires to the display controller.

During standby position, **X+** is grounded, while **X+** and **Y+** are both at high impedance states, and **Y-** is a pull up. The measurement of a 4-wire resistive panel is performed in two stages, the first for one coordinate, then the other. For measuring the **X** coordinate, **X+** is connected to GND, **X-** to VCC, **Y+** is high impedance and **Y-** connected to the ADC. For measuring the **Y** coordinate, **X+** is high impedance, **X-** connected to the ADC, **Y+** to GND and **Y-** to VCC. The (x, y) position is then given by the following equations:

$$x = \frac{V_{Y+}}{V_{Vcc}} \times Length_{Screen} \quad (2)$$

$$y = \frac{V_{X+}}{V_{Vcc}} \times High_{Screen} \quad (3)$$

The resistive type sensors are being progressively replaced by capacitive sensors, as it is necessary to exert a higher pressure to produce the expected feat, thus showing less sensitivity. However, they continue to be used in devices subject to harsh environments, as they are typically more resistant.

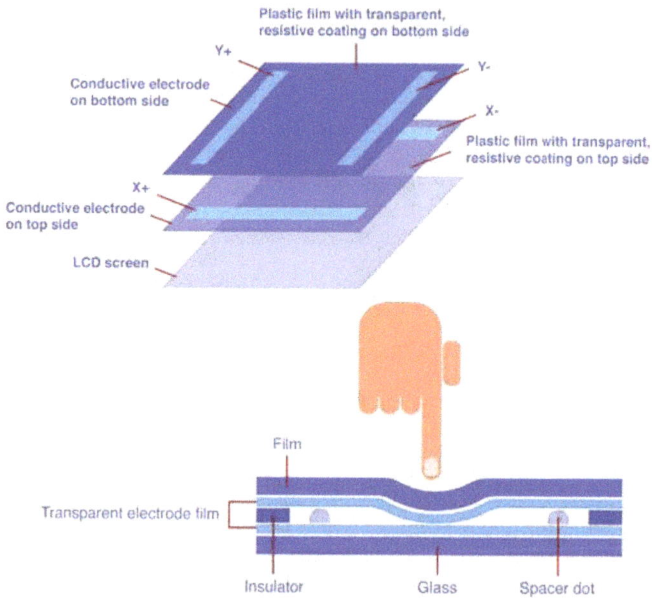

Figure 3. Model representing a resistive panel [22].

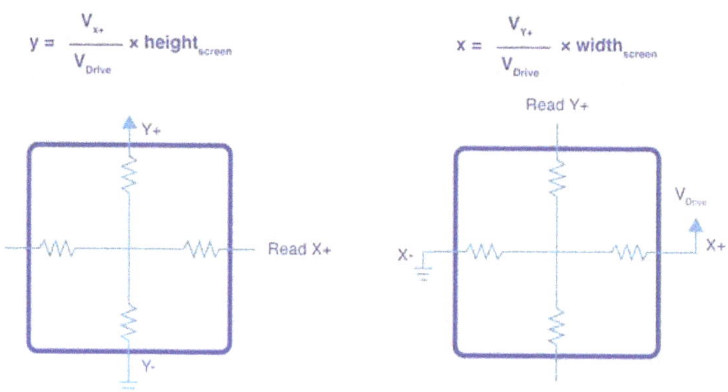

Figure 4. Electrical models of a resistive sensor [23].

2.1.5. Capacitive Sensors

Capacitive sensors are currently widely used in portable devices, such as phones, tablets, and computers, as well as for a wide variety of home appliances and other electronic devices. There are many advantages, including: durability, robustness, and the possibility of building electronic devices with a more attractive design.

The operating principle is analogous to that of a simple capacitor. Two conductive plates are separated by an insulating material where the C capacitance translates to:

$$C = \varepsilon_0.\varepsilon_r.A/d, \tag{4}$$

where ε_0 represents free space permittivity, ε_r is the relative permittivity or dielectric constant, A the capacitive plate area, and d the distance in between.

By analyzing Equation (4), it can be seen that, in the installed system ε_0, ε_r and A are constant. Thus, the capacitance changes are inversely proportional to the distance between the plates as depicted in Figure 5.

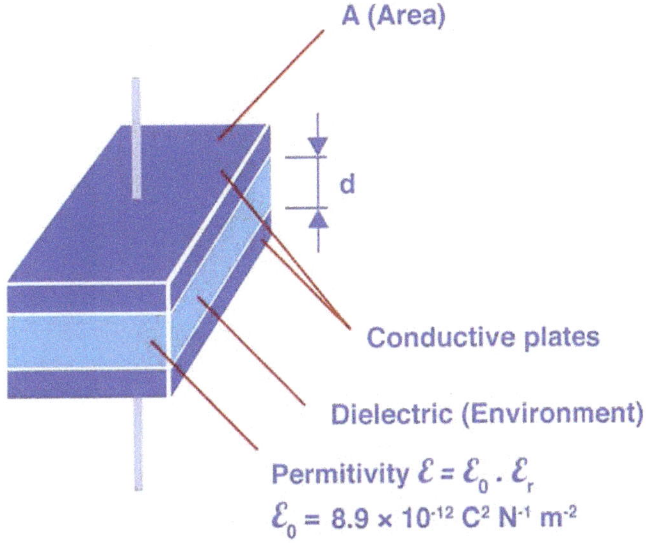

Figure 5. The model representing a capacitive sensor [24].

In a capacitive sensor, the sensing electrode is one of the plates. The second electrode (second plate) consists of two factors: the first is C_0, referring to a parasitic capacitance of the surrounding environment and the other factor is the human finger, which, being conductive, creates a capacitance C_T (please see Figure 6). This way, C_0 and C_T form a parallel, increasing the total capacitance (capacitive devices in parallel, add up their effect). The inspection of the model in Figure 6 shows that the capacitance is lower in the absence of a finger, and higher in the presence of the human part. Additionally, in the presence of the finger, the total capacitance is the sum of the capacitance C_0 and C_T.

The ε_r of plastic materials, which varies between 0.95 and 1.00, is an important aspect for further validation of the applicability of this technology [25]. These parameters are important, as they affect the sensor calibration. Subsequently, the sensor electrode is connected to an electronic circuit that periodically reads its capacity.

There are several analog measurement methods for detecting the presence of an object/finger while using a capacitive sensor: resonance frequency change; frequency modulation; amplitude modulation; charge time measurement; and, duty cycle. Most of the methods require intensive analog circuits that inherit problems, like crosstalk, coupling, and noise sensitivity. However, a digital approach, with dedicated integrated circuits, consumes less area and energy when compared to analog solutions.

2.2. *Evaluation of Processes and Materials*

This section discusses the advantages and limitations of the distinct sensing technologies introduced above. In particular, it analyzes and evaluates their differentiating characteristics, allowing for conscious decision making, which is being supported in the best interest of the *SmartPlastic* project.

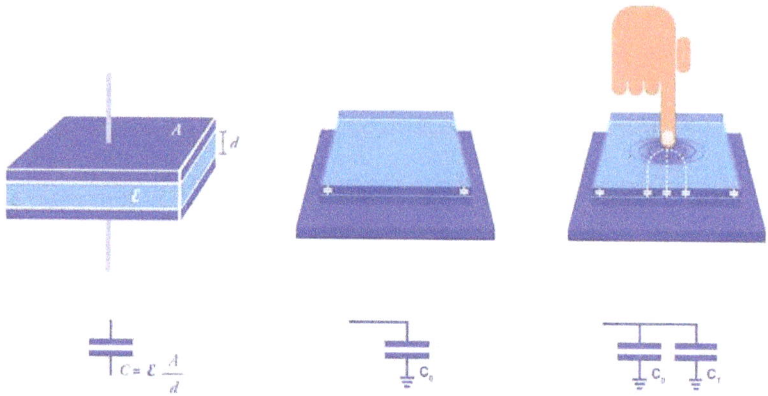

Figure 6. The model representing a capacitive sensor activated by a human finger [26].

2.2.1. Inductive Sensors

Advantages: inductive touch sensors show strong advantages in the following areas:

- Touch and validation: systems where wireless contact and data validation are required. This is one of the great advantages of inductive touch sensors and, thus, these technologies are typically used in ATM cards (i.e., contactless), in tickets and public transportation passes, or in car keys.
- Metal detection: inductive sensors are also widely used in metal detection and their first use was precisely to replace mechanical limit switches in industry. This application is also used for traffic accounting. Often, on roads and car parks, there are inductive loops for automatic vehicle detection.
- Nuclear applications: inductive sensors are also used in nuclear applications, detecting variations in the electromagnetic field of a system that would have previously been stable.

Disadvantages: they show the following negative aspects:

- Additional electronics: they need an exciter or metallic circuit on one side of the system and a resonant circuit on the other.
- Complexity: the implementation of an inductive sensor has several stages, which implies some additional electronics components and the need for proper signal processing conditioning.

2.2.2. Infrared Sensors

Advantages: these sensors are quite popular in a broad range of applications. The following aspects of this technology are highlighted:

- Object detection: the use of infrared sensors is very common and it has been widespread in the area of object detection. The most common example of this application relates to garage gates.
- Large touch screens: the principle used to detect a car at the garage door can be miniaturized to the point of achieving a high density of beams and, thus, can be applied to touch screens. They are primarily used in screens of enormous dimensions, which are generally located outdoors.
- Counting systems: there are systems for counting people or vehicles that use infrared sensors.
- Intrusion alarms: some of these systems also use infrared sensors to detect intruders.

Disadvantages: they present the following weaknesses:

- Dimension: it is necessary that the sensors (as well as the application) occupy a considerable volume to work properly.

- Limitation of the application on screens (I): it is only used in large screens. The incorporation of this technology is justified due to its reach (panels of several square meters).
- Limitation of the application on screens (II): it can only be applied to flat screens.
- Cost: the cost of an infrared sensor for a home gate is very low. However, the manufacture of a panel that uses this technology is much higher, because it uses an array of sensors of this type, thus increasing the number of devices and the construction complexity.

2.2.3. Ultrasound Sensors

Advantages: among the main advantages, we highlight:

- 3D recognition: this type of sensor aims for three-dimensional recognition of touch. It is possible to measure the depth of a touch, and the relief of the object that is touching the sensor.

Disadvantages: the main disadvantages are:

- Cost: the cost per device is high and its application also entails high costs.
- Applicability: for the intended purpose in the context of the current automotive application, the most interesting features that are associated with the use of this type of sensor are not observed.

2.2.4. Resistive Sensors

Advantages: The utilization of resistive sensors in touch panels shows several advantages.

- Robustness: resistive panels are highly resistant and robust, which is why they are widely used in screen panels in the industry.
- Use: they can be touched with objects, such as pens and gloves, which confers them an additional advantage in industrial use, namely regarding the use of personal protective equipment that is often used by operators.
- Cost: they have a lower cost when compared to other technologies.

Disadvantages: however, they also show important disadvantages:

- Sensitivity: touch sensitivity is reduced; therefore, it becomes necessary to exert some force to produce an effect on the screen, which makes the contact a little forced.
- Manufacturing: the manufacture of this type of panels requires meticulous and high precision processes to produce the exact gap between the different layers of the sensor, which makes it difficult to carry out personalized experiments.
- Cost: as mentioned in the previous section, resistive panels exist in different configurations. The most sensitive and accurate ones are in the form of 8 wire resistive panels, and this variant is expensive.
- The cost of implementing ON/OFF buttons: The cost of implementing this technology to produce ON/OFF buttons is high. For the particular case of the present work, it would imply the use of a small panel/screen or a larger panel for controlling several ON/OFF buttons.
- Applicability: bearing in mind that the present objective is to couple electronic sensor components near or inside the plastic after/during overmoulding, this solution would present reduced sensitivity, as it would take more force to execute a command. This issue also adds complexity to the mechanics of fixing the resistive panel.

2.2.5. Capacitive Sensors

Advantages: capacitive sensors have numerous advantages:

- Ease of use and integration: the implementation of a capacitive sensor is relatively less difficult, as the utilization of a conductive material with an appropriate shape suffices. This way, there is a large number of possible shapes and implementations of a capacitive sensor, either through the use of stickers, the positioning of a conductor, or the utilization of a material with a proper shape that transmits mechanical properties, such as elasticity and resistance, which can change an electric field when activated.

- Simple electronics: in view of the existing microchips for performing signal conditioning and capacitance measurement, the use of this technology and the consequent development of new components for human interaction are increasingly expanding.
- Projected capacitance: it represents an evolution of capacitive screens that permit multi-touch and superior sensitivity. This technology consists of the manufacture of a panel in the middle, of which is a layer of any dielectric, and then on its two faces it has conductive surfaces with a matrix design (X, Y, or other personalized shape). These matrix conductors are constantly monitored by an electronic circuit that is sensitive to the change in the capacitance of a given coordinate (X, Y) on the panel.
- Cost: this technology leads to the simplest and cheapest way of implementing ON/OFF buttons.

Disadvantages: the disadvantages of capacitive sensors are mainly related to the difficulty of detecting contact under certain circumstances:

- Glove contacts: procedures of this type affect the capacitive panels/buttons, since gloves change sensitivity.
- False triggers: the triggering of this type of sensors originated from external issues that do not represent a human finger (for example, a drop of water) is undesirable, but possible to mitigate. To minimize this problem, filters and the processing of dedicated algorithms are typically used, producing good results.

2.2.6. Comparison of Touch Sensor Technologies

The following table indicates the main characteristics that are listed by different technologies. All of the evaluations were conducted in relation to the objectives of the SmartPlastic project.

The distinction between the "On/Off Button" and "Panel" columns in the table refers to a simple two-state button and a flat 2D panel with X/Y dimensions. It allows a set of functionalities that vary with the precision of the panel.

Project applicability assesses the suitability of this sensor for the project's objectives. Cost represents the general assessment of the cost of technology, including all of the electronics attached. Robustness evaluates the robustness of a sensor that was developed using these technologies. Integration evaluates the ease with which this technology can be integrated in different media/materials. The requirements to be met must meet dimension, additional electronics, the need for microcontrollers and programming, among others. The inspection of Table 1 seems to indicate that a large set of properties can be more easily found in sensors of capacitive type.

Table 1. Summary of the characteristics of the different technologies covered and investigated under the context of project *SmartPlastic*.

Sensor	On/Off	Panel	Robustness	Integration (ON-OFF/Panel)	Cost (ON-OFF/Panel)	Proj. Applicabil. (ON-OFF/Panel)
Induct.	Yes	No	Good	Good/ ×	High – ×	–/×
Infra.	Yes	Yes	Good	Good/Bad	Medium – High	–/–
Ultra.	Yes	Yes	Good	Bad/Bad	High – High	–/–
Resist.	Yes	Yes	V. Good	Good /V. Good	High – Low	+/–
Capacit.	Yes	Yes	V. Good	V. Good /V. Good	Low – Medium	+/+

2.3. *Integration of Sensors with Microcontrollers*

For the sake of flexibility and scalability in future production, in the development of the prototype we opted for using a microcontroller to perform the reading and aggregation of the signals coming from various touch sensors. There are several methods for connecting switches to a microcontroller and achieving the desired functionality.

2.3.1. Button Directly Connected to the Microcontroller

This is the most direct method, but it is rarely used, since each button uses a digital microcontroller pin, which are limited in number. The signal that is used in this type of assembly is usually digital, and the control can be made through a downward or upward slope (denied). For this, the characteristics of the microcontroller pins must be considered, and the type of "pull" to be inserted in the assembly—pull-up or pull-down, i.e., the device pulls current or supplies current—should match accordingly (see Figure 7). It should be noted that, for the user, there is no difference between the two approaches. They both cause current to flow whenever an event occurs and the process is triggered from there.

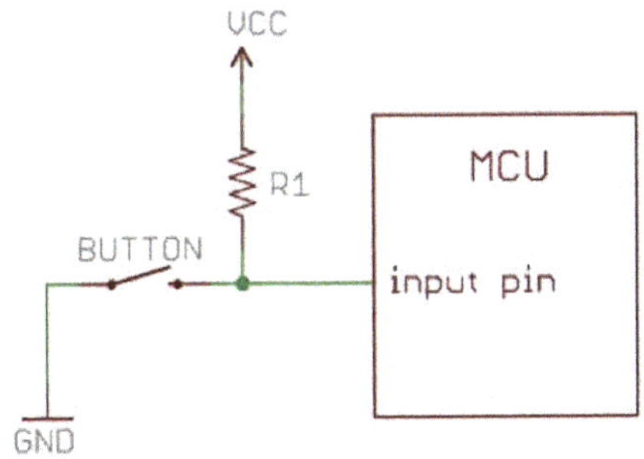

Figure 7. Schematic of connecting a button to a microcontroller with a *pull-up* resistance, that is, normally *High* when not pressed. If the button is pressed, GND forces the logic level '0' (zero) at the input of the microcontroller.

2.3.2. Variable Voltage Drop

This method allows several buttons to be connected to a single pin of the microcontroller, which specifically has to be an analog-digital peripheral (ADC) that converts analog voltage levels into sequences of digital symbols (i.e., sequences of '1's and '0's) .

The inspection of Figure 8 shows that, without any button pressed, the voltage that is read by the ADC is equivalent to VCC. However, by pressing a button, the value read is different. For example, for VCC = 5 V:

- pressing SW1 reads 0 V;
- pressing SW2 reads VCC − (VCC/2) = 2.5 V;
- pressing SW3 reads VCC − (VCC/3) = 3.33 V;
- pressing SW4 reads VCC − (VCC/4) = 3.75 V; and,
- pressing SW5 reads VCC − (VCC/5) = 4 V.

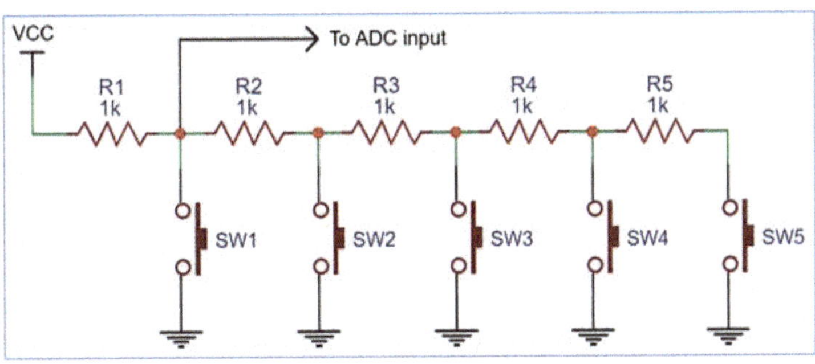

Figure 8. The schematic of several buttons using a circuit with variable resistance to be connected to a single pin of the microcontroller, which, in turn, connects to an analog-to-digital converter (ADC) to sense the voltage measured at the input and convert it to a digital value.

The analysis performed to the example circuit shown in Figure 8 can be used a different number of buttons and other specific resistances. This is the approach employed in the car window lift control buttons provided for analysis and described later in the text.

2.3.3. Matrix with Row and Column Control

This approach is quite common in keyboards and, usually, in panels with many buttons. This technique more efficiently uses the available pins on the microcontroller chip. It allows for a greater number of buttons in relation to the number of occupied pins.

For the matrix topology, the number of buttons relates with the number of available pins on the following way:

- 4 buttons → 2 rows & 2 columns = 4 pins
- 8 buttons → 3 rows & 3 columns = 6 pins
- 16 buttons → 4 rows & 4 columns = 8 pins
- 32 buttons → 8 rows & 8 columns = 16 pins

This implementation requires some automation at the software level and some details at the hardware level. The use of this technique requires scanning rows and columns. In other words, the methodology requires that the columns (in Figure 9 designated by letters) are configured as outputs and the rows (in the figure designated by numbers) configured as inputs. The columns are all set to HIGH ('1'), which is, all of the columns are connected to pins that are at a higher voltage and, one-by-one, scanned at LOW. At the same time, the scan of the rows takes place, and each pin (of the columns) is checked, in order to verify whether any column has the associated LOW value. This way, the pressed column at LOW state is known, together with the corresponding Y row, thus allowing to infer the coordinate of the button being pressed as illustrated in Figure 10.

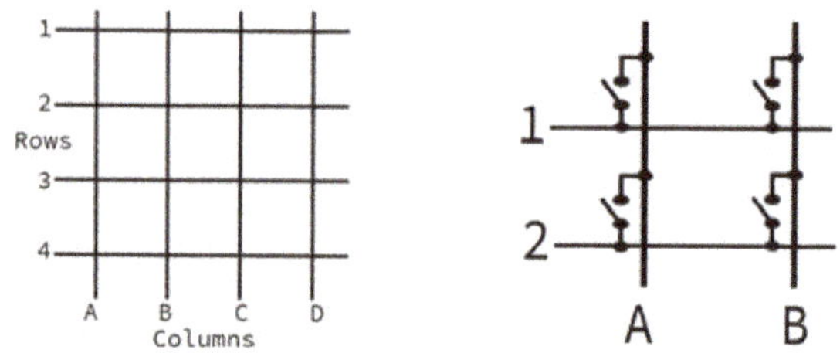

(**a**) (**b**)

Figure 9. Matrix linked buttons. An array of rows and columns is depicted in (**a**). In (**b**), the switches are also represented to illustrate how all possible connections indicate a unique combination.

2.4. Communication Protocol with the Engine Control Unit (ECU)

In this study, a real automobile that was manufactured in 2018 is addressed. The operating mode and electrical functionality of door windows and rear-view mirrors are equally referenced under the context of this work.

2.4.1. Multiplex Door Panels Control System

The functionality of this system consists of the actuation of four window panes with associated functions (descend, ascend, full descend, and full ascend), as well as a button to lock their actuation. There is no substantial amount of technical information available regarding the electrical properties of the system. These are regular buttons developed with polymers, electronic components, and metallic contacts. Nevertheless, there is a wide range of options for this functionality of a car door when looking at other similar products from the same manufacturer. Regarding communications, there are models with CAN, LIN, and "K interface" communication capabilities [27,28]. In terms of electrical characteristics, the supply voltages vary between 5 and 12 V, depending on the model. In the specific case of the vehicle that is used for investigation and development, the analysis of all components to be reproduced observes that the parts have more than one button. The simple command for a window pane is formed by four buttons: one button to raise the window, another to lower it, or to fully raise and fully lower it. Figure 11 depicts such a command obtained from the tested vehicle and the corresponding electrical circuit.

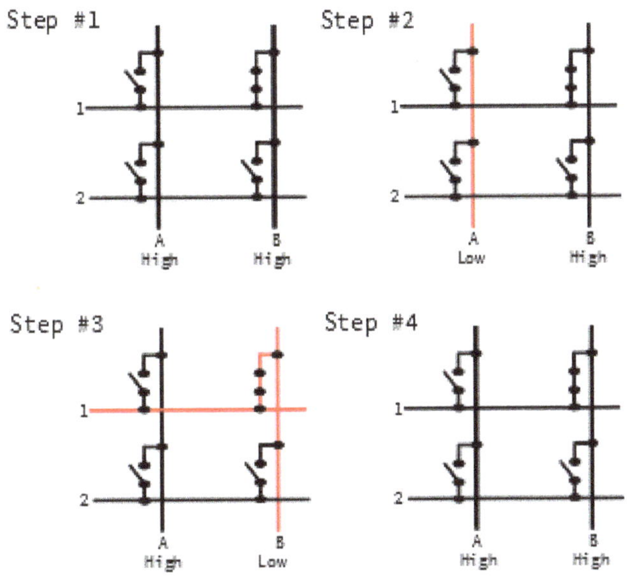

Figure 10. Illustration of the scanning and reading algorithm performed to the buttons that are connected in the matrix.

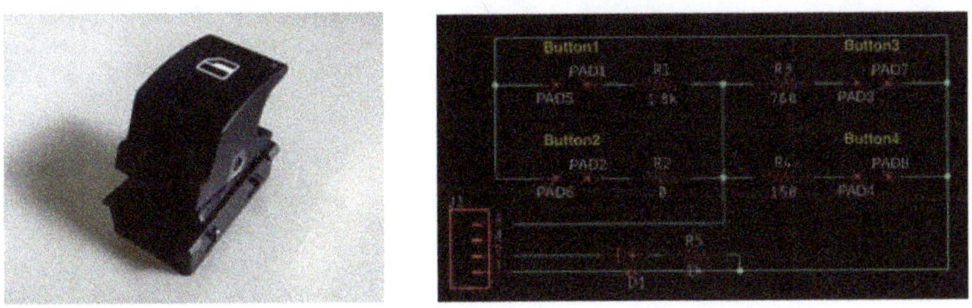

(a) (b)

Figure 11. Passenger window pane elevator button in (**a**). The schematic of the electric button is shown in (**b**).

2.4.2. Disassembling the Communication Protocol between Touch Sensor and ECU

In Figure 12, we can observe the operation of a car's electronic system.

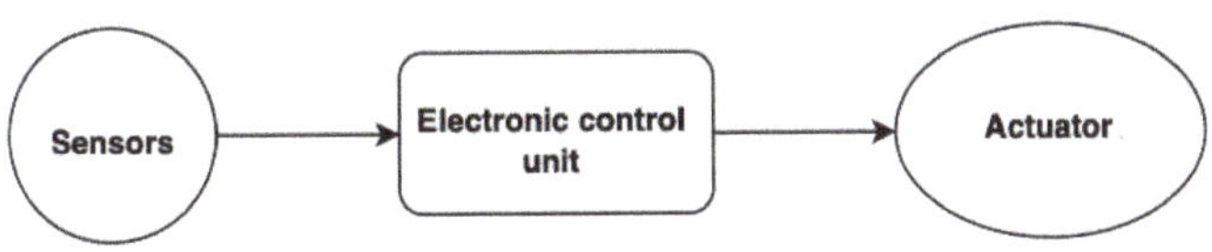

Figure 12. Automobile electronic system.

The ECU is the device that controls all of the electronic components and many mechanical components of an automobile [29].

For the sensors to be read, the control unit has its own communication and protocol capabilities, as well as ADC and DAC ports. The system receives information from different points of the car through the respective sensors and it performs the necessary actions after processing that information. The operations range from raising or lowering the car door window, to injecting more or less fuel into the engine. For this purpose, the control unit also has a processing unit, memory, and internal communication ports. By observing this system, it is thus possible to perceive and extrapolate to the real case, when considering that a car today has several control units, each one performing specific tasks.

In the specific case under investigation, an experimental analysis was carried out by measuring the electrical signals between the door and the control unit under test [30], thus obtaining a detailed pattern of behavior in relation to the observed signals (see Figure 13). The signals do not vary only when the user performs a certain action. It was found that there is always a periodic signal that maintains its shape and frequency over time. Through further analysis, the perception is acquired that this signal is used in the automotive industry to perform diagnoses of the condition of the vehicle and its different parts [31]. The analysis of this paper focuses on the signals measured during the actions of "going up", "going up completely", "going down", and "going down completely", for the left front window, since the method of operation of the remaining windows and other functions is similar.

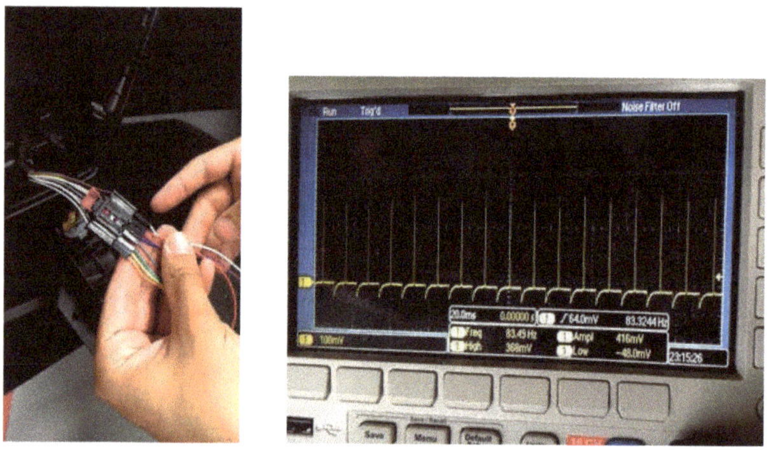

(**a**) (**b**)

Figure 13. Electrical connections performed for measuring the electrical communications signals between the sensor and ECU in (**a**). (**b**) illustrates the measured signal with no action being applied by the user.

In Figure 14, we can observe the signals that were measured by probing the electrical circuits in the door panel.

The electrical values change according to the functionality requested, and they are:

- Minimum: −48 mV;
- Maximum: +368 mV;
- Amplitude: 416 mV; and,
- Frequency: 83 Hz.

When analyzing the board referring to the button on the left front window pane, the electrical diagram shown in Figure 15 is obtained.

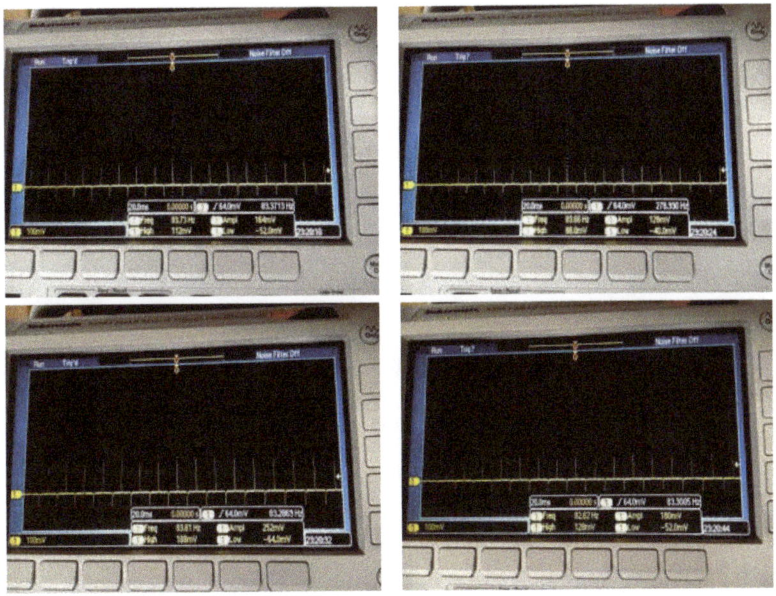

Figure 14. The signals measured by applying all of the functions to the window pane: Descend, Full Descend, Ascend, and Full Ascend. (Left to right, top to bottom)

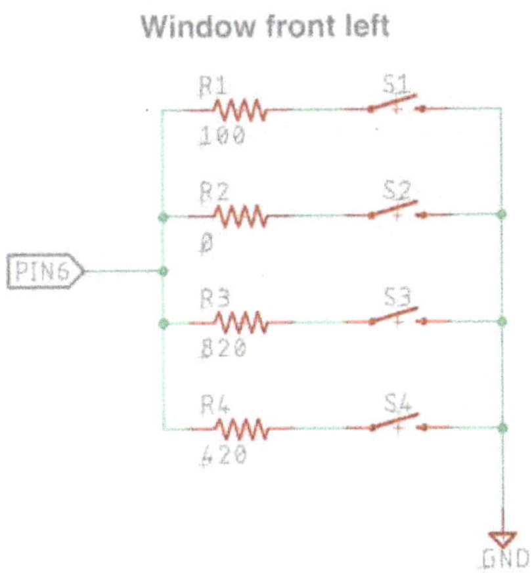

Figure 15. Circuit diagram present on the *PCB* of the tested button assembly.

A mechanical analysis of the button operation was also performed after analyzing the board of the tested button accessory and its circuitry. Thus, it was found that the first "click", be it descending or ascending, is a simple contact button. However, the second "click" forms a parallel between the resistance of the first "click" and the second. Subsequently, we have the following definition for the four possible actions:

- Descend: R1;
- Full Descend: R1 in parallel with R2;
- Ascend: R3; and,
- Full Ascend: R3 in parallel with R4.

When comparing the circuit with the measured signals, there is a relationship and mechanism for actuation by voltage drop where a voltage divider is applied. In this way, it is concluded that, at this moment, the door does not have a communications protocol, but rather a purely analog interaction with the amplitude manipulation of a signal sent by the ECU.

3. Results

By analyzing the characteristics of the sensor technologies that were previously described in Table 1, and when considering the desired use in the context of this project, in particular the criteria of robustness, ease of integration, and cost, the most adequate decision implies adopting sensors of capacitive type.

Capacitive sensors in a system can be used in several fields of application, such as replacing simple buttons, or using touch panels, sliders. Because they present a set of favorable characteristics, whereby their adequate sensitivity to touch and the reduced need to connect electronics that are dedicated to their operation stand out, capacitive type sensors are gaining popularity in several application areas. Cost also represents a very appealing factor, especially if purchased in large quantities. The advantages are numerous and they facilitate the specific implementation of the desired functionalities in the context of the SmartPlastic project.

Regarding the verification of whether some of the indicated disadvantages constitute an impediment to implementation, it appears that the operating environment (e.g., the car's cabin) is sufficiently stable to the point that the normal operation of the device is not affected. In the event of unforeseen scenarios (e.g., an arm inadvertently placed over the sensor), it is always possible to correct this behavior by implementing filters and control algorithms using firmware.

3.1. Testing Capacitive Sensors with Different Polymers and Thickness Specifications

By testing several types of capacitive sensors composed of a simple copper pad on a PCB, and the TTP223B packaging integrated circuit, the experimental results were successful, even for plastic samples (placed near the sensor) that present greater thickness. The sensor proved to be sensitive to touch with a high success rate. The first tests were conducted on hundreds of samples and achieved an approximately 100% success rate in sensing human touch. It should be noted that the hand and touch-based sensor are on opposite sides of the plastic part. For the worst case scenario, plastic pieces as thick as 2.5 mm were perfectly able to detect the hand touch on the other side of the piece, obtaining sensitivity rates above 96.67% and 100% for most cases. These results can be observed from Tables 2–6, where the type and thickness of the polymer tested are depicted for each of the sensors presented in Figure 16 and Table 7.

Table 2. Preliminary tests to test the permissible plastic thickness (for sensor 1).

Polymer	Description	Thickness (mm)	Test Version	Test Number	Effectiveness	Assessment
PC ABS	Taroblend 66	0.8	V1	>10	NA	Approved
PA6GF30	Ultramid B3E2G6	0.8	V1	>10	NA	Approved
PPTD20	Hostacom TRC 352N	0.8	V1	>10	NA	Approved
PP	Sabic PHC3181	0.8	V1	>10	NA	Approved
PA6	Badamid B70S Natur	0.8	V1	>10	NA	Approved
PC ABS	Taroblend 66	1.8	V1	>10	NA	Approved
PA6GF30	Ultramid B3E2G6	1.8	V1	>10	NA	Approved
PPTD20	Hostacom TRC 352N	1.8	V1	>10	NA	Approved
PP	Sabic PHC3181	1.8	V1	>10	NA	Approved
PA6	Badamid B70S Natur	1.8	V1	>10	NA	Approved
PC ABS	Taroblend 66	2.5	V1	>10	NA	Approved
PA6GF30	Ultramid B3E2G6	2.5	V1	>10	NA	Approved
PPTD20	Hostacom TRC 352N	2.5	V1	>10	NA	Approved
PP	Sabic PHC3181	2.5	V1	>10	NA	Approved
PA6	Badamid B70S Natur	2.5	V1	>10	NA	Approved

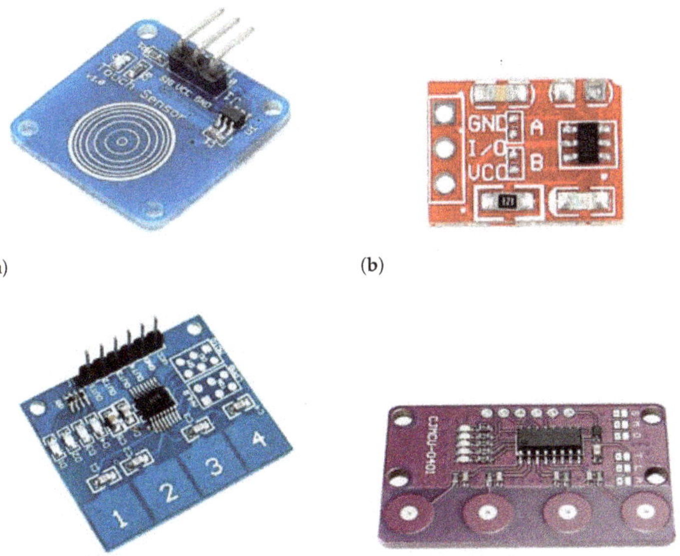

(a) (b) (c) (d)

Figure 16. From (**a**–**d**), the figure depicts the four capacitive sensors tested. The individual electrical properties and datasheet for each sensor can be found in Table 7.

Table 3. Effectiveness tests for sensor 1.

Polymer	Description	Thickness (mm)	Test Version	Test Number	Hits	Effectiveness	Assessment
PC ABS	Taroblend 66	2.5	v2	30	30	100.00%	Approved
PA6GF30	Ultramid B3E2G6	2.5	v2	30	30	100.00%	Approved
PPTD20	Hostacom TRC 352N	2.5	v2	30	29	96.67%	Approved
PP	Sabic PHC3181	2.5	v2	30	29	96.67%	Approved
PA6	Badamid B70S Natur	2.5	v2	30	29	96.67%	Approved

Table 4. Effectiveness tests for sensor 2.

Polymer	Description	Thickness (mm)	Test Version	Test Number	Hits	Effectiveness	Assessment
PC ABS	Taroblend 66	2.5	v2	30	30	100.00%	Approved
PA6GF30	Ultramid B3E2G6	2.5	v2	30	30	100.00%	Approved
PPTD20	Hostacom TRC 352N	2.5	v2	30	30	100.00%	Approved
PP	Sabic PHC3181	2.5	v2	30	30	100.00%	Approved
PA6	Badamid B70S Natur	2.5	v2	30	29	96.67%	Approved

Table 5. Effectiveness tests for sensor 3.

Polymer	Description	Thickness (mm)	Test Version	Test Number	Hits	Effectiveness	Assessment
PC ABS	Taroblend 66	1.8	v2.1	30	29	96.67%	Approved
PA6GF30	Ultramid B3E2G6	1.8	v2.1	30	30	100.00%	Approved
PPTD20	Hostacom TRC 352N	1.8	v2.1	30	29	96.67%	Approved
PP	Sabic PHC3181	1.8	v2.1	30	30	100.00%	Approved
PA6	Badamid B70S Natur	1.8	v2.1	30	30	100.00%	Approved

Table 6. Effectiveness tests for sensor 4.

Polymer	Description	Thickness (mm)	Test Version	Test Number	Hits	Effectiveness	Assessment
PC ABS	Taroblend 66	1.8	v2.1	30	30	100.00%	Approved
PA6GF30	Ultramid B3E2G6	1.8	v2.1	30	30	100.00%	Approved
PPTD20	Hostacom TRC 352N	1.8	v2.1	30	30	100.00%	Approved
PP	Sabic PHC3181	1.8	v2.1	30	29	96.67%	Approved
PA6	Badamid B70S Natur	1.8	v2.1	30	30	100.00%	Approved

Table 7. Electrical characteristics of the tested sensors.

Sensor #	Int. Circuit	Pad	Type	Datasheet (Accessed on 1 April 2019)
Sensor 1	TTP223-B	Circular	1 Channel	https://datasheet.lcsc.com/szlcsc/TTP223-BA6_C80757.pdf
Sensor 2	TTP223N-B	Squared	1 Channel	https://datasheet.lcsc.com/szlcsc/TTP223-BA6_C80757.pdf
Sensor 3	TTP224	Squared	4 Channels	https://download.mikroe.com/documents/datasheets/ttp224.pdf
Sensor 4	0401-8224	Circular	4 Channels	Clone TTP224

All of the pieces injected in relation to the present context of producing polymer components for controlling the opening of door windows and rear-view mirrors are less thick than the values that are indicated in Tables 2–6. Therefore, it is possible to conclude that the objective of the project can be achieved without the need to resort to a necessarily more complex process that involves over-molding the sensors, as tested and illustrated in Figure 17.

3.2. Integration of Capacitive Sensors with Microcontrollers on the PCB

Having validated the touch sensor technology, it was decided to design the sensors themselves on the PCB in order to reduce the costs and area of the circuit. We chose to integrate these sensors that were designed directly on the PCB with a microcontroller that had its own peripheral for conditioning and interpreting the signals emitted from these sensors. The microcontroller chosen was the ATtiny3217 from Microchip, which incorporates a Peripheral Touch Controller that abstracts the functions necessary to deal with capacitive sensors. By using this peripheral, it is advantageous to explore a method for connecting capacitive sensors, which is analogous to the "Matrix with row and column control" method described in Section 2.3.3. This method of measuring the change in capacitance using a matrix electrode configuration is called Mutual Capacitance. This way, where only 10 sensors could be integrated using the "one sensor per pin" or Self Capacitance method, it now becomes possible to support up to 21 sensors using a matrix configuration with 3 rows and 7 columns. Figure 18 illustrates the prototype developed that supports 19 capacitive sensors.

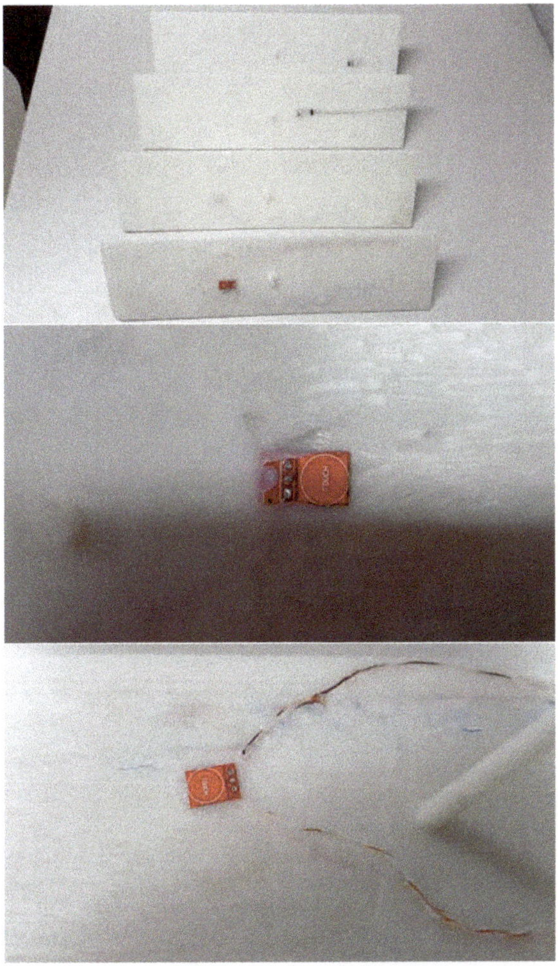

Figure 17. The figure illustrates the tests that were performed by over-moulding embedded sensors in different types of polymers. The tests were conducted for senor 2. Although most of the sensors were functional after moulding, they were subject to high pressure and temperature, which may have consequences related to the robustness and durability of the component during use over time. This the main reason why this path of investigation has been left for future work.

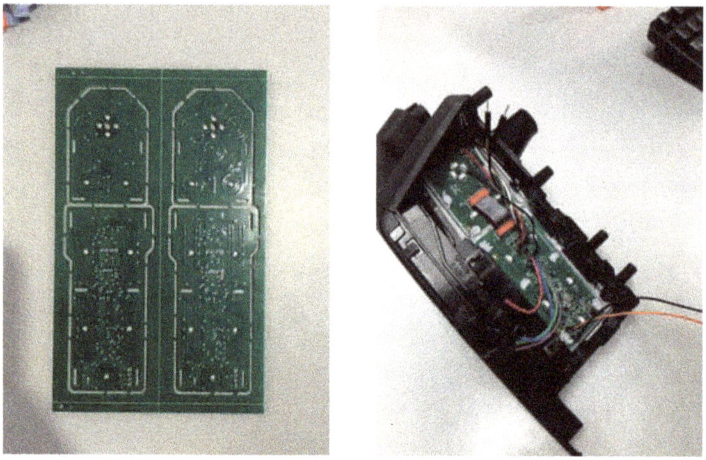

(a) (b)

Figure 18. (a) PCB with 19 capacitive sensors supported. (b) The finished prototype ready to be incorporated in the automobile's door.

4. Analysis and Discussion

The purpose of SmartPlastic was to develop a technological alternative to an automobile part, which is the result of a complex fabrication process that employs several miniature components based on injected plastic, for controlling functionalities inside the car cabin. This part has a limited life-cycle as a result of this process and of the mechanical wear of the components.

The original automobile part that is responsible for the control of the door windows and rear-mirror's movement used in tests is composed of 26 discreet components. Amongst them, there are several pieces made of plastic, rubber, metal, and PCBs with their respective electronic components and connectors as depicted in Figure 19. The intricate design of the side mirror joystick is of particular interest (please see Figure 19c): this component employs a complex electromechanical system that is comprised of metal beads rolling along plastic grooves, closing electrical circuits to move the mirrors. This joystick component is made up of 12 components alone.

Each one of these components has its own production line and they all converge into an assembly line to complete the part mounting. These production lines need to obey specific tolerance limits, defined as an allowance for a specific variation in the size and geometry of a part [32]. These variations in size and geometry add up along the production and assembly, which can result in a shortened life-cycle for the component. Assembly is of particular importance, because it usually is made by hand, which is never error free, takes more time than automation, and can lead to waste if a mistake occurs. There is an inherent cost to the OEM in all of these fault prone processes: the product quality. The shorter the product quality, the shorter its mean time between failures (MTBF) and, thus, the higher the maintenance and replacement costs.

After considering all of the above, the original part mechanism was analyzed to determine which functionality should be incorporated and emulated in the new part. Once those requirements were defined, the development of the new part was made using the least mechanical complexity possible.

The newly developed part is composed of two components: a PCB with its electronic devices and the molded polymer plastic surface. The polymer plastic component is the result of a proven automated process that requires no human intervention during fabrication. The PCB production and its electronic component's assembly are both also automated and, thus, do not require human intervention during the entire process. The only intervention required was to join the PCB and surface components. Such a reduced level of complexity allows for smaller margins of error and faster production times, which results in higher productivity and better quality standards being met.

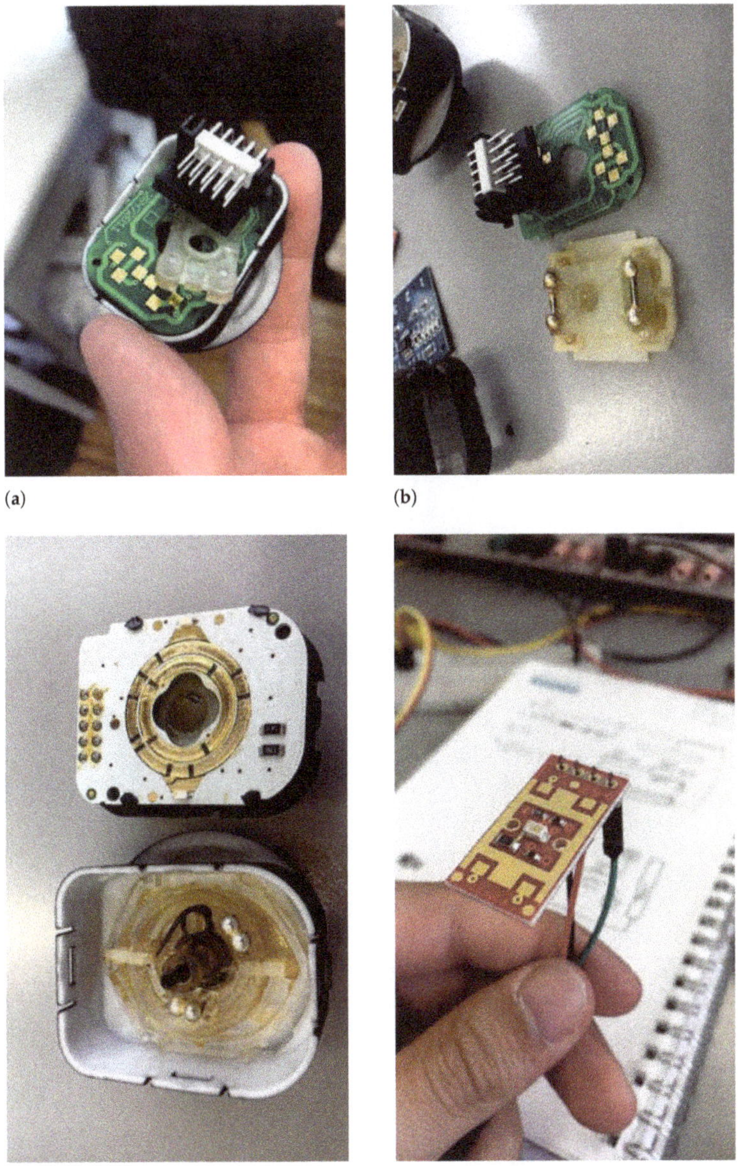

(**a**) (**b**) (**c**) (**d**)

Figure 19. From (**a–d**) the figure illustrates some of the components of joystick rear-view mirrors and buttons used to open/close door windows. (**a–c**) depicts electronic parts of the joystick components and (**d**) illustrates the PCB that interacts with the contacts that are closed/opened by the button.

The plastic polymer surface was designed to allow the user to accurately feel where the hand should be placed in order to activate the controls (see Figure 20a), without the need of looking at the polymer touch surface. This was accomplished by employing grooves along the touch controls and ridges between them during the design process. This requirement was of the utmost importance, as the driver needs to focus on the road while driving, and should avoid looking where the controls are. Therefore, the surface design is important enough to make every other development task, such as the PCBs, gravitate around it. The PCB was designed as a function of the surface limitations and functional requirements, as depicted in Figure 20b.

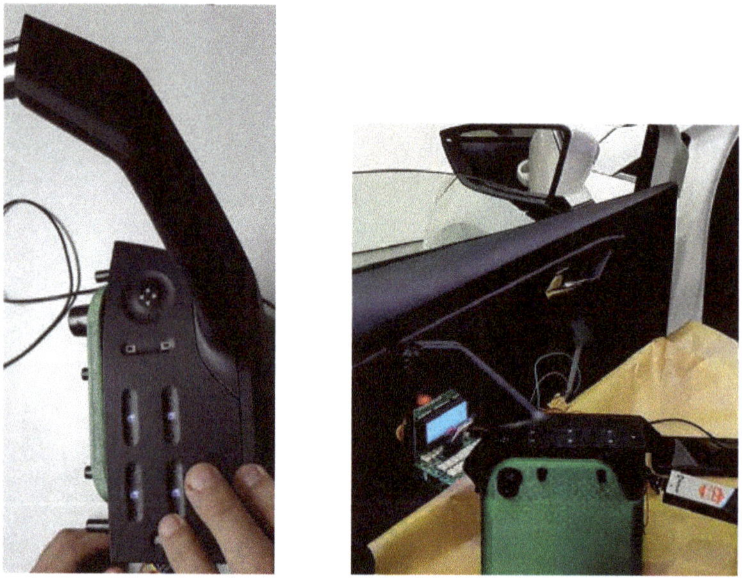

(a) (b)

Figure 20. The functional prototype connected to the door of a real automobile manufactured in 2018. (**a**) depicts the new single-component touch-based polymer panel in detail, while (**b**) shows the panel that was integrated in the vehicle's door, controlling the four windows (front left and right, and rear left and right) and the two rear-view mirrors.

As already noted, all of the development results in a substantial reduction of components production and assembly processes. The lesser amount of human intervention and, thus, probability of error, also contributes to a reduction in the production costs along the road. These factors are of extreme importance to the operator that is responsible for selling the vehicle to the end consumer, because the maintenance costs are a critical factor to be taken into consideration when deciding to buy a new vehicle. There is also a non-monetary cost involved, which is the improvement in the ecological footprint when comparing this newly developed part to the original mechanical part. This new part uses less plastic and it has a longer life-cycle—as a result of less production and assembly faults—which will invariably lead to less waste, even if the OEM decides to replace the part in the event of an operational failure to the detriment of repairing the faulty part. This improvement in reliability is the result of omitting moving mechanical components in the development of this new part. Those moving components are the main contributors to the wear of a part or product. Therefore, by employing no moving components and reducing the assembly complexity, there would be less potential points of failure, which can be measured in a longer MTBF.

There was some level of skepticism in the beginning of the investigation regarding this technology's flexibility when paired to the types of polymers and associated thickness, which some manufacturers use, as well as in regards to the use of painting and surface finishing. This solution can be implemented using different polymers and thickness, which makes it a versatile option for different environments and contexts that require unusual designs and even extreme operating conditions, such as greater thermal resistance or greater resistance to mechanical stress, as shown in Section 3.1. This way, this novel technology meets the requirements of different car manufacturers without the need of any type of physical adaptation.

The next step consists of the endurance validation for the developed part. The first tests allow for concluding that the number of failures is indeed small, being below 0.05%. This process will be mechanized and it is expected to allow the solution to be automatically validated running thousands of operating cycles. With this, the intention is to improve reliability and be able to decide which gestures and interactions are the most appropriate for using this solution in the originally proposed context, which is the control of windows and rear-mirrors on the driver's side in the interior of a car cabin.

Considerations about Incorporation of Touch-Based Technology

There is some debate going on regarding the visual attention that the driver must pay to use touch-based screens and their maintenance as primary control interfaces, being supplemented by secondary devices that are only enabled for certain features [33]. Independently of the chosen path, at later stages of evolution of the car cabin interfaces, the touch-based technology that is proposed in this work can be adopted with high gains in terms of development complexity, integration in the vehicle, ease-of-use, and cost.

5. Conclusions

The paradigm of carbon emissions reduction represents an important goal in the industrial context. In this work, we developed a new technology, which was supported by a case-study for opening/closing electric windows and controlling the rear-view system, which allows the plastic part's number reduction in the manufacturing process through the replacement of a mechanical panel with touch sensors incorporated using low-cost electronics that were attached to a single injected plastic part. The production processes are further reduced, leading to reduced waste of the material, smaller environmental impact at the end of life, and less energy being spent on design.

By eliminating several parts from the current system, we obtain a cheaper component whose functionality is also easier to build and less prone to errors. This conclusion requires quantitative validation. However, the decrease in the number of hours of a human operator largely justifies the reduced cost of the part, rendering the economic study to an economy of scale less relevant.

Regarding the technological purpose of this work, at this stage of development there is a new technology that covers all of the features of its predecessor. However, it shows to be more versatile and able to be reprogrammed to integrate new features in the future, both from the manufacturer and the user, including presets for different vehicle users. This type of personalization, according to the driver, is an add-on that car manufacturers currently pursue with great interest.

What the Future Will Bring

Autonomous self-driving cars will become a reality soon and, supported by new artificial intelligence technology [34], they will provide more functionalities to the passengers. It may be expected that the new concept of a car cabin changes accordingly, namely for the driver who is now going to be able to interact more with other passengers and eventually use the car seat in positions where he/she can face other passengers, thus not necessarily being aligned with the front direction of the vehicle. This may imply that control buttons will have to be integrated into the passenger's seat, roof, or arm rest, and, in that case, buttons will have to become simpler and probably not formed by dozens of plastic parts each. Multi-function, re-programmable, and touch-based sensors, such as the solution proposed in this article, are excellent candidates that are ready to be adopted by these new cars and systems [35].

Author Contributions: Conceptualization, C.A., T.C., P.S., J.S., C.R., R.L., R.P., F.M., R.M., G.T. and G.F.; methodology, C.A., T.C., P.S., J.S., G.T. and G.F.; software, C.A. and T.C.; validation, C.A., T.C. and G.F.; formal analysis, P.S. and G.F.; investigation, C.A., T.C., P.S., J.S. and G.F.; resources, P.S., G.T. and G.F.; data curation, P.S. and J.S.; writing—original draft preparation, C.A., T.C. and G.F.; writing—review and editing, P.S., C.R., R.L., F.M., G.T. and G.F.; visualization, J.S., R.P., R.M.; supervision, P.S. and G.F.; project administration, P.S.; funding acquisition, P.S. and G.T. All authors have read and agreed to the published version of the manuscript.

Funding: This work has been financially supported by project SPaC (POCI-01-0247-FEDER-038379), co-financed by the European Community Fund FEDER through POCI—Programa Operacional Competitividade e Internacionalização. It has also been financially supported by Instituto de Telecomunicações and Fundação para a Ciência e a Tecnologia under grants UIDB/50008/2020 and UIDP/50008/2020.

Acknowledgments: The authors would like to acknowledge the support provided by CJR Motors and Leiribéria.

Conflicts of Interest: The authors declare no conflict of interest.

Abbreviations

The following abbreviations are used in this manuscript:

SPaC	Smart Plastic Cover
LCD	Liquid-Crystal Display
ADC	Analog-to-Digital Converter
DAC	Digital-to-Analog Converter
CAN	Controller Area Network
LIN	Local Interconnect Network
PCB	Printed Circuit Board
ECU	Engine Control Unit
MTBF	Mean Time Between Failures
OEM	Original Equipment Manufacturer
MDPI	Multidisciplinary Digital Publishing Institute
FEDER	Fundo Europeu de Desenvolvimento Regional (Portugal)

References

1. Krenicky, T.; Ruzbarsky, J. Alternative Concept of the Virtual Car Display Design Reflecting Onset of the Industry 4.0 into Automotive. In Proceedings of the IEEE 22nd International Conference on Intelligent Engineering Systems (INES), Las Palmas de Gran Canaria, Spain, 21–23 June 2018; pp. 407–412.
2. Shaygi, M.; Wunderle, B.; Arnold, J.; Pflügler, N.; Pufall, R. Button Shear Fatigue Test: Fracture-Mechanical Interface Characterisation under Periodic Subcritical Mechanical Loading. In Proceedings of the 25th International Workshop on Thermal Investigations of ICs and Systems (THERMINIC), Lecce, Italy, 25–27 September 2019; pp. 1–12.
3. Nunes, J.S.; Castro, N.; Gonçalves, S.; Pereira, N.; Correia, V.; Lanceros-Mendez, S. Marked object recognition multitouch screen printed touchpad for interactive applications. *Sensors* **2017**, *12*, 2786. [CrossRef] [PubMed]
4. Kim, J.; Song, W.; Jung, S.; Kim, Y.; Park, W.; You, B.; Park, K. Capacitive Heart-Rate Sensing on Touch Screen Panel with Laterally Interspaced Electrodes. *Sensors* **2020**, *14*, 3986. [CrossRef] [PubMed]
5. Yeh, S.-K.; Hsieh, M.-L.; Fang, W. CMOS-based tactile force sensor: A review. *IEEE Sens. J.* **2021**. [CrossRef]
6. Bae, J.; Hwang, Y.; Park, S.J.; Ha, J.-H.; Kim, H.J.; Jang, A.; An, J.; Lee, C.-S.; Park, S.-H. Study on the Sensing Signal Profiles for Determination of Process Window of Flexible Sensors Based on Surface Treated PDMS/CNT Composite Patches. *Polymers* **2018**, *10*, 951. [CrossRef] [PubMed]
7. Park, J.K.; Lee, C.J.; Kim, J.T. Analysis of multi-level simultaneous driving technique for capacitive touch sensors. *Sensors* **2017**, *9*. [CrossRef] [PubMed]
8. Brasseur, G. Design rules for robust capacitive sensors. *IEEE Trans. Instrum. Meas.* **2003**, *52*, 1261–1265. [CrossRef]
9. Seo, S.-H.; Park, J.-H.; Hwang, S.-H.; Jeon, J.W. 3-D Car Simulator for Testing ECU Embedded Systems. In Proceedings of the SICE-ICASE International Joint Conference, Busan, Korea, 18–21 October 2006; pp. 550–554.
10. Morales-Alvarez, W.; Sipele, O.; Léberon, R.; Tadjine, H.H.; Olaverri-Monreal, C. Automated Driving: A Literature Review of the Take over Request in Conditional Automation. *Electronics* **2020**, *9*, 2087. [CrossRef]

11. Kouba, S.; Dickson, N. Intuitive Touch Technologies—Semiconductor-Based Electronic Components and their Integration. *Auto Tech Rev.* **2016**, *5*, 38–43. [CrossRef]
12. Fleming, B. Advances in Automotive Electronics [Automotive Electronics]. *IEEE Veh. Technol. Mag.* **2015**, *10*, 4–96. [CrossRef]
13. Miedl, F.; Tille, T. 3-D surface-integrated touch-sensor system for automotive HMI applications. *IEEE/ASME Trans. Mechatron.* **2015**, *21*, 787–794. [CrossRef]
14. MOBINOV, An Aggregating Platform of Knowledge and Skills. Available online: https://mobinov.pt/index.php/en/ (accessed on 12 January 2021).
15. Vojtech, L.; Lopez, A.M.F.; Neruda, M.; Lokaj, Z. Embedded system with RFID technology and inductive proximity sensor. In Proceedings of the 36th International Conference on Telecommunications and Signal Processing (TSP), Rome, Italy, 2–4 July 2013; pp. 213–217.
16. Han, D.; Choi, H.B.; Kim, Y. Design of Road Surface Lighting System for Rear Lamp using Automotive Ultrasonic Sensor. In Proceedings of the International SoC Design Conference (ISOCC), Daegu, Korea, 12–15 November 2018; pp. 249–250.
17. Rodríguez-Machorro, J.C.; Ríos-Osorio, H.; Águila-Rodríguez, G.; Herrera-Aguilar, I.; González-Sánchez, B.E. Development of capacitive touch interfaces. In Proceedings of the International Conference on Electronics, Communications and Computers (CONIELECOMP), Cholula, Mexico, 22–24 February 2017; pp. 1–8.
18. Wang, H.; Jones, D.; de Boer, G.; Kow, J.; Beccai, L.; Alazmani, A.; Culmer, P. Design and Characterization of Tri-Axis Soft Inductive Tactile Sensors. *IEEE Sens. J.* **2018**, *18*, 7793–7801. [CrossRef]
19. Liu, Y.; Liu, J.; Lin, H.; Zeng, L. Research on infrared signal processing circuit of large size infrared touch screen with interference rejection. In Proceedings of the IEEE International Conference on Signal Processing, Communications and Computing (ICSPCC), Macau, China, 21–24 August 2020; pp. 1–5.
20. EE | Times, Ultrasound Sensor Turns Any Surface into a Touch Button. Available online: https://www.eetimes.com/ultrasound-sensor-turns-any-surface-into-a-touch-button (accessed on 18 December 2019).
21. Calpe-Maravilla, J.; Medina, I.; Martínez, M. J.; and Carbajo, A. Dual touch and gesture recognition in 4-wire resistive touchscreens. In Proceedings of the IEEE SENSORS, Valencia, Spain, 2–5 November 2014; pp. 787–790.
22. Touch Screen Technology. Available online: http://inst.eecs.berkeley.edu/~ee16a/sp15/Lecture/Lecture_11.pdf (accessed on 2 March 2020).
23. AVR341: Four and Five-Wire Touch Screen Controller. Available online: http://www.lysator.liu.se/~kjell-e/embedded/doc8091.pdf (accessed on 8 March 2020).
24. TOUCHsemi, Capacitive Touch Sensor. Available online: http://www.touchsemi.com/index.php?Show=10_Touch_Sensors_e/40_Theory_of_operation_e.php (accessed on 4 April 2020).
25. Vanderbuilt. Available online: http://www.vanderbilt.eu/docs/Trebla_Services_Emissie_Tabel.pdf (accessed on 27 May 2020).
26. Fujitsu, Capacitive Touch Sensors. Available online: https://www.fujitsu.com/downloads/MICRO/fme/articles/fujitsu-whitepaper-capacitive-touch-sensors.pdf (accessed on 8 May 2020).
27. Coanda, H.-G.; Ilie, G. Design and implementation of an embedded system for data transfer in a car using CAN protocol. In Proceedings of the 12th International Conference on Electronics, Computers and Artificial Intelligence (ECAI), Bucharest, Romania, 25–27 June 2020; pp. 1–4.
28. Elshaer, A.M.; Elrakaiby, M.M.; Harb, M.E. Autonomous Car Implementation Based on CAN Bus Protocol for IoT Applications. In Proceedings of the 13th International Conference on Computer Engineering and Systems (ICCES), Cairo, Egypt, 18–19 December 2018; pp. 275–278.
29. Ayres, N.; Deka, L.; Paluszczyszyn, D. Continuous Automotive Software Updates through Container Image Layers. *Electronics* **2021**, *10*, 739. [CrossRef]
30. Sharma, P.; Moller, D.P.F. Protecting ECUs and Vehicles Internal Networks. In Proceedings of the IEEE International Conference on Electro/Information Technology (EIT), Rochester, MI, USA, 3–5 May 2018; pp. 465–470.
31. El-Said, M. ECU Counterfeit Mitigation Using Holistic Approach in Modern Automotive Echo System. In Proceedings of the IEEE 17th Annual Consumer Communications & Networking Conference (CCNC), Las Vegas, NV, USA, 10–13 January 2020; pp. 1–2.
32. Puncochar, D.E. In *Interpretation of Geometric Dimensioning and Tolerancing*; Industrial Press Inc.: New York, NY, USA, 1997.
33. Large, D.R.; Burnett, G.; Crundall, E.; Lawson, G.; Skrypchuk, L.; Mouzakitis, A. Evaluating secondary input devices to support an automotive touchscreen HMI: A cross-cultural simulator study conducted in the UK and China. *Appl. Ergon.* **2019**, *78*, 184–196. [CrossRef] [PubMed]
34. Eliot, L. In *AI Self-Driving Cars Consonance: Practical Advances in Artificial Intelligence and Machine Learning*; LBE Press Publishing: New York, NY, USA, 2020.
35. Liu, S.; Li, L.; Tang, J.; Wu, S.; Gaudiot, J.-L. *Creating Autonomous Vehicle Systems*, 2nd ed.; Morgan & Claypool Publishers: San Rafael, CA, USA, 2020.

electronics

Article

Deep and Transfer Learning Approaches for Pedestrian Identification and Classification in Autonomous Vehicles

Alex Mounsey, Asiya Khan * and Sanjay Sharma

School of Engineering, Computing and Mathematics (SECaM), University of Plymouth, Plymouth PL4 8AA, UK; alex.mounsey@postgrad.plymouth.ac.uk (A.M.); Sanjay.sharma@plymouth.ac.uk (S.S.)
* Correspondence: Asiya.khan@plymouth.ac.uk; Tel.: +44-1752-585123

Abstract: Pedestrian detection is at the core of autonomous road vehicle navigation systems as they allow a vehicle to understand where potential hazards lie in the surrounding area and enable it to act in such a way that avoids traffic-accidents, which may result in individuals being harmed. In this work, a review of the convolutional neural networks (CNN) to tackle pedestrian detection is presented. We further present models based on CNN and transfer learning. The CNN model with the VGG-16 architecture is further optimised using the transfer learning approach. This paper demonstrates that the use of image augmentation on training data can yield varying results. In addition, a pre-processing system that can be used to prepare 3D spatial data obtained via LiDAR sensors is proposed. This pre-processing system is able to identify candidate regions that can be put forward for classification, whether that be 3D classification or a combination of 2D and 3D classifications via sensor fusion. We proposed a number of models based on transfer learning and convolutional neural networks and achieved over 98% accuracy with the adaptive transfer learning model.

Keywords: pedestrian identification; classification; autonomous vehicles; CNN; transfer learning

Citation: Mounsey, A.; Khan, A.; Sharma, S. Deep and Transfer Learning Approaches for Pedestrian Identification and Classification in Autonomous Vehicles. *Electronics* **2021**, *10*, 3159. https://doi.org/10.3390/electronics10243159

Academic Editors: Abdeldjalil Ouahabi, Bogdan Ovidiu Varga, Felix Pfister and Calin Iclodean

Received: 8 November 2021
Accepted: 15 December 2021
Published: 18 December 2021

1. Introduction

Autonomous vehicles are becoming increasingly prevalent on roadways around the world; a study conducted in 2020 by Mordor Intelligence [1] reports that "the autonomous (driverless) car market was valued at USD 20.97 billion in 2020" and is projected to increase by 22.75%, to USD 61.93 billion by 2026. While consistent and significant technological advancements are being made in related fields, confidence in autonomous systems for use on roadways is declining. AAA reported in 2018 [2] that 73% of adults in the United States claim to be "too afraid" of allowing a vehicle to autonomously control itself—this is an increase of 10% from a similar study conducted one year prior.

Eliminating the human element of vehicular control, of course, subsequently eliminates the risk of traffic collisions resulting from human error. Furthermore, the occupants of autonomous vehicles are free to spend travel time recreationally or occupationally; this is especially beneficial considering the increasing congestion on roadways within major settlements, alongside a growing world population.

In the event that most, if not all, vehicles on roadways possess fully autonomous capabilities, it would be possible for a system to be implemented wherein these vehicles communicate with one another by sharing information on hazards ahead and manoeuvres they wish to perform. The resulting improvements to travel efficiency would likely have a cascading effect through iterative increases to speed limits.

Additionally, the Mobility-as-a-Service (MAAS) market is likely to see an increase in potential as autonomous vehicles acquire mass-adoption. Fully autonomous MAAS would hypothetically enable individuals to, rather than owning a personal vehicle, lease a vehicle for each journey they embark upon, similar to how companies such as Uber and Lyft currently operate, however in this case, without the need for a driver. Alternatively, those who do opt to purchase their own autonomous vehicle would have the opportunity to lease the vehicle out when not in use, providing an additional stream of income.

In an autonomous driving system, the safety of vehicle occupants, as well as individuals in the surrounding environment, should be guaranteed. One recent study conducted by Najada and Mahgoub [3] revealed that approximately 80% of casualties resulting from vehicle-related accidents were pedestrians.

The safety of vehicle occupants and pedestrians can be achieved through collision avoidance warning systems (CAWS). A key component of CAWS is vehicular situational awareness, which can be facilitated through the use of different types of sensors. These sensors gather data pertaining to the vehicle's surroundings, which can then be processed, with useful information being extracted. LiDAR, RADAR, and camera sensors are the three most prominent sensors currently in use.

In this paper, the use of cameras and computer vision in the scope of pedestrian detection and classification, touching on methods by which LiDAR can be used to improve such a system through sensor fusion are investigated. Here, pedestrian detection can be defined as the process of determining whether an image, generally a frame extracted from a video sequence, contains pedestrian instances. A successful system should be able to leverage computer vision technologies in order to extract the specific locations of any pedestrians in the frame [4], the results of which are usually in the form of bounding boxes encapsulating individual pedestrian instances.

Machine learning models are generally bespoke, designed with a specific use-case in mind. While the performance of these models can be exceptional, the training process requires a substantial amount of labelled data, which can be incredibly time-consuming. ImageNet [5] is an example of such a dataset, consisting of over 14 million images across thousands of classes. Models trained using ImageNet may be exceptional at differentiating between a wide variety of classes, however, applying such a model to a more specific use-case would likely result in a significant loss of performance. Hence, the motivation to make use of transfer learning in this paper to reduce computer complexity and enable the transfer of learning from a previously trained model.

Identifying and localizing pedestrian shapes in images has, perhaps, been one of the greater challenges facing computer vision researchers over the past decades [6], largely due to the variable appearance of the human body and variations in illumination, occlusion, and poses [7]. Recently, however, with the advent of increasingly powerful and compact hardware, pedestrian detection systems have taken great strides in terms of efficiency and accuracy [8–10].

There are two primary methods of achieving pedestrian classification through computer vision: deep learning [11] and machine learning [12] based methods; both approaches follow similar computational pipelines. First, candidate regions must be identified—this can be achieved through the application of either a sliding window, or some more complex region proposal algorithm [13,14]. Once candidate regions are identified, feature extraction is applied to these regions to obtain an accurate classification on the basis of subsequent classification algorithms.

In 1999, Lowe proposed a visual recognition system [15] which makes use of local features which are scale-invariant and partially invariant to changes in illumination. This publication is indicative of researchers' shift in focus at the time, from attempts to reconstruct objects as three-dimensional objects [16], to feature-based object recognition. Soon after, Viola and Jones published a real-time facial recognition framework [17] in the form of a binary classifier consisting of numerous, weaker, classifiers which are trained using Adaboost [18]. Viola and Jones later went on to propose a pedestrian detection algorithm which used motion and appearance information in order to detect a moving person [19]. Dalal and Triggs expanded this work [20] and proposed the use of Histogram of Oriented Gradients (HOG) as a feature extractor, with the resulting HOG features being fed into a linear Support-Vector Machine (SVM) [21] classifier. This HOG-SVM combination is capable of differentiating between regions which contain pedestrians and those which do not. The resulting reduction in the number of false positives was over an order of magnitude, compared to the best performing Haar wavelet detector at the time [22]. While HOG-SVM offers exceptional performance in classification tasks, it fails to achieve a low mean average precision [23].

In 2008, Felzenswalb et al. utilised the HOG-based detector in their multiscale Deformable Part Model (DPM) [24] which deconstructs objects into groups based on pictorial models [25]. The DPM was suggested to be state-of-the-art at the time, outperforming other methods of object detection, such as template matching.

McCulloch and Pitts first proposed the McCulloch-Pitts (MCP) model in 1943 [26], which is widely accepted as the genesis of Artificial Neural Networks. In 1980, Fukushima introduced Neocognitron, a hierarchical, multilayer Artificial Neural Network which was designed for use in handwritten character recognition and similar pattern recognition tasks. The model consisted of several pooling and convolutional layers, which provided the ability to learn how to identify visual patterns in images. LeCun et al. inspired from Neocognitron proposed the concept of Convolutional Neural Networks (CNNs) which utilize error gradient, yielding impressive results in a range of pattern recognition applications [27–29].

CNNs [30] are perhaps the most prevalent application of deep learning for computer vision tasks, as they have proven to be exceptionally well-suited for tackling object detection problems, in part due to their ability to extract discriminative features. CNNs are composed of three different types of neural layers: convolutional layers, pooling layers, and fully connected layers. In the context of computer vision tasks, Yosinski et al. [31] deduced that the lower layers (i.e., convolutional and pooling) act in a similar manner to conventional computer vision-based feature extractors such as edge detectors, while the final, fully connected layers, are more task-specific. In [32] authors showed that CNNs outperformed both HOG descriptor and Haar-classifier.

As discussed in earlier sections, deep learning and machine learning models require significant volumes of data for use during training. This was identified in 2001 in a research report published by Gartner [33], which alluded to the impending surge of big data. An on-board pedestrian detection system is proposed in [34] based on 2D and 3D cues. Just under a decade later, the ImageNet database was introduced by Deng et al. in 2009 [5]. Authors in [35] propose a dataset that includes challenges related to dense urban traffic, based on their dataset they propose a fusion framework for multi-object detection. The advent of larger datasets such as ImageNet required more capable deep learning models and, in 2012, Krizhevsky et al. introduced AlexNet [36]: a breakthrough in CNN architecture which makes use of the Rectified Linear Units (ReLU) activation function which provided a sixfold reduction in training time, compared to the TanH activation function which, at the time, was standard. Additionally, AlexNet has the capability of being trained across multiple GPUs simultaneously, which enabled more complex models to be produced and was a key enabler of the significant reduction in training time.

Transfer learning (TF) aims to provide a middle ground, where knowledge acquired from larger datasets can be used in conjunction with smaller, domain-specific, datasets in order to improve performance in subsequent domain-specific tasks. In this context, prior knowledge can be model weights or low-level image features which describe what is being classified such as edges, shapes, corners, pixel intensity, etc.

Therefore, the models produced in this work make use of transfer learning as it enhances the performance of the proposed CNN model with the VGG-16 architecture, proposed by Simonyan and Zisserman in 2014 [37]. The VGG-16 CNN model improves upon the work carried out for AlexNet by switching the 11×11 and 5×5 kernel-sized filters with two consecutive 3×3 filters in the first two convolutional layers.

The main contributions of this paper are threefold—(i) a review of CNNs in pedestrian classification, (ii) classification models trained on CNNs and transfer learning and (iii) a pre-processing system with LiDAR point cloud with applications in a 3D object classification model.

The rest of the paper is organized as follows. Section 2 presents the review of CNNs. The models developed are explained in Section 3. Results and discussions are elaborated in Section 4 with a conclusion in Section 5.

2. Review of CNNs for Pedestrian Recognition

R. Hecht-Nielsen [38] described neural networks as "a computing system made up of a number of simple, highly interconnected processing elements, which process information by their dynamic state response to external inputs". The review presented here expands on the CNN and deep learning principle presented in [39,40] in the context of AlexNet [36]. They are modelled to mimic the human brain in order to recognize patterns. This is achieved through numerical input vectors that describe real-world information such as images and text, from which an output response can be generated. In the context of pedestrian classification, once a candidate region has been recognized through recognition techniques, it can be classified through the use of a neural network which allows for an appropriate response to be made by the vehicle.

Convolutional Neural Networks (CNNs) mostly used in the computer vision field. CNNs are structured in a three-dimensional layers and processes information first through "convolution", layer where small portions of data are analysed in order to create a "feature map", before passing it to "pooling" layer. Here, each feature of the data set has its dimensionality reduced while retaining the most relevant information. This next section covers the related theory behind CNNs.

2.1. Single Layer Perceptrons

Perceptrons, sometimes referred to as "linear binary classifiers", are a form of supervised classification algorithm that can be used to determine the classification of a given input. If neural networks are considered to be computational representations of the human brain the perceptrons act as individual neurons, which take the form of a single-layer neural network and consist of four key elements: input values, weight and bias, the net sum, and an activation function.

Input values are multidimensional vector values that are fed into the perceptron in order to be processed. The input values are multiplied by a weighting parameter, which is indicative of an individual input's influence over the output value. The sum of the weighted input values is referred to as the "net sum" or "weighted sum", and can be calculated with the following equation:

$$s = \sum_{i}^{m} w_i I_i, \tag{1}$$

where s is the weighted sum, m is the number of inputs, w represents the weight for each input, and I represents the value of each input. Once a weighted sum has been calculated, it is then applied to the activation function, which normalizes it. In simple perceptron models, the activation is a step function. See Figure 1.

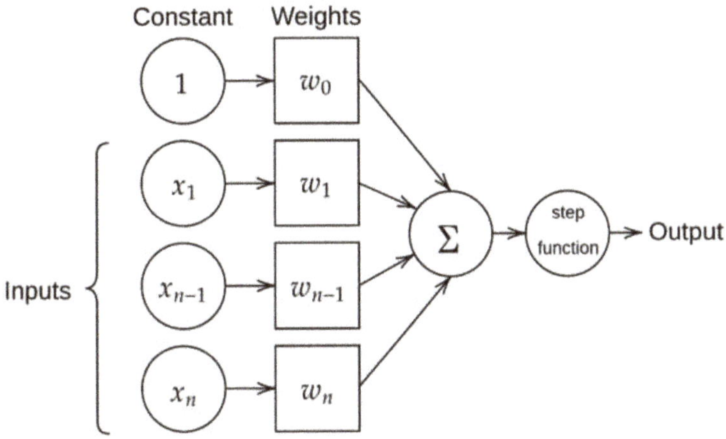

Figure 1. A simple perceptron model.

2.2. *Multi-Layer Perceptrons*

Multi-layer perceptron (MLP) is simply another way of referring to a neural network and consists of a collection of individual single-layer perceptrons (SLP) arranged into distinct "layers". The most basic form of MLP consists of three layers: an input layer, output layer, hidden layer. The input and output layers serve the same purpose as in an SLP, the hidden layer is where most of the MLP's computation is performed.

MLPs allow for non-linear classification, such as XOR functions, which is not possible with SLPs as they are not capable of modelling feature hierarchy. It is for that reason that SLPs generally only find use as building blocks for MLPs, which have been shown to approximate non-linear functions. Furthermore, SLPs simply use the step function as an activation function, whereas MLPs can use more complex activation functions which enable the classification of items into multiple labels as well as to provide probability-based prediction.

2.3. *Activation Functions*

Activation functions are mathematical equations that are not only used to determine the output of the individual perceptrons, but also the accuracy, computational efficiency during training, and the output of a deep learning model in its entirety. Additionally, selecting an appropriate activation function for the task the neural network is attempting to perform is crucial, as they have a significant influence over the network's ability to converge and the speed at which it can converge. Examples of activation functions include the binary step, linear, sigmoid, TanH, rectified linear unit (ReLU), SoftMax and back propagation. ReLU is the most frequently used activation function due to their simplicity where positive values are treated linearly, and negative values are assigned a value of zero [41].

2.4. *Hyperparameters*

2.4.1. Hidden Layers and Units

Hidden layers are layers within a neural network that lie between the input and output layers. Increasing the number of hidden layers has the potential of increasing the accuracy of a model, however as more hidden layers are added, computational requirements will increase yet will yield diminishing returns on the error function.

Not having an adequate number of hidden layers on the other hand will result in poor generalization and unreliable predictions, so it is important to strike a balance in the selection of number of hidden layers.

2.4.2. Dropout

Dropout is a technique used during the training of a model in which certain nodes are deactivated so that it does not become overwhelmed with information, which can isolate nodes that may not be contributing to an improved error function, which in turn should produce a more efficient model.

2.4.3. Activation Function

Activation functions, which have been discussed earlier in this section, can be added to any point of a neural network and there is no limit to the amount that can be added which again results in the process of determining a suitable balance between the number of activation functions and the overall efficiency of the model.

2.4.4. Learning Rate

The learning rate determines the strength of changes made to weights during the process of backpropagation. A lower learning rate results in smoother convergence at the cost of an increase in training time and a higher learning rate will have opposing effects, which means the appropriate learning rate will be model-specific.

2.4.5. Epochs and Batch Size

The number of epochs represents the number of instances that the training dataset is fed into a neural network during the training process. Increasing the number of epochs will increase the accuracy to a certain extent, after which overfitting will start to occur, and training accuracy can decrease. Batch size controls the percentage of the dataset to be exposed to the network through each iteration (epoch) of the training process, which can reduce the over generalization of the model.

2.4.6. Optimisation Algorithm

Optimisation algorithms are those that attempt to minimize the error function of a model. There are two subcategories of optimization algorithms: "first-order" (e.g., gradient descent), which adjust the loss function with respect to given parameters, and "second-order", which use what is known as the 'second order derivative' or "Hessian" to adjust the loss function.

First order optimizations are easier to compute and require less computational time to converge reasonably well with larger data sets. Second order derivatives are only able to outperform first order optimizations when a second order derivative is known, otherwise they are more computationally intensive and take longer to execute.

2.4.7. Momentum

Momentum is the process of tracking changes made to a model and the direction of those changes, which can be used to influence subsequent changes so that they follow the same direction, towards a lower error function.

3. Proposed Pedestrian Classification Models

Pedestrian detection can be described as a binary classification problem, where the goal is to predict whether candidate regions contain an instance of a pedestrian (positive sample, denoted as 1) or not (negative sample, denoted as 0).

All models and software produced for this project were developed using Python 3.8. TensorFlow 2.3.0, Keras 2.4.3, and scikit-learn 0.24.2 were used in the development of deep learning models. Scikit-learn also found use in the development of our 3D pre-processing system, in the creation of the DBSCAN clustering algorithm and RANSAC regressor. TensorBoard was used during the training of CNN models in order to monitor training progress in real-time.

The overall concept of our proposed method is presented in Figure 2.

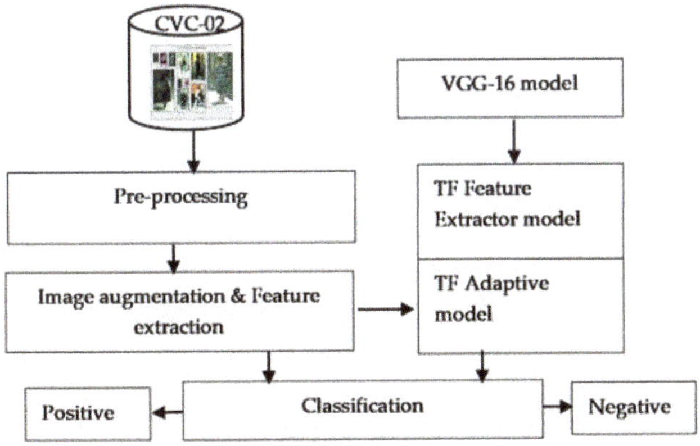

Figure 2. Concept of our proposed method.

In Figure 2, at the pre-processing stage the training and validation images have their pixel values scaled from (0 to 255) to (0 to 1) upon import. In the next block, image generators augment the imported images before supplying them to our models. This includes minimal random rescaling, as well as random zooming and shear-transformations by a factor of 0.3. In the TF adaptive model block, we do not use bottleneck features as we use image generators, whereas the TF feature extractor model makes use of the VGG bottleneck features.

This section presents the datasets used, data preparation and image augmentation parameters.

3.1. Datasets

3.1.1. CVC-02 Dataset

The CVC-02 [34] dataset was used during the development of the pedestrian classification systems for an autonomous vehicle. Images provided by the CVC-02 dataset have been recorded in urban scenarios in and around Barcelona, Spain. Images have been recorded using a colour stereo camera with a resolution of 640 × 480 pixels and 6 mm focal length. Specifically, the classification subset which consists of 3140 positive and 6175 negative samples (in which pedestrians are present, and are not present, respectively) were prepared for this paper; these images have also been rescaled to a size of 64 × 128 pixels and have been split into training, validation, and testing dataset. Table 1 presents the overview of the CVC-02 data splits used in the development of pedestrian classification models.

Table 1. Overview of CVC-02 data splits.

Data Split	Total Samples	No. of Positive Samples	No. of Negative Samples
Training	3000	1500	1500
Validation	1000	500	500
Testing	1000	500	500

3.1.2. NuScenes Dataset

The NuScenes [42] dataset is a large-scale dataset provided for use in autonomous driving research and has been used here in the development of a 3D LiDAR pre-processing system. The entire database consists of 1000 20-s scenes recorded in Boston, United States and Singapore under challenging driving conditions. Each scene contains data captured using 6 cameras, 1 LiDAR sensor, 5 RADAR sensors, a GPS, and an IMU. The 10 scenes which is a subset of the main dataset are used here to show the effectiveness of this approach.

3.2. Image Augmentation

When training deep learning models such as CNNs, the quantity of available data may be a limiting factor for model performance—this can be somewhat alleviated through the application of image augmentation, which artificially creates a new data samples based on original samples. While there are appropriately sized datasets available for the training of pedestrian classification models, image augmentation can also be used to apply artificial transformations to input images such as rotation, scale, shearing, zoom, etc. These augmentations are employed in an effort to mimic noise, variations in illumination, and variations in fundamental image properties (e.g., scale, rotation).

It is preferable that the types of augmentation, and the intensity to which they are applied, strike a balance between providing a suitable amount of noise and variation, which can increase the complexity of the data, and preserving the image features, which is crucial in a model being able to generate accurate predictions. Figure 3a,b demonstrates examples of suitable and unsuitable image augmentation. The images can be improved by pre-processing techniques such as wavelet-based de-noising to remove the noise. Note that in the unsuitable data, parts of the pedestrian have been cropped out of the image—this, of course, will not provide the model with representative information. Furthermore, the augmentations should make sense in the context of the problem being solved; applying a vertical flip to a pedestrian dataset is ill-suited as it is unlikely that a deployed model will encounter a pedestrian in such a position, although edge cases will exist.

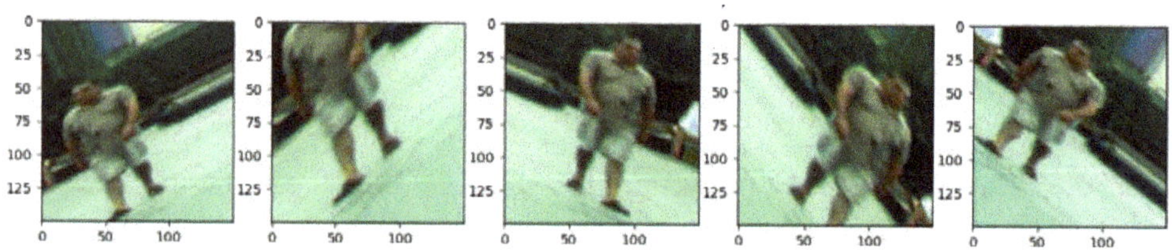

(a)

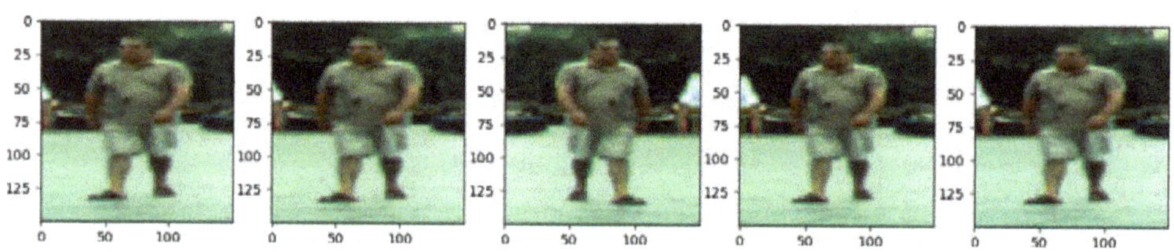

(b)

Figure 3. (**a**) Unsuitable image augmentation. (**b**) Suitable image augmentation.

All models trained during this project made use of identical image preparation techniques. Augmentation was applied to all models, except for the model described in Section 3.5.1, which takes the VGG-16 bottleneck features as input.

Image dimensions are first changed to 150×150, with three channels corresponding to the RGB colour space. Image pixel values are then normalized, from values between 0 and 255, to values between 0 and 1. The normalization of pixel values is to increase the rate at which models learn; large pixel values can disrupt or slow down the process. The original sample labels, which are strings ("positive" or "negative") are encoded to numerical values (1 or 0, respectively), which is more suitable for machine learning techniques.

Once the data has been prepared, image augmentation is applied. Keras' Image-Data Generator() method was used to facilitate this augmentation. The parameters of this augmentation are as follows: a shear range of 0.1, a zoom range of 0.1 with the fill mode set to "nearest" (new pixels are set to the nearest neighbouring pixel values), with horizontal flipping enabled. Figure 3b illustrates the effects of these parameters on training data.

During the development of the models, it was found that prior configurations for image augmentation resulted in models that were unable to identify pedestrians in the majority of samples (REF RESULTS). After investigation, it was found that the augmentation being applied was too intense, resulting in images which contained pedestrians who could not be identified by the model. The offending parameters included: A zoom range of 0.3, rotational range of 90 degrees, width and height shift ranges of 0.2, and shear range of 0.2. This faulty image augmentation is illustrated in Figure 3a.

3.3. *Rudimentary CNN Classifier*

The first CNN model produced in this paper was developed to serve as a benchmark for comparison against transfer learning-based models. The architecture of this model is relatively simple, consisting of three blocks of convolutional layers, the output of which are flattened into feature maps of shape 17 × 17 × 128, which are then processed by two dense layers as shown in Figure 4.

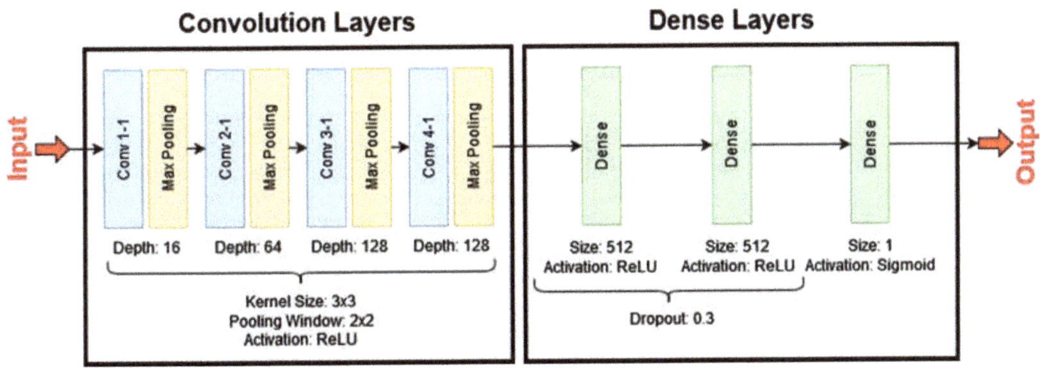

Figure 4. Rudimentary CNN model architecture.

In the convolutional layers, a kernel size of 3 × 3 has been used—this is based on the kernel filter size used in the VGG-16 architecture which, in-part, resulted in an improvement of performance over the AlexNet architecture (see Section 3.4). The final output of the dense layers is a confidence score for each class, indicating how confident the model is that a sample does or does not contain a pedestrian instance.

Training of the first rudimentary model spanned 100 epochs, with a batch size of 30. 100 iterations per epoch was used to cover the training data consisting of 3000 samples (Table 1), with each sample being used to generate 30 additional samples via image augmentation. As the augmentation of validation data results in only 20 images per original sample (scaling is the only augmentation applied to the validation set), the number of validation steps per epoch has been set to 50, such that the model is able to validate on all available samples.

As observed in Figure 5, during training, the model began to overfit after the fourth epoch, as illustrated by the increase in loss on validation data. This means that the model was not capable of generalizing unseen data. However, decent results with a final validation accuracy of 96% were obtained.

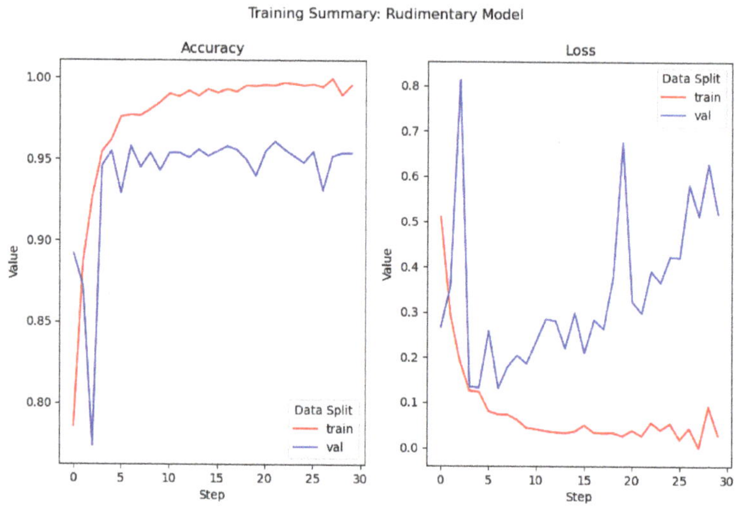

Figure 5. Training summary of the rudimentary CNN model.

Figure 5 is composed of two training summary plots. The left plot details the accuracy of the model during training, i.e., what percentage of data samples can it accurately identify; the red line corresponds to the model accuracy on training data, while the blue line corresponds to the model accuracy on validation data. The right plot details the model loss during training. Loss is a measure of how far an estimated value is from its true value. In all models described in this paper, binary cross entropy is used as a loss function.

3.4. *Rudimentary CNN Classifier with Regularization*

In an attempt to improve upon the performance of the previous model, a second model has been developed which incorporates regularization in the fully connected layers. This is achieved via the addition of two dropout layers, one after each of the first two dense layers. A fourth convolutional block, consisting of a convolution and max-pooling layer is also added. The parameters of this new block (depth, filter size, and activation function) are identical to those in the preceding layer (Figure 6). The reasoning behind the addition of a fourth convolutional block is to enable the model to extract more features from input samples, resulting in an improvement of performance.

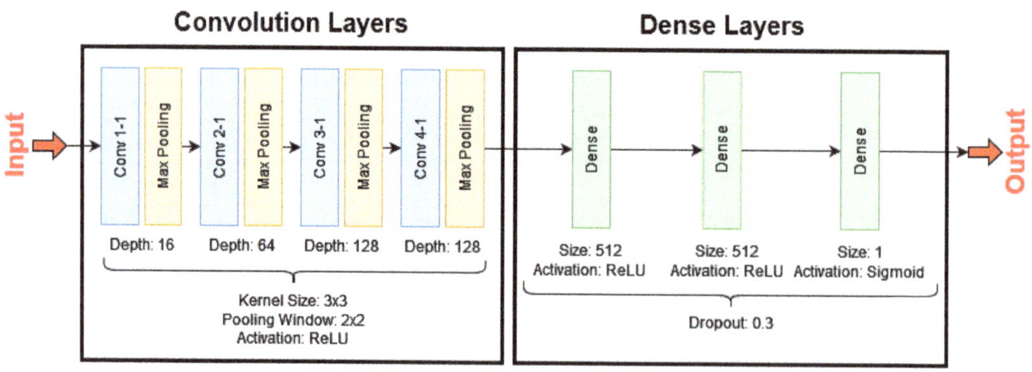

Figure 6. Rudimentary CNN model architecture with regularization.

Dropout layers enable regularization in deep learning models, which simulates the use of multiple architectures during training. Dropout layers randomly mask a fraction of units' output by setting their values to zero. Dropout is a computationally inexpensive and effective technique of reducing the rate of overfitting in a model and improves its generalization error.

Here, a second rudimentary model which makes use of regularization through the addition of dropout layers is presented. In this case, a dropout of 0.3 is applied to the output of the first two dense layers. The result of this is that 30% of the units in these dense layers are masked (i.e., 30% of the units in each layer have their output nullified; set to zero). Additionally, a fourth convolutional block has been added. This enables the model to extract more features from input samples, with the goal of an increase in performance over the original model.

The regularized model is trained with the same hyper parameters as in the original model, spanning 100 iterations across 30 epochs. The addition of regularization and a fourth convolutional block, while resulting in similar validation accuracy (96%), reduces the validation loss considerably (approx. 0.51%, down to approx. 0.40%). Furthermore, the model begins overfitting later in the training cycle, and the rate of overfitting is less intense than that in the previous model. The training data indicates that this model is more suitable for use on unseen data than the former model, suggesting that this model is more capable of generalizing better (Figure 7).

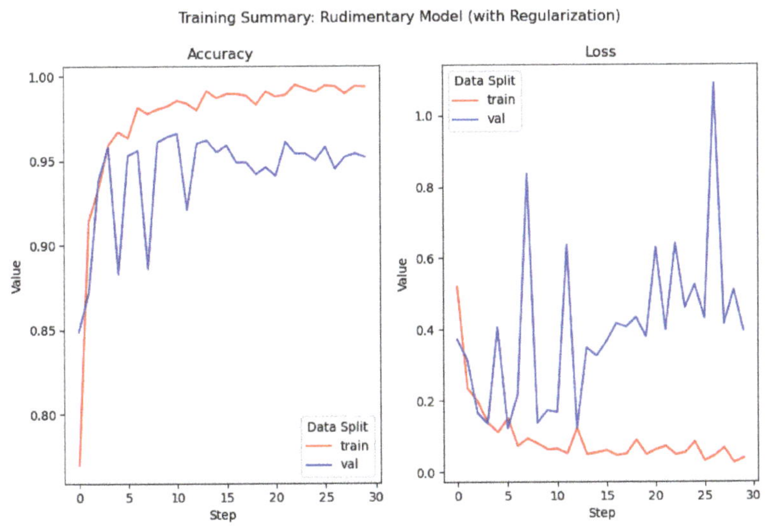

Figure 7. Training summary of the rudimentary CNN model with regularization via dropout layers.

3.5. Transfer Learning Classification Models

VGG-16 [36] is a CNN architecture which specializes in computer vision recognition tasks and has been trained on the ImageNet dataset [5]. It takes 224 × 224 RGB images as input. The "16" in its name refers to the 16 composite layers, which are split into five convolutional blocks and a single fully connected block, that make up the architecture (Figure 8).

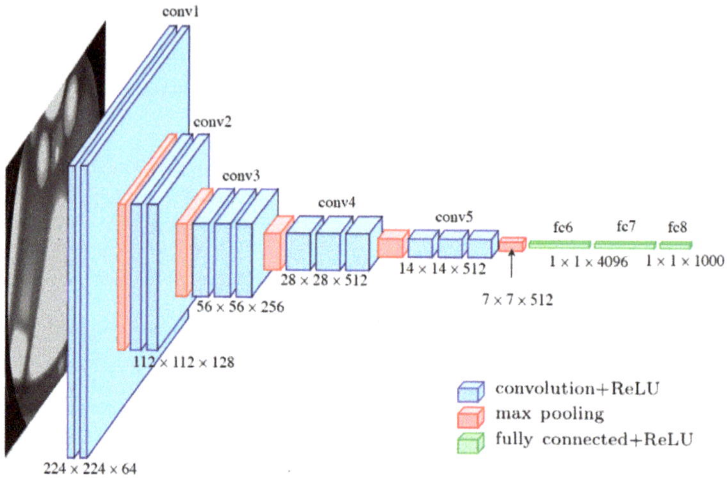

Figure 8. VGG architecture diagram [41].

The convolutional layers all make use of 3 × 3 convolution filters, with each convolutional block paired with a max pooling layer. The max pooling layers are applied over a 2 × 2 window with a stride of 2, for down sampling purposes. Note that, in blocks 3 and 4, VGG-16's use of 3 × 3 filters resulted in an improvement of performance over AlexNet and ZF-Net, which used 11 × 11 and 7 × 7 filters, respectively. It is also worth highlighting that 3 × 3 filters are the smallest dimensions that allow a model to learn directional features (e.g., up, left, centre, etc.).

Convolutional block 5 produces a feature map of shape 7 × 7 × 512, which is flattened into a feature vector containing bottleneck features. These bottleneck features are then processed by a block of fully connected layers consisting of two 4096-channel layers and a single 1000-channel layer (with one channel for each class in the ImageNet dataset). Finally, a SoftMax activation layer normalizes the classification vector, ensuring that the sum of all probabilities is equal to zero.

In this sub-section, we describe three models developed via transfer learning: the first two of which "freezes" the convolutional blocks in order to preserve the original VGG-16 model weights, which are not updated during training. The original dense layers are replaced with bespoke dense layers which are more suited to the specific task of pedestrian classification. The second and third models make use of previously discussed image augmentation methods, while the first does not. Instead, aforementioned bottleneck features are provided to the model during training. These features must also be processed for subsequent classifications on new data.

The third model enables the updating of weights in convolutional blocks 4 and 5. These weights are initialized as those in the vanilla VGG-16 model. The dense layers are also replaced with those used in the previously described model.

In all three approaches outlined above, the dense layers are replaced. This is in-line with the findings of Yosinski et al. [31], described in the literature review: convolutional layers act as conventional computer vision feature extractors, which can be applied to a wide scope of tasks. The outputs of the final convolutional block take the form of 'bottleneck features' (Figure 9) which are the final activation maps prior, processed by the fully connected dense layers. The blurriness in Figure 9 can be improved by pre-processing the image.

Figure 9. Visual representation of bottleneck features from the VGG-16 model.

3.5.1. VGG-16 Feature Extractor Model

This section describes the first of three models with the VGG-16 architecture used in the transfer learning model. In this model, all convolutional layers of the original architecture have been frozen; their weights will remain the same throughout the training process, not being updated as in the traditional process of training CNN models. The dense layers have been replaced: the new dense layers follow the same form as in the rudimentary models proposed in Section 3.3. Furthermore, no image augmentation has been applied to the data samples provided during training– this is to compare the results of image augmentation and will be compared to a subsequent model which does incorporate image augmentation. Instead, the vanilla VGG-16 model is used to generate bottleneck features for all training and validation data samples, these bottleneck features are then flattened for use as inputs for our model.

The hyper parameters used in the training of this model are identical to those used in the training of the rudimentary models described previously: 30 epochs of 100 iterations in training, with 50 iterations during validation. The model performance appears to have significantly improved over the aforementioned rudimentary models (Figure 10): validation loss, while still indicative of overfitting at the fifth epoch, has been reduced by approximately 30%. Furthermore, the training and validation accuracies are more closely correlated (a variation of approximately 2%) which suggests that the transfer learning model is more capable of generalizing on unseen data, than the rudimentary models.

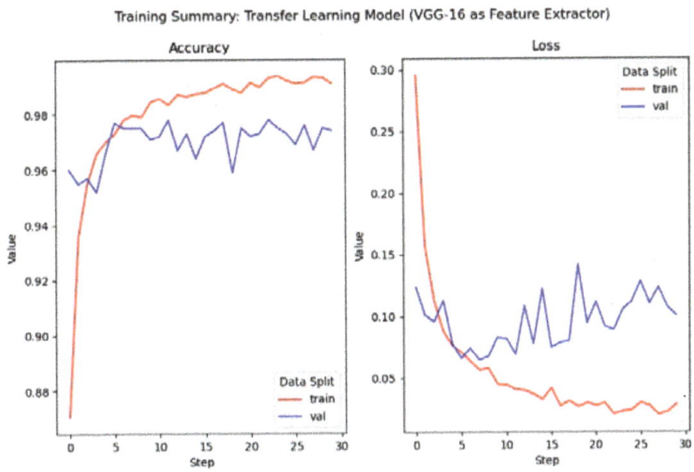

Figure 10. Training summary of the VGG-16 feature extraction model.

3.5.2. Applying Image Augmentation

In this model, image augmentation is applied to input samples—the process of which is identical to that in the rudimentary models (Section 3.3). The model has been trained for 100 epochs of 100 iterations, with 50 iterations for validation steps. Additionally, the learning rate is reduced (from 1×10^{-4} to 2×10^{-5}) in order to avoid rapid and abrupt changes to weight values during backpropagation which may adversely affect model performance.

The training of this model spanned 100 epochs of 100 iterations, with a batch size of 32: the default batch size for Keras. The training summary of this model (Figure 11) is similar to that of the previous model, suggesting that the application of image augmentation did not provide significant improvements to model performance in this case.

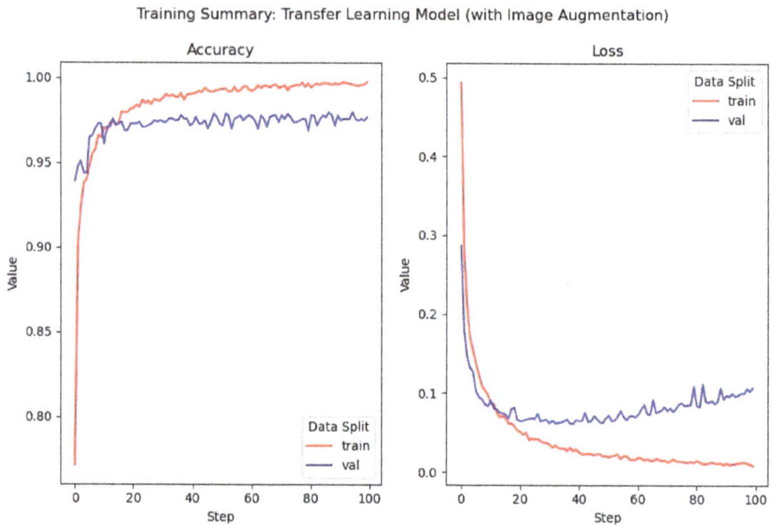

Figure 11. Training summary of the VGG-16 transfer learning model with image augmentation.

3.5.3. Adaptive VGG-16 Model

This section describes the third and final transfer learning model developed during this project. This model enables the updating of weights in convolutional blocks 4 and 5; blocks 1–3 remain frozen, preventing the weights of these blocks from being updated during backpropagation. The learning rate is further reduced from 1×10^{-4} to 1×10^{-5}, again to prevent adverse effects on model performance as a result of intense adjustments to weight values and avoid the model from getting stuck in local minima. This model has been trained with the same image augmentation as in previously discussed models.

The training of this model, again, spanned 100 epochs of 100 iterations, with a batch size of 32. The training summary of this model (Figure 12) illustrates that, while validation loss appears to have increased from previous models, the validation accuracy has improved (approx. 98%). This is likely due to the model's ability to better understand the provided data though the adjustment of weight values during backpropagation.

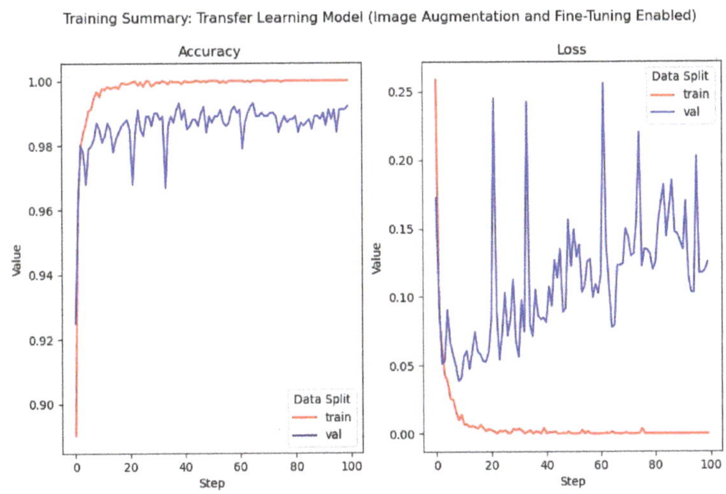

Figure 12. Training summary of the VGG-16 transfer learning model with convolution blocks unfrozen.

3.6. 3D Point Cloud Pre-Processing

In the development of the 3D pre-processing system described in this section, a number of tools and software were used in this paper to produce results. Most notably: Python 3.8, NumPy, Pandas, and the NuScenes devkit. Furthermore, Scikit-Learn facilitated the implementation of our RANSAC regressor and DBSCAN clustering algorithm.

3.6.1. Identification and Removal of Ground Points

LiDAR point cloud data is inherently noisy and can contain large amounts of data which may not be pertinent to the task at hand. Examples of such points are those which correspond to the ground on which a vehicle is travelling. While information about the ground can be useful in the identification of road markings, it does not provide meaningful information in the context of pedestrian detection (there are specific use-cases, an example of which being the identification of regions of high pedestrian activity such as crossings, however). Removing ground points not only removes points which may skew the results of further operations, it also significantly improves the computational time and effort required to process the entire dataset.

3.6.2. Data Preparation

A crucial and effective method of removing irrelevant points from the provided point clouds is to identify and remove points which reside outside of the target camera's field of view (FOV). The NuScenes dataset provides images and point cloud information for each "snapshot" (i.e., frame) within each 20-s scene. The point cloud data retrieved from the dataset resides in the point sensor frame: in order to determine which points lie within the camera FOV and which points do not; the point cloud must be transformed from the point sensor frame to the image frame. Once the driving data have been collected, well-synchronized keyframes are sampled (image, LIDAR, RADAR) at 2 Hz. These samples are annotated using expert annotators and multiple validation steps in order to produce highly accurate annotations. All objects in the nuScenes dataset comes with a semantic category, as well as a 3D bounding box and attributes for each frame they occur in.

First, the point cloud is rotated and transformed from the point sensor frame to the vehicle ego frame for the timestamp of the relevant LiDAR sweep. This same process is subsequently used to transform the data from the vehicle ego frame to the global frame, from the global frame to the ego vehicle frame, and finally, the ego vehicle frame to the camera plane.

Next, a "snapshot" of the LiDAR point cloud is taken. The resulting 2D matrix is multiplied by the camera intrinsic matrix, and renormalization is applied. Points which lie outside of the camera FOV can now be removed using logical AND functions. A margin of 1 pixel is applied for aesthetic purposes, and we ensure that all points are positioned at least 1 m in front of the sensor in order to exclude points which pertain to the camera casing. This process returns a 2D mask of points, whose IDs are matched with the original 3D point cloud in order to identify and remove the points which lie outside of the camera FOV. Figure 13 illustrates the significance of removing points which fall outside of the camera FOV.

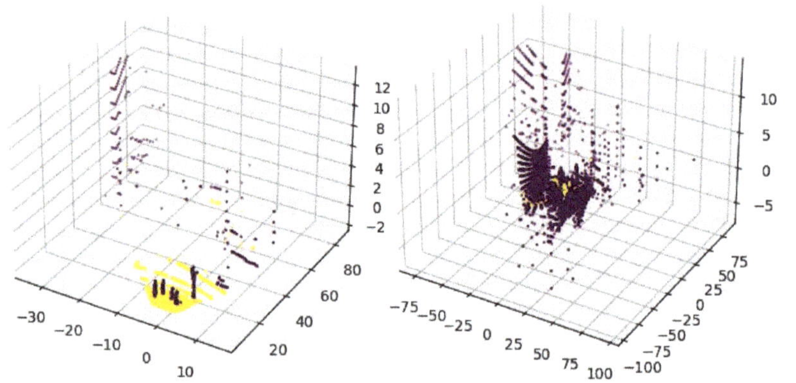

Figure 13. RANSAC regression applied to 3D point cloud data.

3.6.3. RANSAC Regression

RANSAC (random sample consensus) is a simple, yet highly effective, method of handling outliers through trial and error. It separates data into inliers and outliers, which can be used in further processing techniques. In the context of this paper, a plane on which the majority of points lie is identified with an that this plane is likely to be the ground plane, these points can be removed.

In 3D space, RANSAC functions by first selecting three data points at random from a given point cloud—three points are chosen as this is the minimum number of points required in order to define a plane in 3D space. These identified points are assumed to be inliers and, once the plane has been defined, the number of data points which lie on this plane is tallied. This tally, alongside the points used to define the plane, is stored. This process is repeated for n number of trials or until a plane, on which x% of points lie, has been defined.

Here, a residual threshold of 0.4 is used. The algorithm is halted when a plane is identified which consists of 30% of the total point cloud data, or 50 trials have been conducted. The result is a mask of inlier points, which correspond to those lie on the ground plane; these points are removed from the original point cloud data. Figure 13 illustrates the effectiveness of RANSAC. Yellow points on Figure 13 distinguish the identified ground points and indicate the RANSAC-identified ground plane.

Note that it would be possible for the algorithm to misidentify objects as ground points. Consider the case where a vehicle approaches a T-intersection with a large office building directly ahead; it is likely that the points which make up the office building far outnumber those which lie on the ground and would subsequently be identified as 'ground'. This can be alleviated through the definition of an angle threshold, relative to the vehicle, which must be respected by proposed ground planes (e.g., if a plane is angled at >5° from the vehicle, it is ignored).

3.6.4. Clustering Objects

DBSCAN is a widely used clustering technique. It makes use of two parameters: epsilon and minimum points. It works by randomly selecting a point within the dataset from which a potential cluster can be defined. The epsilon is a distance parameter which forms a radius around the selected data point; all other points which fall within this radius are considered "core" points. If the number of core points exceeds the defined number of minimum points, a cluster is initialized. Once a cluster has been initialized, all points which lie within the epsilon of the core points are added to the cluster-these are known as "border" points. Border points are those which are considered to be part of the cluster, but do not lie within the epsilon of the starting point. These border points then project their own radius, which gathers further related points and adds them to the cluster-this process is repeated until no further points are identified. Once a cluster is finalized, and no further points have been identified in an iteration, a new starting point is randomly selected from the remaining points which do not belong to an existing cluster, and the entire process is repeated.

The algorithm developed during this project made use of DBSCAN, with an epsilon of 0.3 and a minimum point's value of 2. The results can be seen in Figure 14a. Individual objects in Figure 14 are distinguished by colour. Each colour indicates a separate DBSCAN-identified cluster. Those on the left are pedestrians, and the yellow cluster to the right is a false detection of a traffic light pole. Figure 14b shows the actual image.

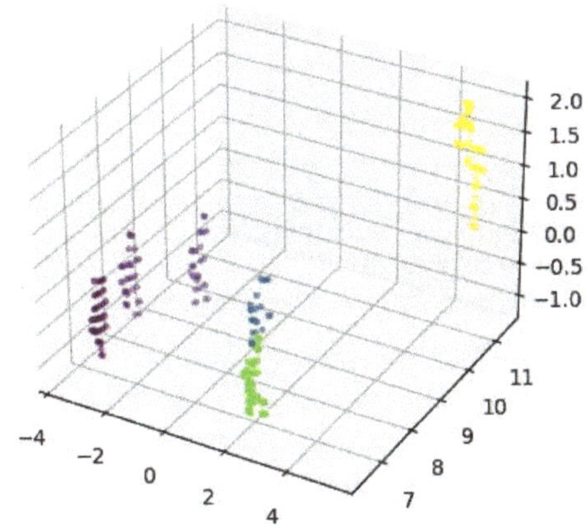

(**a**)

(**b**)

Figure 14. (**a**) Clustered objects defined via DBSCAN clustering. (**b**) Actual image.

While the system described here is incapable of classifying pedestrians, it serves as a foundation from which a system, more accurate than the classification models discussed in previous sections, can be developed through sensor fusion.

4. Results and Discussion

When summarizing the performance of classification algorithms, a simple metric to use is classification accuracy, which is the total number of correct predictions divided by the total predictions. While this allows for a general overview of how well a model might be performing, it lacks details which can be used to better understand the performance of a model and diagnose where it might be failing. A confusion matrix allows for a better understanding of model performance, directly illustrating the ratio of correct and incorrect predictions for each class. In this section, the use of many confusion matrices—one for each model produced— are utilized to show the effectiveness of the approach.

4.1. Rudimentary CNN Models

The two rudimentary models that produced reasonably well results, covered in detail at the start of Section 3, are used as a benchmark to compare the further improved transfer learning models.

The most basic model achieved a classification accuracy of 96% on the test set, with the majority of incorrect classifications belonging to the positive sample class (i.e., pedestrians). Interestingly, an identical model with regularization applied via dropout layers produced almost identical results: It can be seen that the model with regularization performed slightly better in predicting negative samples (i.e., no pedestrian present), however it did misidentify one more positive sample than the most basic model.

It may appear that the addition of dropout resulted in a questionable increase in performance, referring back to the training summary for each model (Figure 4 for the most basic model 6 for the dropout model), there is a significant difference in overfitting-indicated by the disparity between training and validation loss over time. This may be the result of using the same core dataset to compose the training, validation, and testing data splits. It would likely prove beneficial to produce a second testing dataset in order to further investigate the differences between the two models described here.

4.2. Transfer Learning Models

Three transfer learning models were developed and discussed below with results presented on the test data.

4.2.1. Transfer Learning Feature Extractor Model

Moving on to Transfer Learning-based models, starting with the VGG-16 feature extraction model. It is seen that this model underperforms compared to the previous two models (Section 4.1). Referring back to the training summary for this feature extractor model, the loss during training is significantly lower than the best performing rudimentary model (Figure 7). This is indicative that the feature extractor model is less overfit than the rudimentary model—the rudimentary model performs better on the test set, however it may be the case that the feature extractor model performs better on data derived from completely new datasets. Furthermore, the fact that convolutional blocks have been frozen during the training of this feature extractor model should also be taken into account; it was unable to adapt to the provided dataset.

4.2.2. Transfer Learning Models with Image Augmentation

As discussed in Section 3.3, it is imperative that suitable image augmentation is applied prior to the training of any classification model. Here, two models which make use of image augmentation at varying intensity are used.

A model trained using overly intense image augmentation significantly underperforms, compared to all other models. Approximately 90% of positive samples (i.e., those that contain pedestrians) are misidentified as not containing a pedestrian. The model trained with suitable image augmentation, shows a substantial improvement in the model's ability to identify pedestrians in data samples, slightly outperforming the feature extraction model.

4.2.3. Transfer Learning Adaptive Model

The adaptive model is the best-performing model described in this paper. It yields an accuracy of approximately 98% on testing data (100% for negative samples, 96.6% on positive samples). Of course, this is due to the 'unfreezing' of convolutional blocks 4 and 5 during training. The ability of this model to adjust its weights during training enabled it to better understand the dataset and adapt accordingly.

Table 2 shows the accuracy of each model developed. All models, with the exception of that which has been trained using flawed image augmentation, perform admirably. The caveat is that, as shown in the training summaries under Section 3.3 all models appear to be overfitting to varying degrees. Further, Table 3 presents the confusion matrix compares the models in terms of additional metrics of precision, recall and F1 score.

Table 2. Classification results on the models proposed.

Model	Percentage Accuracy on Positive Samples	Percentage Accuracy on Negative Samples	Percentage Accuracy on All Samples
Rudimentary CNN model	92.6	99.2	96
Rudimentary CNN model with regularization	99.8	92.4	96.4
TF model with feature extractor	100	68.2	84.1
TF model with flawed image augmentation	100	8.2	54.1
TF model with suitable image augmentation	100	66.4	83.2
Adaptive TF model	100	96.6	98.3

Table 3. Confusion matrix.

Model	Precision		Recall		F1 Score	
	Positive	Negative	Positive	Negative	Positive	Negative
Rudimentary CNN model	0.99	0.93	0.93	0.99	0.96	0.96
Rudimentary CNN model with regularization	1	0.93	0.92	1	0.96	0.96
TF model with feature extractor	1	0.76	0.68	1	0.81	0.86
TF model with flawed image augmentation	1	0.52	0.08	1	0.15	0.69
TF model with suitable image augmentation	1	0.75	0.66	1	0.80	0.86
Adaptive TF model	1	0.97	0.97	1	0.98	0.98

The adaptive transfer learning model, the best performing model proposed in this paper, suffers from minimal overfitting. Furthermore, it appears to perform exceptionally well on training data. Rudimentary models, while offering acceptable performance in classification of test data, have not been trained on a dataset as large as those which make use of transfer learning. Comparing the results in [43] where they used three different datasets and trained models using transfer learning, the results in this paper are very similar. The authors achieved an accuracy of 96.71% with 2000 training samples and 99.52% with 5000 training samples using SVM classifiers on the PRID database. In this paper, an accuracy of 98.3% on 3000 training samples is achieved.

As transfer learning models proposed in this paper use the VGG-16 model weights, they can be considered more reliable; while not trained directly on the ImageNet database, the inherent knowledge acquired from the VGG-16 model's training on ImageNet provided a foundation from which a domain-specific (i.e., pedestrian classification) understanding can be cultivated.

4.3. Sensor Fusion

In Sections 3.1–3.5, a description how a CNN might be trained for use in a pedestrian classification system is explained, and in Section 3.6, a process for object clustering on LiDAR point cloud data can be applied to segment points which represent pedestrians in the environment surrounding a road vehicle. While CNN methods can provide acceptable classification results for pedestrian detection, the combination of visual and spatial data holds the potential to improve the efficiency and effectiveness of a pedestrian detection system through sensor fusion. There are two categories of sensor fusion: early fusion and late fusion.

Sensor fusion compiles the outputs of multiple sensors, such as LiDAR, RADAR, cameras, etc. The goal of sensor fusion is to create a model which leverages the strengths of each sensor type in the hopes of mitigating their weaknesses. An example of this is the use of LiDAR point cloud data to identify pedestrians in low-light conditions, in which a camera would likely underperform. Conversely, consider a situation where a pedestrian is standing next to a set of traffic lights: a 3D classification system may determine that the pedestrian and traffic lights make up the same object-a camera would be able to differentiate between the two.

In early fusion, the raw data produced by sensors is fused together. LiDAR point cloud data produced by LiDAR sensors can be projected onto 2D images gathered by cameras, for example. 2D object detection on images can be combined with region of interests (ROIs) generated through this point cloud projection through ROI matching, which validates candidate detections proposed by the 2D detection system.

In late fusion, the results of independent detections are fused. LiDAR point clouds are fed into 3D object detection systems, while images are fed into 2D object detection systems. A 3D projection can be created from these 2D detections, which are then cross-referenced with the LiDAR detections via intersection over union (IOU) matching.

5. Conclusions

In this paper, a discussion of the existing work pertaining to pedestrian classification through machine learning and deep learning techniques for an autonomous vehicle is presented and a review of convolutional neural networks and how they can be applied in the scope of pedestrian classification is included.

A number of classification models have been proposed, trained on the CVC-02 dataset. It was found that the regularization did not lead to significant improvements in accuracy, however, it did result in a less-overfitting model which is able to better generalize unseen data. Additionally, the image augmentation must be appropriately applied to training data prior to the training of a classification model. Failure to do so will produce a model which significantly underperforms and is unsuitable for use in an autonomous driving system.

The advantage of VGG-16 architecture with a transfer learning model is discussed and shown to have better performance than the models trained using traditional methods. Furthermore, it is concluded that allowing convolutional layers to update their weights during training is beneficial and can lead to exceptional results when compared to models trained with their convolutional layers frozen.

Additionally, a pre-processing system, whereby LiDAR point cloud data is prepared for use in a 3D object classification model, making use of RANSAC regression and DBSCAN clustering, and methods by which visual and spatial data can be combined via sensor fusion in order to boost the results of a pedestrian classification system, are proposed.

Author Contributions: Conceptualization, A.K. and A.M.; methodology, A.M.; software, A.M.; validation, A.M.; formal analysis, A.M.; investigation, A.M.; resources, A.M. and A.K.; data curation, A.M.; writing—original draft preparation, A.M.; writing—review and editing, A.M., A.K. and S.S.; visualization, A.M. and A.K.; supervision, A.K.; project administration, A.K.; funding acquisition, A.K. and S.S. All authors have read and agreed to the published version of the manuscript.

Funding: This research received no external funding. The submitted research is an outcome of MSc dissertation which was commissioned to produce preliminary results in support of the EU European Regional Development Fund (ERDF) funded project Marine-i. The publication costs will be met by EU ERDF funded Marine-i project.

Conflicts of Interest: The authors declare no conflict of interest.

References

1. Mordor Intelligence. 2021. Available online: https://www.mordorintelligence.com/industry-reports/autonomous-driverless-cars-market-potential-estimation (accessed on 29 August 2021).
2. Edmonds, E. AAA: American Trust in Autonomous Vehicles Slips. 2021. Available online: https://newsroom.aaa.com/2018/05/aaa-american-trust-autonomous-vehicles-slips/ (accessed on 30 August 2021).
3. Al Najada, H.; Mahgoub, I. Big vehicular traffic Data mining: Towards accident and congestion prevention. In Proceedings of the 2016 International Wireless Communications and Mobile Computing Conference (IWCMC), IEEE, Paphos, Cyprus, 5–9 September 2016; pp. 256–261. [CrossRef]
4. Li, H.; Wu, Z.; Zhang, J. Pedestrian detection based on deep learning model. In Proceedings of the 2016 9th International Congress on Image and Signal Processing, BioMedical Engineering and Informatics (CISP-BMEI), IEEE, Datong, China, 15–17 October 2016; pp. 796–800. [CrossRef]
5. Deng, J.; Dong, W.; Socher, R.; Li, L.J.; Li, K.; Li, F.F. Imagenet: A Large-Scale Hierarchical Image Database. In Proceedings of the 2009 IEEE Conference on Computer Vision and Pattern Recognition, Miami, FL, USA, 20–25 June 2009; pp. 248–255.
6. Polana, R.; Nelson, R. Detecting activities. In Proceedings of the IEEE Conference on Computer Vision and Pattern Recognition, Boston, MA, USA, 7–12 June 2015. [CrossRef]
7. Tuzel, O.; Porikli, F.; Meer, P. Pedestrian Detection via Classification on Riemannian Manifolds. *IEEE Trans. Pattern Anal. Mach. Intell.* **2008**, *30*, 1713–1727. [CrossRef] [PubMed]
8. Nguyen, D.T.; Li, W.; Ogunbona, P.O. Human detection from images and videos: A survey. *Pattern Recognit.* **2016**, *51*, 148–175. [CrossRef]
9. Li, B.; Yao, Q.; Wang, K. A review on vision-based pedestrian detection in intelligent transportation systems. In Proceedings of the 2012 9th IEEE International Conference on Networking, Sensing and Control, IEEE, Beijing, China, 11–14 April 2012, pp. 393–398.
10. Bali, S.; Tyagi, S.S. A Review of Vision-Based Pedestrian Detection Techniques. 2018. Available online: https://papers.ssrn.com/sol3/papers.cfm?abstract_id=3315411 (accessed on 13 October 2021).
11. Hao, X.; Zhang, G.; Ma, S. Deep Learning. *Int. J. Semant. Comput.* **2016**, *10*, 417–439. [CrossRef]
12. Kim, H.; Kim, J.; Kim, Y.; Kim, I.; Kim, K.J. Design of network threat detection and classification based on machine learning on cloud computing. *Clust. Comput.* **2018**, *22*, 2341–2350. [CrossRef]
13. Lei, J.; Chen, Y.; Peng, B.; Huang, Q.; Ling, N.; Hou, C. Multi-Stream Region Proposal Network for Pedestrian Detection. In Proceedings of the 2018 IEEE International Conference on Multimedia & Expo Workshops (ICMEW), IEEE, San Diego, CA, USA, 23–27 July 2018; pp. 1–6. [CrossRef]
14. Sun, R.; Wang, H.; Zhang, J.; Zhang, X. Attention-Guided Region Proposal Network for Pedestrian Detection. *IEICE Trans. Inf. Syst.* **2019**, *E102.D*, 2072–2076. [CrossRef]
15. Lowe, D. Object Recognition from Local Scale-Invariant Features. In Proceedings of the Seventh IEEE International Conference on Computer Vision, Corfu, Greece, 20–25 September 1999; IEEE Computer Society: Washington, DC, USA, 1999; Volume 2, pp. 1150–1157.
16. Marr, D. *Vision: A Computational Investigation into the Human Representation and Processing of Visual Information*; The Mitt Press: Cambridge, MA, USA, 1982. [CrossRef]
17. Viola, P.; Jones, M. Rapid object detection using a boosted cascade of simple features. In Proceedings of the 2001 IEEE Computer Society Conference on Computer Vision and Pattern Recognition, CVPR 2001, Kauai, HI, USA, 8–14 December 2001. [CrossRef]
18. Freund, Y.; Schapire, R. A Short Introduction to Boosting. *J. Jpn. Soc. Artif. Intell.* **1999**, *1612*, 771–780.
19. Viola, P.; Jones, M. Snow Detecting pedestrians using patterns of motion and appearance. In Proceedings of the 9th IEEE International Conference on Computer Vision, IEEE, Nice, France, 13–16 October 2003; Volume 2, pp. 734–741.
20. Dalal, N.; Triggs, B. Histograms of Oriented Gradients for Human Detection. In Proceedings of the IEEE Computer Society Conference on Computer Vision and Pattern Recognition (CVPR'05), San Diego, CA, USA, 20–25 June 2005. [CrossRef]
21. Mathias, M.; Cheriet, A.; Cheriet, M. Support Vector Machine. In *Encyclopedia of Biometrics*; Springer: Berlin/Heidelberg, Germany, 2015; pp. 1504–1511. [CrossRef]
22. Mohan, A.; Papageorgiou, C.; Poggio, T. Example-based object detection in images by components. *IEEE Trans. Pattern Anal. Mach. Intell.* **2001**, *23*, 349–361. [CrossRef]

23. Nafiah, F.; Sophian, A.; Khan, R.; Hamid, S.B.A.; Abidin, I.M.Z. Image-Based Feature Extraction Technique for Inclined Crack Quantification Using Pulsed Eddy Current. *Chin. J. Mech. Eng.* **2019**, *32*, 26. [CrossRef]
24. Felzenszwalb, P.; McAllester, D.; Ramanan, D. A discriminatively trained, multiscale, deformable part model. In Proceedings of the IEEE Conference on Computer Vision and Pattern Recognition, Anchorage, AK, USA, 23–28 June 2008; p. 1984.
25. Fischler, M.; Elschlager, R. The Representation and Matching of Pictorial Structures. *IEEE Trans. Comput.* **1973**, *C-22*, 67–92. [CrossRef]
26. McCulloch, W.S.; Pitts, W. A logical calculus of the ideas immanent in nervous activity. *Bull. Math. Biol.* **1943**, *5*, 115–133. [CrossRef]
27. Lecun, Y.; Bottou, L.; Bengio, Y.; Haffner, P. Gradient-based learning applied to document recognition. *Proc. IEEE* **1998**, *86*, 2278–2324. [CrossRef]
28. Rawat, W.; Wang, Z. Deep Convolutional Neural Networks for Image Classification: A Comprehensive Review. *Neural Comput.* **2018**, *2733*, 2709–2733. [CrossRef] [PubMed]
29. Tygert, M.; Bruna, J.; Chintala, S.; LeCun, Y.; Piantino, S.; Szlam, A. A Mathematical Motivation for Complex-Valued Convolutional Networks. *Neural Comput.* **2016**, *28*, 815–825. [CrossRef] [PubMed]
30. LeCun, Y.; Boser, B.; Denker, J.S.; Howard, R.E.; Habbard, W.; Jackel, L.D.; Henderson, D. Handwritten digit recognition with a back-propagation network. In *Advances in Neural Information Processing Systems 2*; Morgan Kaufmann Publishers Inc.: Burlington, MA, USA, 1990; pp. 396–404. ISBN 1558601007.
31. Yosinski, J.; Clune, J.; Bengio, Y.; Lipson, H. How transferable are features in deep neural networks? *arXiv* **2014**, arXiv:1411.1792.
32. Day, C.; McEachen, L.; Khan, A.; Sharma, S.; Masala, G. Pedestrian Recognition and Obstacle Avoidance for Autonomous Vehicles Using Raspberry Pi. In *Advances in Intelligent Systems and Computing*; Springer: Berlin/Heidelberg, Germany, 2019; pp. 51–69.
33. Laney, D. 3D data management: Controlling data volume, velocity and variety. *META Group Res. Note* **2001**, *6*, 1.
34. Gerónimo, D.; Sappa, A.; Ponsa, D.; López, A. 2D–3D-based on-board pedestrian detection system. *Comput. Vis. Image Underst.* **2010**, *114*, 583–595. [CrossRef]
35. Mimouna, A.; Alouani, I.; Ben Khalifa, A.; El Hillali, Y.; Taleb-Ahmed, A.; Menhaj, A.; Ouahabi, A.; Ben Amara, N.E. OLIMP: A Heterogeneous Multimodal Dataset for Advanced Environment Perception. *Electronics* **2020**, *9*, 560. [CrossRef]
36. Krizhevsky, A.; Sutskever, I.; Hinton, G.E. ImageNet Classification with Deep Convolutional Neural Networks. *Commun. ACM* **2017**, *60*, 84–90. [CrossRef]
37. Simonyan, K.; Zisserman, A. Very deep convolutional networks for large-scale image recognition. *arXiv* **2014**, arXiv:1409.1556.
38. Dishashree Gupta and Dishashree. Activation Functions: Fundamentals of Deep Learning. 2020. Available online: https://www.analyticsvidhya.com/blog/2020/01/fundamentals-deep-learning-activation-functions-when-to-use-them/ (accessed on 29 August 2021).
39. Arbaoui, A.; Ouahabi, A.; Jacques, S.; Hamiane, M. Concrete Cracks Detection and Monitoring Using Deep Learning-Based Multiresolution Analysis. *Electronics* **2021**, *10*, 1772. [CrossRef]
40. Arbaoui, A.; Ouahabi, A.; Jacques, S.; Hamiane, M. Wavelet-based multiresolution analysis coupled with deep learning to efficiently monitor cracks in concrete. *Frat. Integrità Strutt.* **2021**, *15*, 33–47. [CrossRef]
41. Ferguson, M.; Ak, R.; Lee YT, T.; Law, K.H. Automatic localization of casting defects with convolutional neural networks. In Proceedings of the 2017 IEEE International Conference on Big Data, Boston, MA, USA, 11–14 December 2017; pp. 1726–1735. [CrossRef]
42. Caesar, H.; Bankiti, V.; Lang, A.H.; Vora, S.; Liong, V.E.; Xu, Q.; Krishnan, A.; Pan, Y.; Baldan, G.; Beijbom, O. nuScenes: A multimodal dataset for autonomous driving. *arXiv* **2019**, arXiv:1903.11027.
43. Wang, J.-T.; Yan, G.-L.; Wang, H.-Y.; Hua, J. Pedestrian recognition in multi-camera networks based on deep transfer learning and feature visualization. *Neurocomputing* **2018**, *316*, 166–177. [CrossRef]

electronics MDPI

Article

A Change of Paradigm for the Design and Reliability Testing of Touch-Based Cabin Controls on the Seats of Self-Driving Cars

Tiago Custódio [1,2,†], Cristiano Alves [1,2,†], Pedro Silva [3,*], Jorge Silva [3], Carlos Rodrigues [3], Rui Lourenço [3], Rui Pessoa [3], Fernando Moreira [3], Ricardo Marques [3], Gonçalo Tomé [3] and Gabriel Falcao [1,2]

1 Instituto de Telecomunicações, 3030-290 Coimbra, Portugal; tiagocustodio1@gmail.com (T.C.); cristianoalves1717@gmail.com (C.A.); gff@co.it.pt (G.F.)
2 Department of Electrical and Computer Engineering, University of Coimbra, 3030-290 Coimbra, Portugal
3 CIE Plasfil, Zona Industrial Da Gala, Lote 6, 3090-380 Figueira da Foz, Portugal; jorge.silva@plasfil.pt (J.S.); carlos.rodrigues@plasfil.pt (C.R.); rui.lourenco@plasfil.pt (R.L.); rui.pessoa@plasfil.pt (R.P.); fernando.moreira@plasfil.pt (F.M.); ricardo.marques@plasfil.pt (R.M.); goncalo.tome@plasfil.pt (G.T.)
* Correspondence: pedro.silva@plasfil.pt; Tel.: +351-233-401-218
† These authors contributed equally to this work.

Abstract: The current design paradigm of car cabin components assumes seats aligned with the driving direction. All passengers are aligned with the driver that, until recently, was the only element in charge of controlling the vehicle. The new paradigm of self-driving cars eliminates several of those requirements, releasing the driver from control duties and creating new opportunities for entertaining the passengers during the trip. This creates the need for controlling functionalities that must be closer to each user, namely on the seat. This work proposes the use of low-cost capacitive touch sensors for controlling car functions, multimedia controls, seat orientation, door windows, and others. In the current work, we have reached a proof of concept that is functional, as shown for several cabin functionalities. The proposed concept can be adopted by current car manufacturers without changing the automobile construction pipeline. It is flexible and can adopt a variety of new functionalities, mostly software-based, added by the manufacturer, or customized by the end-user. Moreover, the newly proposed technology uses a smaller number of plastic parts for producing the component, which implies savings in terms of production cost and energy, while increasing the life cycle of the component.

Keywords: cabin control functions; touch sensor; touch button; capacitive sensor; injected plastic; plastic button; embedded lighting; reliability testing; automotive industry; self-driving car

Citation: Custódio, T.; Alves, C.; Silva, P.; Silva, J.; Rodrigues, C.; Lourenço, R.; Pessoa, R.; Moreira, F.; Marques, R.; Tomé, G.; et al. A Change of Paradigm for the Design and Reliability Testing of Touch-Based Cabin Controls on the Seats of Self-Driving Cars . *Electronics* **2022**, *11*, 21. https://doi.org/10.3390/electronics11010021

Academic Editors: Calin Iclodean, Bogdan Ovidiu Varga and Felix Pfister

Received: 13 November 2021
Accepted: 13 December 2021
Published: 22 December 2021

1. Introduction

Recent advances in self-driving cars are expected to translate into a significant number of new vehicles circulating using this new paradigm in the coming years [1,2]. Many works state that by 2050 self-driving cars will dominate, which creates new opportunities but also new challenges [3]. In this context, the human driver will likely be released from functions, offloading that responsibility to the machine and artificial intelligence algorithms [4,5]. Once relieved from driving duties [6], the driver will benefit from new services inside the car cabin, including those related to using multimedia and communications such as games, work meetings, movies, music and Internet browsing, to name a few. Furthermore, there may no longer exist a clear separation of the passengers the way we have it today due to the transmission tunnel on the cockpit's floor. In this context, it is possible that a brand differentiator may well consist of the interior design of the car cabin, namely the seat arrangement and orientation, as well as the floor of the car cabin [7,8].

Therefore, the drivers' seat, as well as the seats of the other passengers no longer need to be aligned with the moving direction. Moreover, the seats should be able to rotate on their axis instead of simply sliding back and forth to adjust proper leg positioning as they do today.

To serve this purpose, the passengers no longer will be aligned with the driving direction, possibly turning to face each other, which creates the need to add control buttons to operate general functionalities from their seats, as opposed to having these control buttons on the doors and central panel.

The development of this technology will imply redesigning the seats to incorporate these controls. Today's standard for implementing funcional controls in the automotive industry is to build buttons which are composed of mechanical devices comprised of several plastic polymer parts [9]. The actual standard presents several disadvantages as compared to robust alternatives that can provide a functionality within a single plastic part.

Touch-based technologies [10] can pave the way towards the integration of sensors coupled to a single plastic part attached to the seat. In this case, the touch sensors lying below the plastic part will be able to sense the interaction of the passenger to control a certain functionality. This is important as it allows a reduced assembling time and a larger Mean Time Between Failures (MTBF). Moreover, it allows the same part design to be incorporated into many models of the brand, since the different funcionalities are controlled by software.

In this work, we developed the proposed technology and designed a new car seat that incorporates a viable design for controlling the opening and closing of car side windows and seat positioning. We were able to develop and test the technology on a real car used in the market and also on a new seat with an innovative design able to be incorporated in self-driving cars.

This article presents the following contributions:

- Performance analysis of capacitive touch sensors for the automotive industry;
- Integration of microelectronics to control capacitive sensors with injected polymer plastic to be coupled to the seat;
- Development and implementation of a testing and reliability assessment system, by automatically applying and monitoring thousands of touch interactions per plastic part;
- Design of a futuristic car seat with integrated controls;
- Development, implementation and testing of a functional prototype to control the opening and closing of car side windows and the position of the car seats.

This work is also part of the 'Collective Efficiency Strategies', totally aligned with the 'Mobinov—Automobile Cluster Association', which identified as one of its main goals "to contribute to making Portugal a reference in research, innovation, design, development, manufacture and testing of products and services of the automotive industry" and "strengthen the competitiveness of a fundamental sector of the economy, promoting an increase in exportation" [9].

This article is structured as follows. Section 2 describes the materials and methods used, analyzing in depth the capacitive touch sensor technology selected for the current context, and validating its use over thousands of cycles of operation. Section 3 addresses implementation and results. It analyzes the design of the newly developed car seat and the methods used to integrate injectable plastic with capacitive touch sensors in it. The discussion and future research directions in Section 4 close the article.

2. Materials and Methods

Distinct sensing technologies such as inductive, infrared, ultrasound, resistive and capacitive-based ones were analyzed and tested in depth [9], and their characteristics compared. The decision for the most appropriate technology to use in this context befell upon capacitive touch sensors depicted in Figure 1a,b. The comparison criteria ranged from ease of use, reliability, cost, ease of integration and smallest design footprint [11,12].

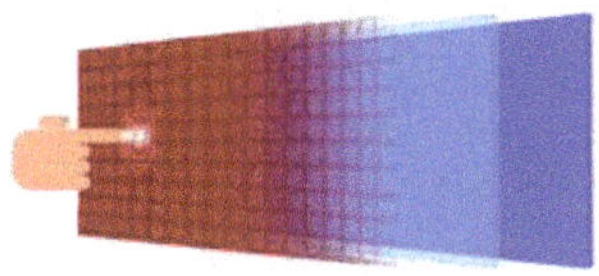

(a)

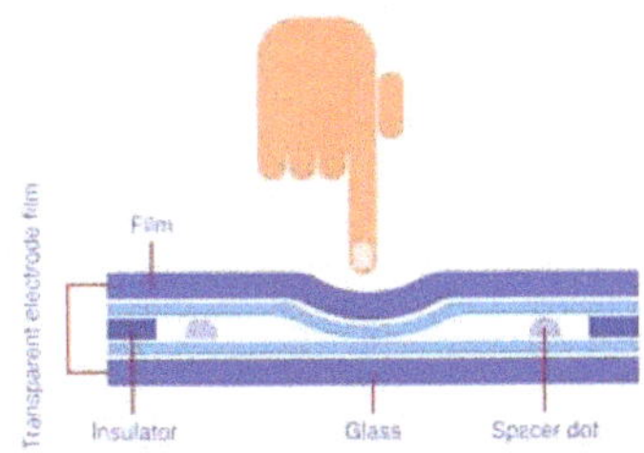

(b)

Figure 1. Touch-based technology. (**a**) General illustration that resembles the use of mobile phones and (**b**) technical detail of the touch surface.

2.1. Touch Sensors for Activating Functions in Polymers

Capacitive touch sensing is a low-cost and low-complexity technology [13] ubiquitous in today's devices. Thus, in a way, there is already a precedent on how people expect these interfaces to work and how to interact with them [14]. Several capacitive sensors (four examples are depicted in Figure 2) were tested with plastic samples of varying thickness and composition as exhaustively described in Subsection 3.1 of [9]. The sensors experimented offered two different designs and two different channel counts. The experimental results were successful to the extent where the touch detection rate nearly reached 100% for every sensor tested, as presented in Table 1.

After validating the technology and method, in order to reduce the cost and the design footprint, it was decided to conceive the sensors directly on the Printed Circuit Board (PCB) (see Figure 3). The way these capacitive sensors work is based on a method known as self-capacitance [15], where each sensor is read using a single input of the system. Self-capacitance touch sensors use a single sensor electrode to measure the apparent capacitance between the electrode and the ground of the touch sensor. This method offers good immunity to noise induced from neighbouring sensors and circuitry.

To build a proof-of-concept design using this sensor, a microcontroller is required for reading sensing data and acting accordingly. A microcontroller consists of an embedded processor with auxiliary peripherals, I/O and electronics to interface with external sensors and other hardware. The microcontroller necessary for this prototype should ease the job of integrating all the technology by having peripherals to interact with capacitive sensors and potentially interact with an automobile's Electronic Control Unit (ECU) [16], as depicted in Figure 4.

(a) (b) (c) (d)

Figure 2. The figure depicts the capacitive sensors tested for the current design from (**a**–**d**). The electrical properties and datasheets for each sensor can be found in Table 1. The effectiveness tests performed these sensors are described in Table 2.

Table 1. Electrical characteristics of the tested sensors in Figure 2.

Sensor #	Int. Circuit	Pad	Type	Datasheet
Sensor 1	TTP223-B	Circular	1 Channel	https://datasheet.lcsc.com/szlcsc/TTP223-BA6_C80757.pdf (accessed on July 13 2020)
Sensor 2	TTP223N-B	Squared	1 Channel	https://datasheet.lcsc.com/szlcsc/TTP223-BA6_C80757.pdf (accessed on 13 July 2020)
Sensor 3	TTP224	Squared	4 Channels	https://download.mikroe.com/documents/datasheets/ttp224.pdf (accessed on 13 July 2020)
Sensor 4	0401-8224	Circular	4 Channels	Clone TTP224

Table 2. Effectiveness tests for sensors 1–4 depicted in Figure 2.

Sensor	Polymer	Description	Thickness (mm)	Test Number	Hits	Effectiveness	Assessment
Sensor 1	PC ABS	Taroblend 66	2.5	30	30	100.00%	Approved
Sensor 1	PA6GF30	Ultramid B3E2G6	2.5	30	30	100.00%	Approved
Sensor 1	PPTD20	Hostacom TRC 352N	2.5	30	29	96.67%	Approved
Sensor 1	PP	Sabic PHC3181	2.5	30	29	96.67%	Approved
Sensor 1	PA6	Badamid B70S Natur	2.5	30	29	96.67%	Approved
Sensor 2	PC ABS	Taroblend 66	2.5	30	30	100.00%	Approved
Sensor 2	PA6GF30	Ultramid B3E2G6	2.5	30	30	100.00%	Approved
Sensor 2	PPTD20	Hostacom TRC 352N	2.5	30	30	100.00%	Approved
Sensor 2	PP	Sabic PHC3181	2.5	30	30	100.00%	Approved
Sensor 2	PA6	Badamid B70S Natur	2.5	30	29	96.67%	Approved
Sensor 3	PC ABS	Taroblend 66	1.8	30	29	96.67%	Approved
Sensor 3	PA6GF30	Ultramid B3E2G6	1.8	30	30	100.00%	Approved
Sensor 3	PPTD20	Hostacom TRC 352N	1.8	30	29	96.67%	Approved
Sensor 3	PP	Sabic PHC3181	1.8	30	30	100.00%	Approved
Sensor 3	PA6	Badamid B70S Natur	1.8	30	30	100.00%	Approved
Sensor 4	PC ABS	Taroblend 66	1.8	30	30	100.00%	Approved
Sensor 4	PA6GF30	Ultramid B3E2G6	1.8	30	30	100.00%	Approved
Sensor 4	PPTD20	Hostacom TRC 352N	1.8	30	30	100.00%	Approved
Sensor 4	PP	Sabic PHC3181	1.8	30	29	96.67%	Approved
Sensor 4	PA6	Badamid B70S Natur	1.8	30	30	100.00%	Approved

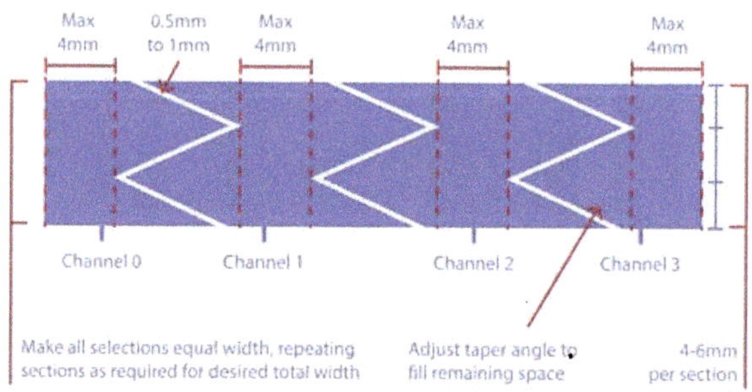

(a)

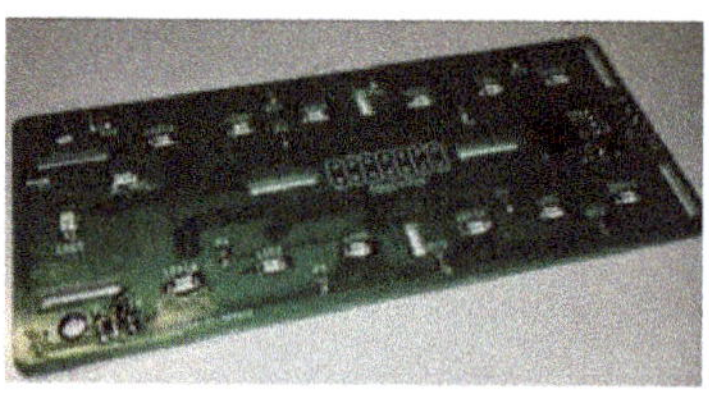

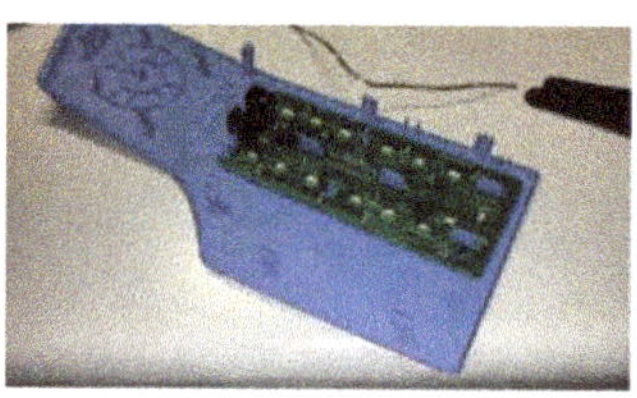

(b) (c)

Figure 3. First prototype of a touch sensor with self capacitance technology. The resulting touch sensor design is shown in (**a**). The bottom PCB is shown in (**b**). The PCB assembled on a plastic part is depicted in (**c**).

With these criteria in mind, the ATSAMC21J18A microcontroller from Microchip was adopted. It features a 32-bit processor architecture, a processor speed of 48 MHz, 32 KB of RAM, 264 KB of non-volatile (Flash) storage memory and 52 general-purpose input/output (GPIO) pins. It also features two important peripherals needed to ease the integration: a Peripheral Touch Controller (PTC) and a Controller Area Network (CAN) [17].

The PTC peripheral makes the process of sampling and validating the capacitive touch sensors more reliable and robust as all of these steps are performed automatically and periodically by the hardware itself. The CAN peripheral [18] was planned as a viable means of integrating this prototype with the automobile's ECU since it implements a standard bus of communications used by the automobile industry. The microcontroller development board and the developed prototype are depicted in Figures 4 and 5, respectively.

Figure 4. PCB and plastic part connected to the sensors and the ATSAMC21J18A microcontroller development board.

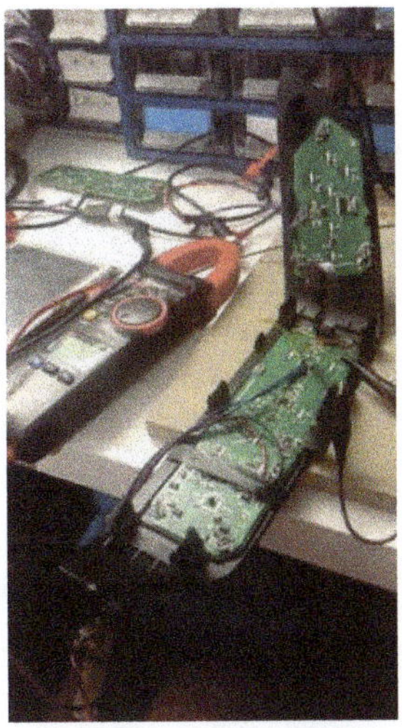

Figure 5. PCB with sensors based on a mutual-capacitance design assembled on a plastic part.

After prototyping this first version and validating the integration method, further design changes were implemented in order to reduce cost and footprint even more. Where previously there was a need to match the amount of GPIOs of the microcontroller to the number of sensors employed using the "Self-Capacitance" [15] design previously explained how we further reduced the number of GPIOs needed using a design called "Mutual-Capacitance" [19]. By using this alternative approach, if sensors are arranged with the electrical connections in a matrix-like array, it would only require M × N GPIOs on the microcontroller, M being the number of rows and N being the number of columns.

Compared against the previous version of the prototype, where 19 GPIOs were needed in order to read the 19 sensors using the "Self-Capacitance" design, in this case only 10 GPIOs are required using a configuration of 3 rows and 7 columns. This design improvement allowed for a greater reduction in cost by making it viable to select a cheaper but powerful microcontroller at the cost of a slightly more complex PCB design (with negligible effect in final cost). The microcontroller chosen was the ATTINY3217 [20], which allows a saving of nearly 71% of the cost when compared to the previously chosen microcontroller.

2.2. Touch Surface Backlighting

Another challenge tackled is the need for adequate backlighting on the touch surface. Backlighting can be used to appropriately guide the user to the touch control, to illuminate pictograms depicting the control function, or both functions.

Integrating backlighting poses a technical challenge as the light needs to be concentric with the touch sensor but cannot interfere with the touch sensing, neither by constraining the surface area for the sensor employment nor by inducing noise on the sensor line. The employed method consisted of drilling a hole in the middle of the sensor that is large enough to allow proper dispersion of the light on the control pictograms. The LEDs were assembled on the underside of the PCB, but shining upwards as suggested in Figure 6.

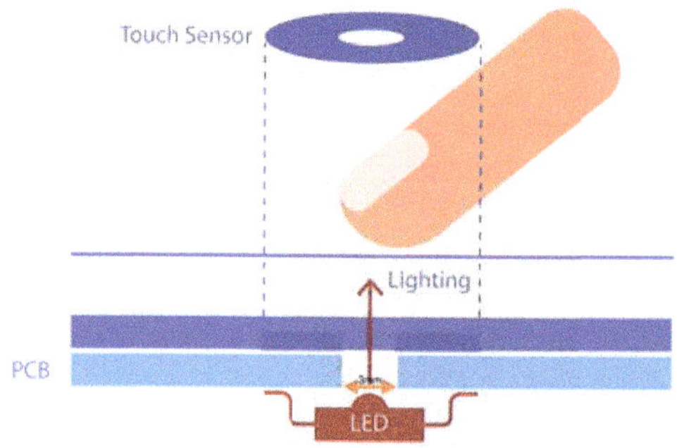

Figure 6. Illustration on how to couple a LED back light to a capacitive touch sensor. The hole where the LED scatters light has a diameter equal to 3 mm.

This approach allows the LED not to interfere electrically with the sensor while properly illuminating the control surface.

After carefully iterating over several drilling diameters, a proper dispersion was achieved using 3 mm, as depicted in Figure 7 and this development was employed in the final prototype. The backlighting effect is illustrated in the intermediate prototype developed shown in Figure 8.

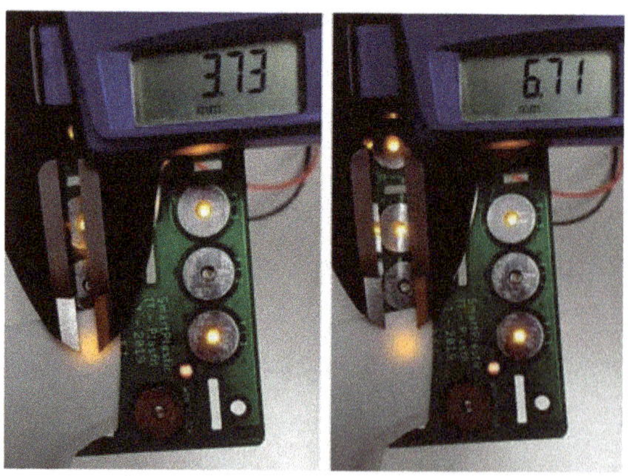

Figure 7. PCB with LEDs shining from behind the sensors and the measuring of the light's hotspot and dispersion diameters.

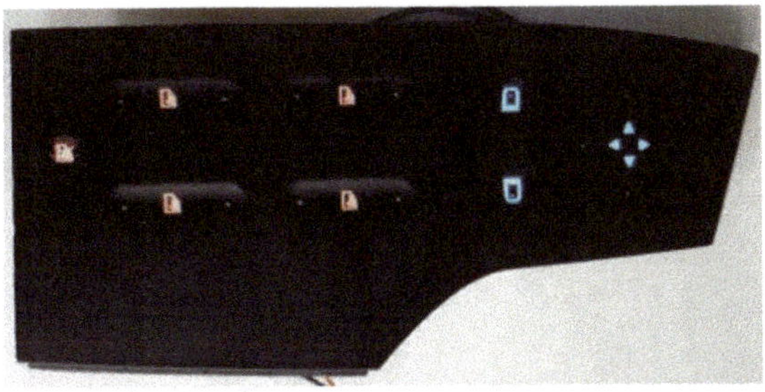

Figure 8. Prototype with backlight testing of two different LED colors.

2.3. Validation and Reliability

The testing of such a piece of hardware can be performed manually or automated with machinery. Testing manually is a time-intensive task and—albeit closer to the real usage of the equipment—requires at least one person to perform the tests by hand and keep track of the success and failure rates. Furthermore there is no way to ensure that the tester performs every test in the same way, with the same motions and adequate finger pressure. Automating the testing by using machinery allows faster tests and a greater degree of accuracy in the motions required. In addition, performing tests faster allows for a greater number of test samples to be acquired in the same time-frame, thus making the whole validation more reliable. Therefore, whenever possible, tests should be automated. An adequate machine for performing this task is a 3-axis Computer Numerical Control (CNC) system [21]. These machines have 3 axis of independent movement and are ubiquitous in several industries (e.g., ranging from 3D printers to assembly lines). They work by having a coupled computer following a precise script telling the machine the movements that have to be done. The adopted setup uses an industrial 3-axis CNC machine [21] with a capacitive "finger" probe (please see Figure 9c) attached in the gripper part as depicted in Figure 9d.

In order to automate the testing procedure we made use of available digital input lines on the CNC controller board in order to allow the CNC to be told by the tested equipment if a given probe touch was detected. All of this was orchestrated using a G-Code script which is the standard scripting language to program a CNC machine to perform a set of movements. The source code and flowchart developed are illustrated in the next page and in Figure 10.

The algorithm employed in our G-Code script can be summarized in a few simple steps:

1. Move the probe to the initial position (2 cm above the sensor);
2. Move the probe down 2 cm (touching the sensor);
3. Wait for a digital signal on the CNC controller's input, warning that a touch was detected;
4. If the signal comes within 5 s, register this iteration as a success; otherwise, mark as a failure;
5. Move the probe up 2 cm (no longer touching the sensor);
6. Repeat the process from 2.

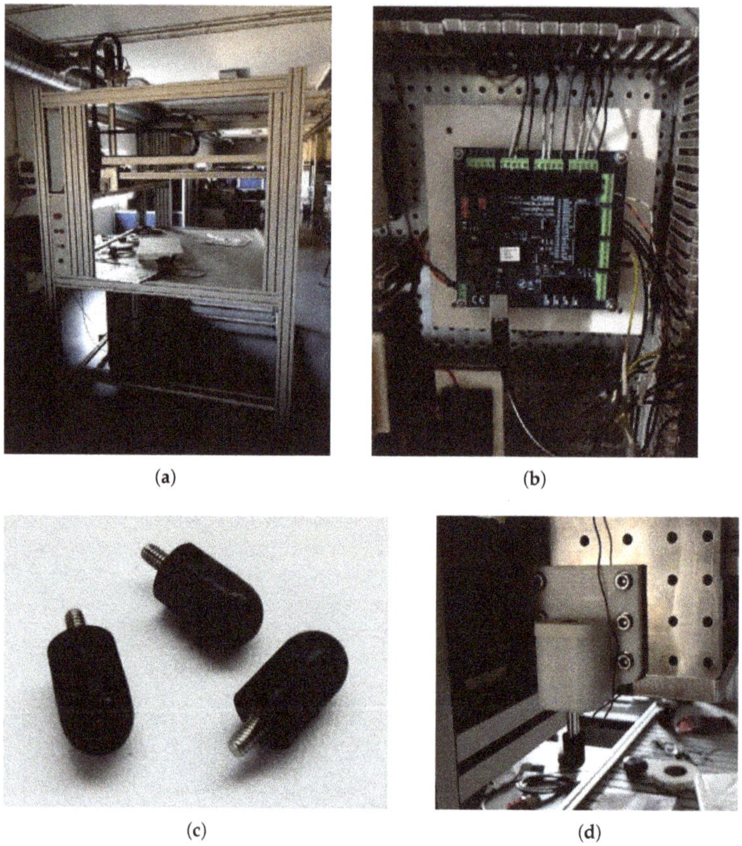

Figure 9. The figure depicts the testing environment specifically designed and developed from scratch to validate the touch-based prototype. In (**a**) the structure of the CNC is shown. (**b**) Depicts the CNC controller board with digital inputs. (**c**) Illustrates the the "finger" probes that emulate a human touch. The gripper in (**d**) was custom-designed and printed using 3D printing technology to allow attachment the probe to the moving part of the CNC strucutre.

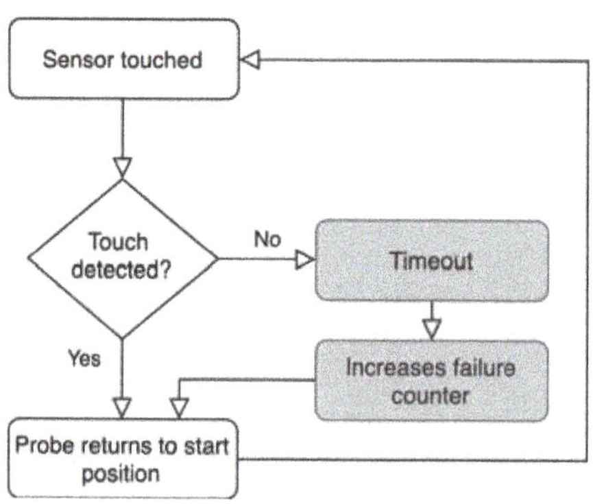

Figure 10. Flowchart of the CNC test script. The G-Code source code is illustrated next.

```
#<loop_count> = 0
#<timeout> = 0
#<timeout_limit> = 50     ;tenths of second
#<successes> = 0
#<misses> = 0
#<t_start> = datetime[]

(logopen,Teste_Touch_Tranca.log)

G00 X338 Y188 Z-430

o100 repeat [10000]     ;repeat ten thousand times
    #<s> = [datetime[] - #<t_start>]

    (print,#<s>, Iteration #<loop_count,0>, Successes: #<successes,0>, Misses: #<misses,0>)
    (log,#<s>, Iteration #<loop_count,0>, Successes: #<successes,0>, Misses: #<misses,0>)

    G00 Z-430     ;move to start position

    o200 while [[#<timeout> LE #<timeout_limit>] AND [#<_hw_input> EQ 1]] ;wait for previous touch release
     G04 P0.1
     #<timeout> = [#<timeout> + 1]
    o200 endwhile
    o300 if [#<timeout> EQ #<timeout_limit>]
     (print,Touch release timeout)
     (log,Touch release timeout)
    o300 endif
    #<timeout> = 0

    G00 Z-440     ;move to touch position

    o400 while [[#<timeout> LE #<timeout_limit>] AND [#<_hw_input> EQ 0]] ;wait for touch signal or
    timeout
     G04 P0.1
     #<timeout> = [#<timeout> + 1]
    o400 endwhile
    o500 if [[#<timeout> LE #<timeout_limit>] AND [#<_hw_input> EQ 1]] ;check if we reached a timeout
     #<timeout> = [#<timeout> / 10]
     (print,Touch signal received after #<timeout,1>s)
     (log,Touch signal received after #<timeout,1>s)
     #<successes> = [#<successes> + 1]     ;increment touch success counter
    o500 else
     (print,Touch signal timeout)
     (log,Touch signal timeout)
     #<misses> = [#<misses> + 1]     ;increment touch miss counter
    o500 endif
    #<timeout> = 0

    #<loop_count> = [#<loop_count> + 1]     ;increment iteration counter
o100 endrepeat

#<t_duration> = [datetime[] - #<t_start>]

G00 Z-420

(print,Took #<t_duration> seconds doing #<loop_count,0> iterations)
(print,#<successes,0> touches detected and #<misses,0> touches missed)

(log,Took #<t_duration> seconds doing #<loop_count,0> iterations)
(log,#<successes,0> touches detected and #<misses,0> touches missed)
(logclose)
```

The complete definition of the testing methodology allowed each one of the 19 slider and on-off sensors on the PCB to be tested and validated in the CNC structure developed, illustrated in Figure 11. Each test was comprised of 10,000 iterations, running from 2 to 3 h. This allowed testing of all the sensors within a week's time.

All tests reported more than 99% reliability, the worst performing sensor achieving 99.45% reliability (see Figure 12 and Table 3) and the best performing one achieving 100% (please refer to Figure 13 and Table 4).

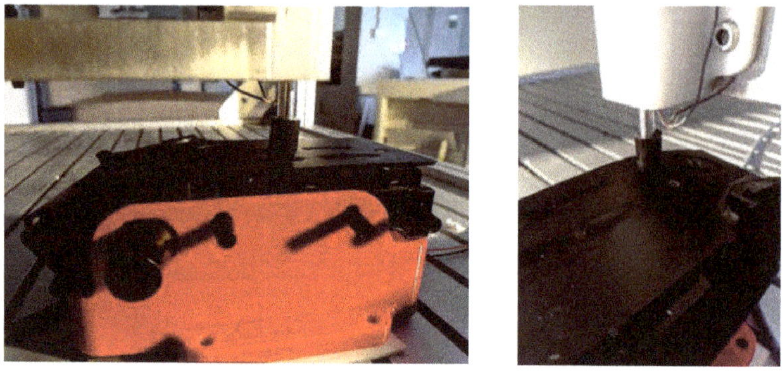

(**a**) (**b**)

Figure 11. (**a**) Testing of a slider-type sensor. (**b**) Testing an on–off sensor.

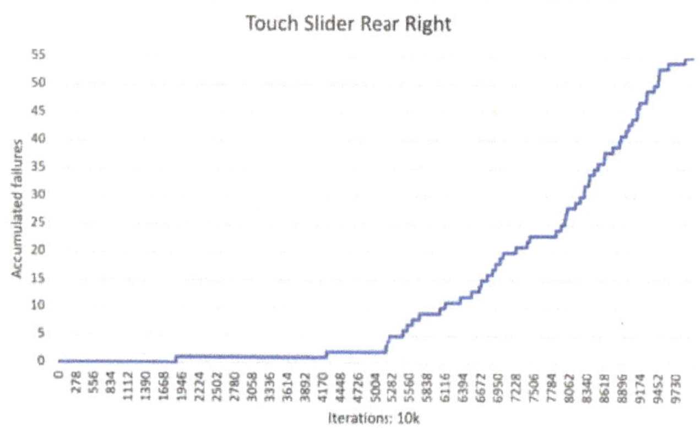

Figure 12. Sensor presenting the worst performance after 10,000 iterations.

Table 3. Statistical data regarding the worst-performing sensor.

Statistical Data	
Number of tests	10,000
Test duration (s)	10,842
Average response time (s)	0.138
Failure count	55
Failure rate	0.55%
Failure distribution, Q1	1
Failure distribution, Q2	1
Failure distribution, Q3	21
Failure distribution, Q4	22

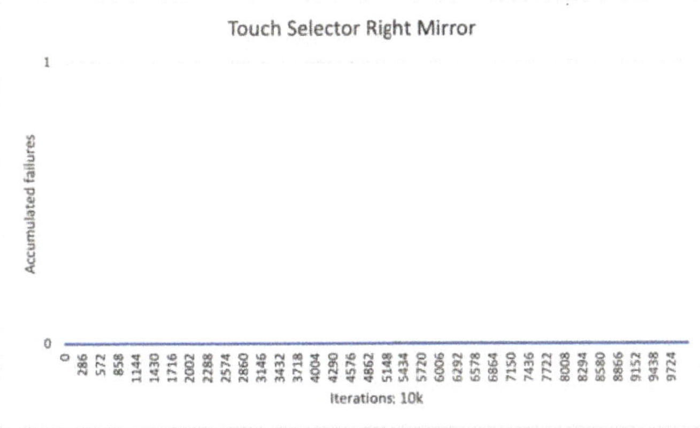

Figure 13. Sensor presenting the best performance after 10,000 iterations.

Figure 12 is a good example of a phenomenon observed where errors tend to show up later during the tests. This might be explained taking into consideration that the integration system calibrates the capacitive sensor baseline over time and, because each one of the 10,000 iterations is performed as quickly as possible (1.1 s for the worst case scenario) the integration system might not have enough time to reliably calibrate back to where the baseline was before, resulting in this accumulated error over time. In a real usage scenario no touch sensor or any other button is used thousands of times during a single trip and not as frequently as in these tests. Considering the strict test conditions, these sensors were also submitted, and considering that the tests show reliability values above 99%, it can be concluded that they are appropriate and reliable.

Another interesting aspect is that the worst performing sensors are positioned within the PCB layout near noise sources such as the power supply circuitry and communication data lines. Therefore, future designs should take care in properly isolating the sensor lines from such circuitry either by properly employing ground planes, by changing the routing of the data and sensor lines or by simply changing the layout of the components. Sensors positioned far from these noise sources have close to no errors as shown in the sensors portrayed in Figures 13–15 and the corresponding Tables 4–6.

Table 4. Statistical data regarding the best performing sensor.

Statistical Data	
Number of tests	10,000
Test duration (s)	6909
Average response time (s)	0.18
Failure count	0
Failure rate	0%
Failure distribution, Q1	0
Failure distribution, Q2	0
Failure distribution, Q3	0
Failure distribution, Q4	0

Table 5. Statistical data regarding the sensor in Figure 14.

Statistical Data	
Number of tests	10,000
Test duration (s)	6820
Average response time (s)	0.172
Failure count	3
Failure rate	0.03%
Failure distribution, Q1	0
Failure distribution, Q2	0
Failure distribution, Q3	2
Failure distribution, Q4	1

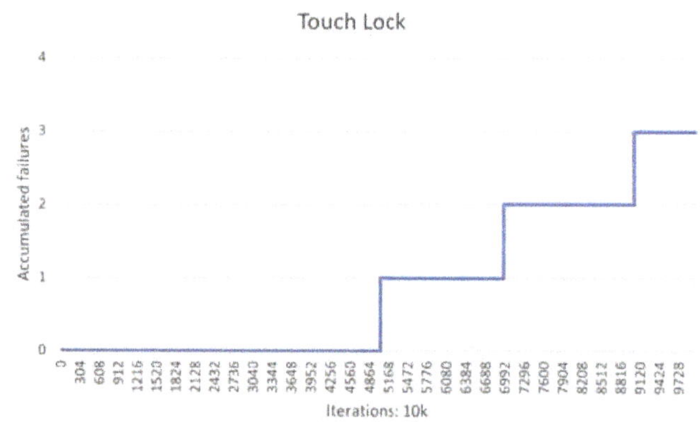

Figure 14. Sensor presenting a typical performance after 10,000 iterations.

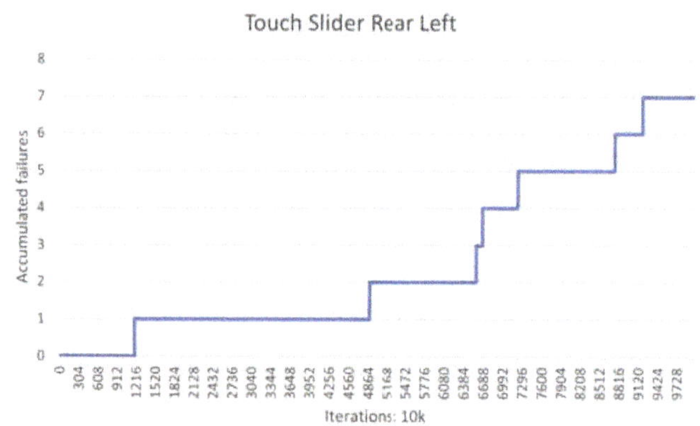

Figure 15. Sensor presenting another typical performance after 10,000 iterations.

Table 6. Statistical data regarding the sensor in Figure 15.

Statistical Data	
Number of tests	10,000
Test duration (s)	10,625
Average response time (s)	0.139
Failure count	7
Failure rate	0.07%
Failure distribution, Q1	1
Failure distribution, Q2	1
Failure distribution, Q3	3
Failure distribution, Q4	2

Finally, one aspect of the technology that deserves consideration is what happens if a sensor becomes out of control. For this, we can assume two possible cases: (1) the sensor starts sending undesired control commands or (2) the sensor freezes and stops working. For both cases there are approaches that must necessarily be foreseen. First, if there is a malfunction of the sensing mechanism, then it may need repair assistance as happens with current systems and cars. However, if the system systematically (or sporadically) assumes an erratic behavior after a certain number of utilizations, then a solution that incorporates a periodic reset of the electrical reference values of the touch system can easily be implemented. For any case, it must be noted that in the thousands of tests conducted through the CNC-based testing system developed, we never experienced a situation where the sensor became out of control. The tests we are currently implementing consist of the utilization of a CNC specifically developed for testing the developed system thousands of times and assessing its response capability and robustness. The manuscript describes these tests, which indicate from 0 (zero) to a maximum of 55 failures detected per 10,000 touches performed, which means the system is robust and reliable (less than 0.55% of failures detected in the worst case scenario, and (0) zero in the best one). Moreover, the newly proposed touch-based system will require less maintenance and become more robust to the presence of water and humidity (e.g., resulting from rain, coming from an open door window) inside the car cabin and near the control buttons, a problem that has been known and described by manufacturers for decades.

2.4. Comparing New and Current Paradigms

The integration tests were developed on a mass-produced automobile (please see Section 4 of [9]). The original vehicle's button responsible for controlling the doors' windows and side mirrors is comprised of 26 discreet components. Amongst these components there are those made of plastic, rubber, metal and PCBs with circuitry as shown in Figure 16. Each one of these components has its own production and/or assembly line which, all together, contribute to a complex sourcing and assembly process that produces a part with many moving components. Moreover, each component fabrication process needs to comply with physical tolerances that may lead to faults. Furthermore, hand assembly is often required and, thus, it is not an error-free process. All this contributes to a part that may often result in waste during fabrication and, having a reduced quality standard due to the tolerances may lead to short MTBF, which produces high maintenance and replacement costs.

The developed prototype has no moving parts as it is made up of a polymer plastic cover and two PCBs. Both the polymer plastic component and the PCBs with their respective circuitry have a fully automated and mature fabrication process that does not require human intervention. Thus, there is little room for tolerance error propagation along the process. This results in larger MTBF which, in turn, represents lower maintenance and replacement costs.

(a)

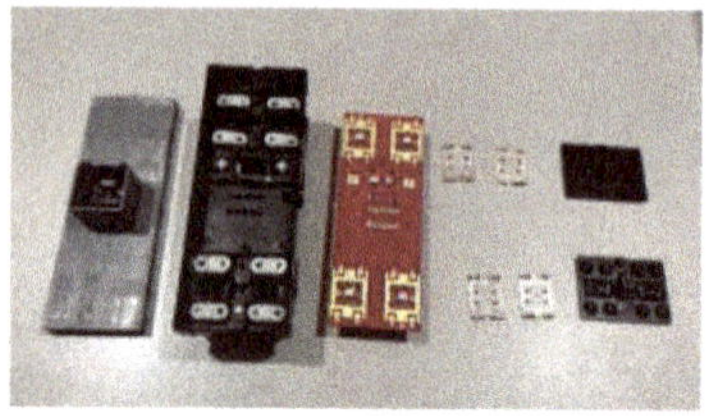

(b)

Figure 16. Example of the number of components and complexity of (**a**) side mirror's and (**b**) windows' control buttons

Of particular interest is the fact that this part is fully programmable and can communicate with the automobile using industry standard protocols. Therefore, it is suitable to be mass-produced and adapted to several automobiles or even different applications, or environments such as aircraft seats, working space seats, etc.

3. Implementation and Results

The first prototype was developed to run a proof of concept with a mass-produced automobile from 2018 [9]. However, the monitor-supported prototype seat (depicted in Figure 17b) was also built to house the SPaC part so that it could be presented to customers and disseminate the new technology. This seat was conceived, developed and built by the company AlmaDesign [22].

In Figure 17a we can see the prototype seat with the integrated SPaC part. The location of the SPaC part in an autonomous vehicle environment of the future will allow the passenger to access different commands for controlling functions regardless of the position of the seat inside the vehicle. In the current case study, these functions are: positioning of the seat, opening and closing of the windows, and control of the rear-view mirrors. In case it becomes necessary to change/add functions, the developed technology has flexibility to allow further development and scaled integration.

This implementation led to the development of a touch system with two PCBs interconnected by a flat-cable, as the curved geometry of the plastic part dictated by the aesthetics conceived by the design team did not allow the electronic system to be implemented using only a single PCB. This is depicted in more detail in Figure 5. There is a possibility of using a flexible PCB, which would allow the PCB to adapt to complex contours of the control surface. However, this would raise costs significantly and consequently this possibility was discarded.

Figure 17a,b illustrates the seat developed under the context of this change of paradigm that are capable of integrating futuristic vehicles' requisites. The new touch-based technology developed adapts to the design of novel seats that integrate functionalities of the car cabin. In fact, as self-driving cars assume more relevance in a global context, it is expected that the passengers no longer have to travel aligned with the driving direction. This implies that they may not be able to reach all parts of the cabin in order to control some functionalities.

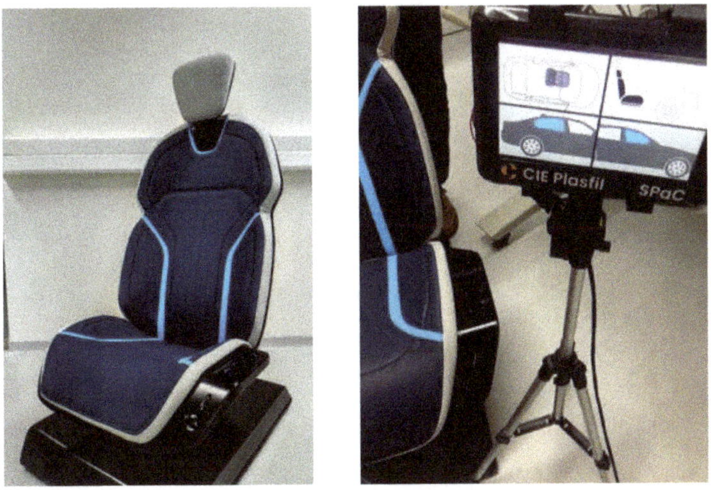

(**a**) (**b**)

Figure 17. (**a**) Button-controlled seating prototype developed for controlling car cabin functionalities. (**b**) Demo screen developed for emulating the car's windows and seat positioning functionalities.

Therefore, the current buttons were designed for the seat to incorporate the car functionalities. They can also be applied in other parts of the vehicle. Moreover, the incorporated functionalities can control the seat positioning and movement, as well as the window closest to that passenger. Figure 18a,b details the controls of window opening and rear mirror positioning. Rear seats, which in many cases may not move, can include advanced multimedia controls that are tipically accessible only by front-seat users.

Another aspect to consider is that unknown (at this point) functionalities may have to be included in the near future to control novel features of self-driving vehicles, which do not exist in current automobiles. The proposed technology makes room for a variety of human–machine interfaces, including the use of other types of sensors, wgucg have never been tried before on a single vehicle. These should allow the passengers of self-driving vehicles to better enjoy the travelling experience towards the destination.

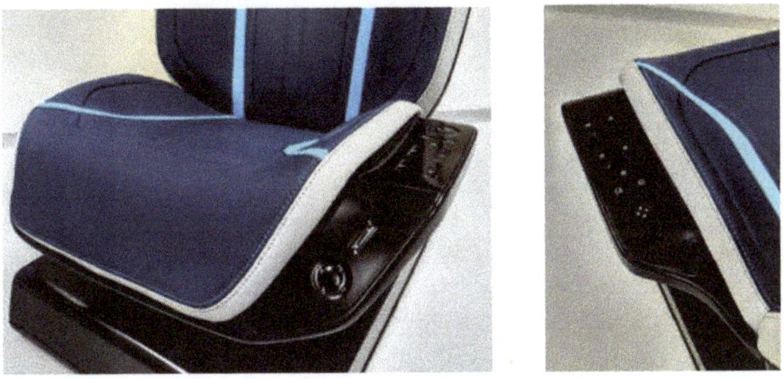

(**a**) (**b**)

Figure 18. (**a**) Front and (**b**) lateral details of the developed car seats for controlling functionalities inside the cabin.

Since the prototype seat was developed with aesthetic concerns in mind, a demo screen was developed to emulate the operation of the previously described car functionalities (shown in Figure 17b). This allows the seat to be demonstrated at conferences, fairs or to any other interested parties while proving the touch-based control concept for the car seat. The demo screen shows animations depicting some functionalities idealized for the seat control panel. The functionalities are: seat rotation (as illustrated in Figure 19) seat height adjustment, seat front and back sliding, seat reclination, and front and back door windows opening and closing.

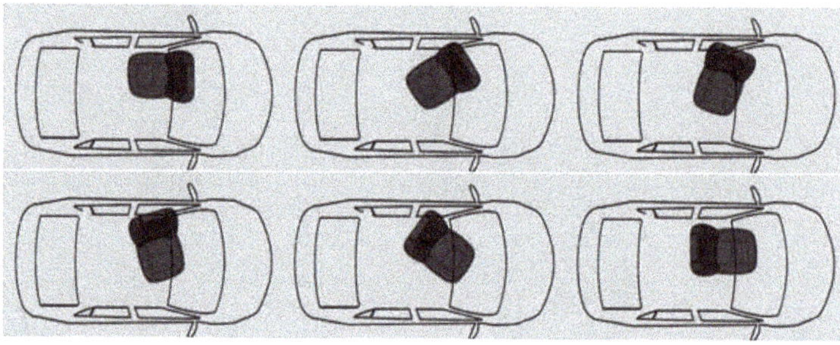

Figure 19. Driver's seat rotation animation as shown on the demo screen.

4. Discussion

The current paradigm of car seats aligned with the driving direction is about to change [23]. This may imply that passengers no longer have access to the usual controls on the front panel or in doors inside the car cabin [24], thus creating the need to incorporate new control functionalities near the seats' surfaces.

4.1. Conclusions of This Study

In this paper we address this change of paradigm in the sense that we propose new low-cost and easy to implement touch-based models for controlling car cabin functionalities. The sensors are based on self capacitance technology coupled to a low-cost microcontroller that connects to the automobile's ECU [16]. To this end, we have developed the necessary electronics and performed thousands of tests on real plastic injected components, to assess reliability and robustness. We have been able to verify that, for the thousands of tests performed to each touch-based sensor, the maximum failure rate achieved was below 0.55% for a behaviour that is far more demanding than real-life utilization on a vehicle. On a conventional utilization, no sensor is consecutively used thousands of times without a reset. This is even more significant as these can be integrated with car seats in order to allow the passenger control over functionalities that otherwise would be no longer accessible when the seat changes position, a possibility that may become a reality when the passenger becomes free from driving duties.

The main **contributions** of this paper can be summarized below:

- Innovative touch-based technology for controlling functionalities inside the car cabin;
- Thousands of cycles automatic reliability test and validation of touch-based sensors;
- Novel and appealing design of the cabin of self-driving cars;
- Moving control buttons and features to other parts of the cabin instead of the doors and central panel;
- More natural and intuitive human-machine interaction similar to the one used in mobile phones;
- Scalable and flexible addition of new functionalities via software technology;
- Retro-compatible technology;
- Supported by versatile communication protocols;
- The same low-cost solution can be incorporated into cars of distinct segments.

4.2. Future Research Directions

There is an active discussion [25] about the attention that touch-based technology requires from the driver and passengers. It is not clear if the feedback obtained by pressing a touch sensor is enough to create the sense of action completed by the user.

As future research directions there is an ample debate that haptic feedback [26] may have to be incorporated in touch-based control [27]. The haptic effect caused by the pressure of a mechanical button needs to be emulated for the new vehicle's control interfaces to act more naturally and give the user the perception that the action has been launched.

Another aspect that is indirect implications to this paper regards the ongoing evolution of deep neural networks and the challenges still to be faced [28] before self-driving cars can be massively adopted. In particular, several aspects of autonomous vehicles still have to undergo strict assessment concerning functional safety [29]. Furthermore, the use of deep learning for predicting decisions [30] based on data captured from cameras [31] and their association to other metrics such as positioning, velocity of the car, traffic or the presence of pedestrians near by, will have to be validated for many more thousands of kilometers to come.

Author Contributions: Conceptualization, T.C., C.A., P.S., J.S., C.R., R.L., R.P., F.M., R.M., G.T. and G.F.; methodology, C.A., T.C., P.S., J.S., G.T. and G.F.; software, C.A. and T.C.; validation, C.A., T.C. and G.F.; formal analysis, P.S. and G.F.; investigation, C.A., T.C., P.S., J.S. and G.F.; resources, P.S., G.T. and G.F.; data curation, P.S. and J.S.; writing—original draft preparation, C.A., T.C. and G.F.; writing—review and editing, P.S., C.R., R.L., F.M. and G.F.; visualization, J.S., R.P., R.M.; supervision, P.S. and G.F.; project administration, P.S.; funding acquisition, P.S. and G.T. All authors have read and agreed to the published version of the manuscript.

Funding: This work has been financially supported by project SPaC (POCI-01-0247-FEDER-038379), co-financed by the European Community Fund FEDER through POCI—Programa Operacional Competitividade e Internacionalização. It has also been financially supported by Instituto de Telecomunicações and Fundação para a Ciência e a Tecnologia under grants UIDB/50008/2020 and UIDP/50008/2020.

Acknowledgments: The authors would like to acknowledge the support provided by CJR Motors and Leiribéria.

Conflicts of Interest: The authors declare no conflict of interest.

Abbreviations

The following abbreviations are used in this manuscript:

SPaC	Smart Plastic Cover
LCD	Liquid-Crystal Display
ADC	Analog-to-Digital Converter
DAC	Digital-to-Analog Converter
CAN	Controller Area Network
LIN	Local Interconnect Network
PCB	Printed Circuit Board
ECU	Engine Control Unit
MTBF	Mean Time Between Failures
OEM	Original Equipment Manufacturer
MDPI	Multidisciplinary Digital Publishing Institute
FEDER	Fundo Europeu de Desenvolvimento Regional (Portugal)

References

1. Ji, K.; Orsag, M.; Han, K. Lane-Merging Strategy for a Self-Driving Car in Dense Traffic Using the Stackelberg Game Approach. *Electronics* **2021**, *10*, 894. [CrossRef]
2. Park, M.; Kim, H.; Park, S. A Convolutional Neural Network-Based End-to-End Self-Driving Using LiDAR and Camera Fusion: Analysis Perspectives in a Real-World Environment. *Electronics* **2021**, *10*, 2608. [CrossRef]
3. Talpes, E.; Sarma, D.D.; Venkataramanan, G.; Bannon, P.; McGee, B.; Floering, B.; Jalote, A.; Hsiong, C.; Arora, S.; Gorti, A. Compute solution for tesla's full self-driving computer. *IEEE Micro* **2020**, *40*, 25–35. [CrossRef]
4. Driverless Cars Are Coming—A Paradigm Shift. Available online: https://www.computer.org/publications/tech-news/neal-notes/driverless-cars-are-coming-a-paradigm-shift (accessed on 17 August 2020).
5. Eliot, L. *AI Self-Driving Cars Consonance: Practical Advances in Artificial Intelligence and Machine Learning*; LBE Press Publishing: New York, NY, USA, 2020.
6. Morales-Alvarez, W.; Sipele, O.; Léberon, R.; Tadjine, H.H.; Olaverri-Monreal, C. Automated Driving: A Literature Review of the Take over Request in Conditional Automation. *Electronics* **2020**, *9*, 2087. [CrossRef]
7. Autonomous Vehicles Will Create a Radical Paradigm Shift in Vehicle Design. Available online: https://www.automotive-fleet.com/159798/autonomous-vehicles-will-create-a-radical-paradigm-shift-in-vehicle-design (accessed on 12 June 2020).
8. Krenicky, T.; Ruzbarsky, J. Alternative Concept of the Virtual Car Display Design Reflecting Onset of the Industry 4.0 into Automotive. In Proceedings of the IEEE 22nd International Conference on Intelligent Engineering Systems (INES), Las Palmas de Gran Canaria, Spain, 21–23 June 2018; pp. 407–412.
9. Alves, C.; Custódio, T.; Silva, P.; Silva, J.; Rodrigues, C.; Lourenço, R.; Pessoa, R.; Moreira, F.; Marques, R.; Tomé, G.; et al. smartPlastic: Innovative Touch-Based Human-Vehicle Interface Sensors for the Automotive Industry. *Electronics* **2021**, *10*, 1223. [CrossRef]
10. Kouba, S.; Dickson, N. Intuitive Touch Technologies—Semiconductor-Based Electronic Components and their Integration. *Auto Tech Rev.* **2016**, *5*, 38–43. [CrossRef]
11. Rodríguez-Machorro, J.C.; Ríos-Osorio, H.; Águila-Rodríguez, G.; Herrera-Aguilar, I.; González-Sánchez, B.E. Development of capacitive touch interfaces. In Proceedings of the International Conference on Electronics, Communications and Computers (CONIELECOMP), Cholula, Mexico, 22–24 February 2017; pp. 1–8.
12. Park, J.K.; Lee, C.J.; Kim, J.T. Analysis of multi-level simultaneous driving technique for capacitive touch sensors. *Sensors* **2017**, *17*, 2016. [CrossRef] [PubMed]
13. Brasseur, G. Design rules for robust capacitive sensors. *IEEE Trans. Instrum. Meas.* **2003**, *52*, 1261–1265. [CrossRef]
14. Kim, J.; Song, W.; Jung, S.; Kim, Y.; Park, W.; You, B.; Park, K. Capacitive Heart-Rate Sensing on Touch Screen Panel with Laterally Interspaced Electrodes. *Sensors* **2020**, *14*, 3986. [CrossRef] [PubMed]
15. Capacitive Touch Sensor Design Guide. Available online: http://ww1.microchip.com/downloads/en/Appnotes/Capacitive-Touch-Sensor-Design-Guide-DS00002934-B.pdf (accessed on 13 July 2020).
16. Seo, S.-H.; Park, J.-H.; Hwang, S.-H.; Jeon, J.W. 3-D Car Simulator for Testing ECU Embedded Systems. In Proceedings of the SICE-ICASE International Joint Conference, Busan, Korea, 18–21 October 2006; pp. 550–554.
17. Coanda, H.-G.; Ilie, G. Design and implementation of an embedded system for data transfer in a car using CAN protocol. In Proceedings of the 12th International Conference on Electronics, Computers and Artificial Intelligence (ECAI), Bucharest, Romania, 25–27 June 2020; pp. 1–4.
18. Elshaer, A.M.; Elrakaiby, M.M.; Harb, M.E. Autonomous Car Implementation Based on CAN Bus Protocol for IoT Applications. In Proceedings of the 13th International Conference on Computer Engineering and Systems (ICCES), Cairo, Egypt, 18–19 December 2018, pp. 275–278.
19. Design with Surface Sensors for Touch Sensing Applications on MCUs. Available online: https://www.st.com/resource/en/application_note/dm00087990-design-with-surface-sensors-for-touch-sensing-applications-on-mcus-stmicroelectronics.pdf (accessed on 10 October 2019).
20. ATtiny3216/3217–Datasheet. Available online: https://ww1.microchip.com/downloads/en/DeviceDoc/ATtiny3216-17-DataSheet-DS40002205A.pdf (accessed on 4 February 2021).
21. Del Guerra, M.; Coelho, R.T. Development of a low cost Touch Trigger Probe for CNC Lathes. *J. Mater. Process. Technol.* **2006**, *179*, 117–123. [CrossRef]
22. AlmaDesign: Managing Process and Culture in a Design Studio. Available online: https://www.almadesign.pt/studio (accessed on 17 December 2019).
23. Fleming, B. Advances in Automotive Electronics [Automotive Electronics]. *IEEE Veh. Technol. Mag.* **2015**, *10*, 4–11. [CrossRef]
24. The Future of the Car: A Paradigm Shift of the Century. Available online: https://autocrypt.io/future-of-car-paradigm-shift-of-the-century/ (accessed on 7 June 2021).
25. Liu, S.; Li, L.; Tang, J.; Wu, S.; Gaudiot, J.-L. *Creating Autonomous Vehicle Systems*, 2nd ed.; Morgan & Claypool Publishers: Ulverston, UK, 2020.
26. Hannaford, B.; Okamura, A.M. Haptics. In *Springer Handbook of Robotics*; Springer: Cham, Switzerland, 2016; pp. 1063–1084.
27. Miedl, F.; Tille, T. 3-D surface-integrated touch-sensor system for automotive HMI applications. *IEEE/ASME Trans. Mechatron.* **2015**, *21*, 787–794. [CrossRef]

28. Rao, Q.; Frtunikj, J. Deep Learning for Self-Driving Cars: Chances and Challenges. In Proceedings of the IEEE/ACM 1st International Workshop on Software Engineering for AI in Autonomous Systems (SEFAIAS), Gothenburg, Sweden, 28 May 2018; pp. 35–38.
29. Xu, J.; Howard, A. How much do you Trust your Self-Driving Car? Exploring Human-Robot Trust in High-Risk Scenarios. In Proceedings of the IEEE International Conference on Systems, Man, and Cybernetics (SMC), Toronto, ON, Canada, 11–14 October 2020; pp. 4273–4280. [CrossRef]
30. Gilpin, L.H. Anticipatory Thinking: A Testing and Representation Challenge for Self-Driving Cars. In Proceedings of the 55th Annual Conference on Information Sciences and Systems (CISS), Baltimore, MD, USA, 24–26 March 2021; pp. 1–2. [CrossRef]
31. Mihalea, A.; Samoilescu, R.; Nica, A.C.; Trăscău, M.; Sorici, A.; Florea, A.M. End-to-end models for self-driving cars on UPB campus roads. In Proceedings of the IEEE 15th International Conference on Intelligent Computer Communication and Processing (ICCP), Cluj-Napoca, Romania, 5–7 September 2019; pp. 35–40. [CrossRef]

electronics

Article

Microphone Array for Speaker Localization and Identification in Shared Autonomous Vehicles

Ivo Marques [1,*], João Sousa [1], Bruno Sá [1], Diogo Costa [1], Pedro Sousa [1], Samuel Pereira [1], Afonso Santos [1], Carlos Lima [1], Niklas Hammerschmidt [2], Sandro Pinto [1] and Tiago Gomes [1]

1 Centro ALGORITMI, Escola de Engenharia, Universidade do Minho, 4800-058 Guimarães, Portugal; a82273@alunos.uminho.pt (J.S.); id10037@alunos.uminho.pt (B.S.); a81176@alunos.uminho.pt (D.C.); a82041@alunos.uminho.pt (P.S.); a81408@alunos.uminho.pt (S.P.); id9490@alunos.uminho.pt (A.S.); carlos.lima@dei.uminho.pt (C.L.); sandro.pinto@dei.uminho.pt (S.P.); mr.gomes@dei.uminho.pt (T.G.)

2 Bosch Car Multimedia, 4705-820 Braga, Portugal; niklas.hammerschmidt@pt.bosch.com

* Correspondence: ivo.marques@dei.uminho.pt; Tel.: +351-2535-10180

Abstract: With the current technological transformation in the automotive industry, autonomous vehicles are getting closer to the Society of Automative Engineers (SAE) automation level 5. This level corresponds to the full vehicle automation, where the driving system autonomously monitors and navigates the environment. With SAE-level 5, the concept of a Shared Autonomous Vehicle (SAV) will soon become a reality and mainstream. The main purpose of an SAV is to allow unrelated passengers to share an autonomous vehicle without a driver/moderator inside the shared space. However, to ensure their safety and well-being until they reach their final destination, active monitoring of all passengers is required. In this context, this article presents a microphone-based sensor system that is able to localize sound events inside an SAV. The solution is composed of a Micro-Electro-Mechanical System (MEMS) microphone array with a circular geometry connected to an embedded processing platform that resorts to Field-Programmable Gate Array (FPGA) technology to successfully process in the hardware the sound localization algorithms.

Keywords: Shared Autonomous Vehicle (SAV); Field-Programmable Gate Array (FPGA); microphone array; sound source localization

Citation: Marques, I.; Sousa, J.; Sá, B.; Costa, D.; Sousa, P.; Pereira, S.; Santos, A.; Lima, C.; Hammerschmidt, N.; Pinto, S.; et al. Microphone Array for Speaker Localization and Identification in Shared Autonomous Vehicles. *Electronics* **2022**, *11*, 766. https://doi.org/10.3390/electronics11050766

Academic Editors: Calin Iclodean, Bogdan Ovidiu Varga and Felix Pfister

Received: 24 January 2022
Accepted: 26 February 2022
Published: 2 March 2022

1. Introduction

In the near future, autonomous vehicles will be sufficiently reliable, affordable, and widespread on our public roads, replacing many current human driving tasks [1]. The emergence of different autonomous applications will not only require accurate perception systems [2], but also vehicles with high-performance processing capabilities with the ability to communicate with cloud services and other vehicles, requiring low communications response time and high network bandwidth [3]. One of the applications of autonomous vehicles will be for shared mobility. This concept, which includes car-sharing or rent-by-the-hour vehicles where passengers can partially or totally share the same trip, tend to become a common practice in modern societies [1]. At the same time, the technological advances in the automotive industry have highly contributed to the future of autonomous vehicles in a sustainable urban mobility scenario [4,5]. The merging between autonomous driving and shared mobility trends resulted in the emergence of the concept of Shared Autonomous Vehicle (SAV) [1,6], which enables unrelated passengers to share the same vehicle during their trips. The adoption of SAV will completely change the current paradigm of shared rides and it will surely contribute to more sustainable and affordable passenger mobility in urban areas [7].

In current ride-share or taxi services, despite the driver moderating the activities inside the common space, some companies, such as Uber, Lyft, and Didi, have reported several safety problems between passengers and drivers [8]. There have been records of harassment, assault and robbing passengers, and unfortunately no strict measures could be taken since the company cannot have full control over the passengers, drivers, vehicles or rides. In the context of an SAV, and since the vehicle will not require the driver's control, these problems can become worse since the absence of a moderator can leave the vehicle vulnerable to misuse and inappropriate behavior between passengers, causing several consequences both for the occupants and the car.

The safety of all occupants being a major concern, it is crucial to develop solutions to ensure a normal ride during shared trips. Current trends aim at equipping an SAV with sensor-based monitoring solutions to analyze and identify several situations inside the vehicle's shared space, for example, driver's and passenger's behavior, violence between occupants, vandalism, assaults, and so forth, to trigger safety measures. Only by ensuring the effectiveness of these triggers will passengers trust SAV solutions, which is essential to bring forward their mass acceptance and adoption [7,9,10]. Current solutions that monitor the activity inside the vehicle are mostly video-based systems [11–13]. In other fields, these video-based solutions tend to be very useful as they can look through facial or object movements to find possible sound sources in the environment [14–17]. However, video-only solutions are not able to capture all the surroundings as they have a limited field of view, making the classification and detection of all human actions inside the vehicle difficult. Thus, it is almost mandatory to collect audio events inside the shared space [18].

Outside the automotive context, current audio-only sensor systems use microphone arrays to localize different sound sources in a wide range of applications, for example, robot and human–robot interactions [19,20], drones direction calculation [21], audio recording for multi-channel reproduction [22], and multi-speaker voice and speech recognition [23]. In such solutions, the accuracy and detection performance is affected by the array geometry, where linear arrays are only able to localize sound sources in a 2D range [24], and circular [19,22,25], spherical [20,26], or other geometries [27,28] allow the system to localize in a 3D space. Besides the geometry, the number of microphones also affects the localization accuracy [27]. Other hybrid approaches combine microphone arrays with video cameras combined with facial recognition techniques to localize and detect audio sources, which can be used to monitor the SAV [29,30]. However, they sometimes require complex sensor fusion systems with high processing capabilities.

In a microphone array solution, the estimation of the Direction of Arrival (DoA) is a well-known research topic. This mechanism can be applied to either a simple scenario where only one sound source is present or to a complex setup with several sound sources to be processed simultaneously. Several solutions allow the estimation of DoA for narrow-band signals including high-resolution subspace algorithms like MVDR [31], MUSIC [32], and ESPRIT [33]. Meanwhile, new research has suggested new directions in this field of study [34–36]. Recently, different solutions that allow obtaining these results both in digital signal processing-based systems [37–39] and in machine learning-based approaches [40,41] have been proposed. Although these proposals are highly accurate, their application is not always feasible in a real-time scenario. This is mainly due to the computational complexity of these approaches being very high, essentially in implementing sound separation mechanisms [42,43].

With the challenge of creating an audio-only sensor system to localize and identify speakers without requiring a video-based solution, this work presents an embedded, cost-effective, low-power, and real-time microphone array solution for speaker localization and identification that can be used inside an SAV. To accelerate the processing tasks, the sensor system resorts to Field-Programmable Gate Array (FPGA) technology to deploy dedicated processing modules in hardware to interface, acquire, and compute data from different microphones [44–49]. Moreover, the processing system provides a Robot Operating System (ROS) interface to make data available to other high-level applications (for the identification and classification of audio events) or to other sensor fusion systems. Finally, and since the system deals with sensitive data, we have deployed the processing systems over a static partitioning hypervisor to guarantee data security and prevent unwanted access of private information. This work was developed in partnership with Bosch Car Multimedia Portugal, S.A., and contributes to the state-of-the-art with:

(1) a microphone array system to monitor sound events inside an SAV, which can be easily integrated with other sensor fusion strategies for automotive;
(2) a hardware-based system with data acquisition and format conversion, i.e., Pulse Density Modulated (PDM) to Pulse Code Modulated (PCM) to interface the microphone array;
(3) hardware-accelerated algorithms to localize different sound sources that can achieve good accuracy and performance metrics with real-time response.

The remainder of this paper is organized as follows: Section 2 describes the sensor system architecture; Sections 3 and 4 detail, respectively, the design and implementation steps to develop the sensor system (these sections are further divided in the microphone array and the processing platform); Section 5 presents the system evaluation, while Section 6 concludes this paper with a summary of our findings; Finally, Section 7 discusses some open issues regarding this research topic, pointing out some future work directions.

2. Sensor System Architecture

The sensor system's architecture with all modules and their respective interactions is depicted in Figure 1. It is mainly divided into three main blocks:

(1) the microphone array;
(2) the processing platform; and
(3) the ROS environment.

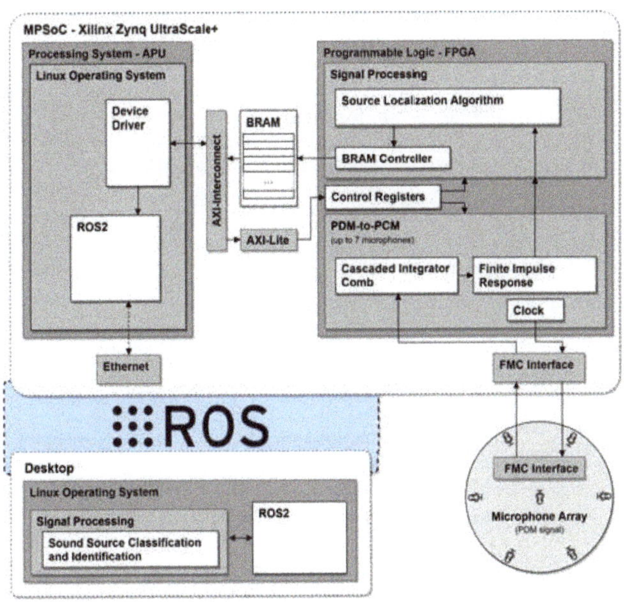

Figure 1. Sensor system architecture.

Microphone Array

The microphone array includes seven microphones in a Uniform Circular Array (UCA) geometry. The output of each microphone is a PDM signal containing the necessary information to localize and separate the sound sources. The microphone's board is connected to the processing platform through an FPGA Mezzanine Card (FMC) interface, which is used to power the board, obtain data from the microphones, and provide the clock signal to generate the data output.

Processing platform

The processing platform is responsible for acquiring and processing, in real-time, the data retrieved from the microphone array module. It consists of the Xilinx Zynq UltraScale+, which includes a MultiProcessor System on a Chip (MPSoC) with Programmable Logic (PL) FPGA technology. This allows the acceleration in hardware of the microphone array interface and the source localization algorithms. On the PL side, the ***PDM-to-PCM*** module is responsible for converting the audio signal from the PDM to PCM format and for applying filtering steps to prevent signal aliasing and spatial-aliasing [50]. The conversion from PDM to PCM is required by the sound processing algorithm in the next hardware block. This step is performed in the ***Signal Processing*** module, where a set of calculations and bit operations in the FPGA are executed to estimate the DoA of the sound sources. The processed data, which contain the estimated DoA and the acquired signal, are sent to the Processing System (PS) through the Advanced eXtensible Interface (AXI) protocol in the Advanced Microcontroller Bus Architecture (AMBA) bus. On the PS side, the data are collected through a standard device driver supported by a virtualized embedded Linux Operating System (OS).

ROS Environment/Interfaces

On top of the embedded Linux, we run the ROS environment to provide the processing data (audio source data and localization) to higher-level applications that perform the identification and classification of the audio events inside the vehicle.

Figure 2 depicts the ROS architecture of the microphone array for sound identification and localization. Upon the arrival of new audio data, the ***MicArray Node*** reads and processes the output from the localization algorithm in the PL, that is, the DoA and the source audio sample in the WAV format, and publishes it to multiple topics according to the audio source (one topic for each source). Finally, each audio source topic is subscribed to by the ***Identification and Characterization Node***, which applies the classification and identification algorithms and forwards data to higher-level applications for further processing. Moreover, the system provides a collection of services, which act as an interface to execute several actions, for example, performing hardware initialization, change parameters, configure the microphone array, and so forth.

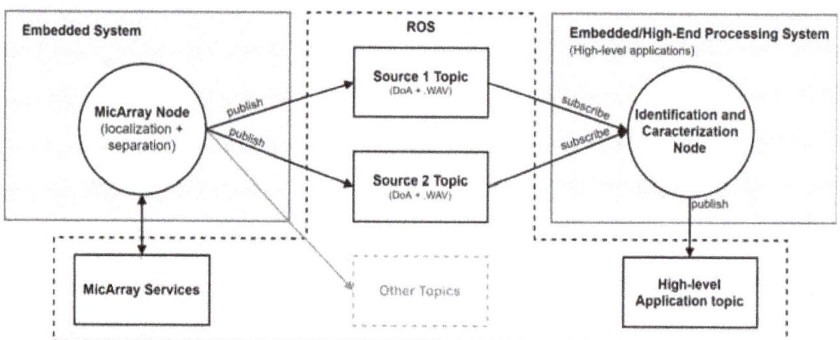

Figure 2. ROS architecture/interface between the sound source localization and separation system and the characterization algorithm.

3. Microphone Array System

To develop the sensor system, and to comply with the project's requirements, the microphone array must provide specific features such as: (1) have an UCA geometry with a maximum diameter of 10 cm; (2) use up to seven low-power MEMS microphones; and (3) include an FMC interface. In contrast to linear arrays, a UCA geometry allows the use of sound source localization in a 3D space, and the FMC interface allows the Printed Circuit Board (PCB) to be plugged into other platforms that support this connector. All microphones are spaced five centimeters apart, providing a UCA with six microphones placed around the circumference (60° between microphones). Additionally, one microphone is placed in the center to be used as reference during the computation of the algorithms. The PCB layout also has six LEDs, placed around the array and parallel to the microphones, to indicate the calculated localization of the detected sound source. The FMC connector, placed at the bottom side, allows us to make the direct connection between the processing platform and the microphones and LEDs.

Figure 3 shows the PCB layout developed for the microphone array. For testing different microphone devices, this board supports three kinds of omnidirectional and low-power MEMS microphones: INMP621 and ICS-51360 (from TDK InvenSense) [51,52], and SPK0641HT4H-1 (from Knowles) [53]. Among other features, they all provide low-power modes, data output in PDM format, and they all work with a clock signal around 2.4 MHz. These features allow the utilization of the same controller for any board configuration. However, and to simplify the sound source localization process, on each PCB prototype only one type of microphone in the array can be used. Due to this geometry, the maximum frequency that the system can handle is 3430 Hz, which is the spatial aliasing frequency, $f_{SpatialAliasing}$, as shown in Equation (1), defined by [50]:

$$f_{SpatialAliasing} = \frac{c}{2d}, \quad (1)$$

where c is the sound speed (in this case it was considered the sound speed in the air at 20 °C, 343 m·s^{-1}), and d is the smaller distance between microphones, 0.05 m.

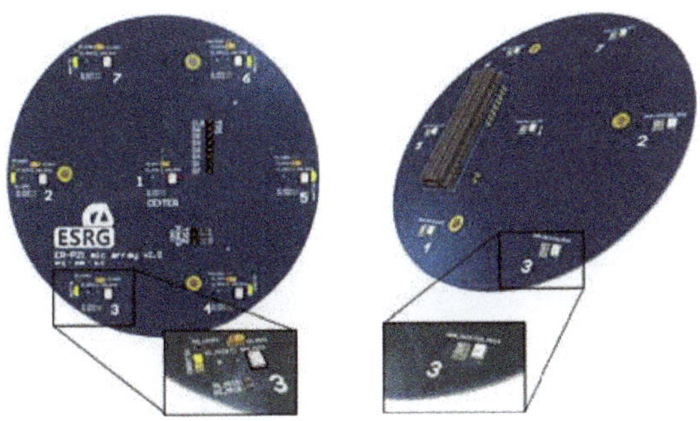

(**a**) (**b**)

Figure 3. Three-dimensional (3D) view of the PCB microphone array with the microphone position (numbered 1 to 7). (**a**) Top-view. (**b**) Bottom-view.

4. Processing Platform

The processing platform includes a PL system with FPGA technology, and a PS with an Application Processing Unit (APU), which are both the main units used in this prototype. The communication between both systems is achieved via the AXI protocol through the AMBA bus and by resorting to the available Direct Memory Access (DMA) controller, and the communication with the microphone array PCB is done through the FMC interface. The processing platform deploys on the PL, the Signal Acquisition module (responsible of acquiring and converting the signal of each microphone) and the Signal Processing module (which includes the algorithm to localize the sound source). By its turn, the PS provides support to the hypervisor, Linux OS, device drivers, and the ROS interfaces.

4.1. Signal Acquisition Module

The data acquisition module is responsible for generating a clock signal of 2.4 MHz to interface the microphones and to collect and convert each microphone data output from the PDM to the PCM format. Figure 4 displays the data flow between each microphone and its corresponding acquisition block. Since the microphone output is a PDM signal with a switching frequency of 2.4 MHz, the PDM to PCM converter block deployed in the FPGA is responsible for: (1) generating a clock signal at 2.4 MHz; (2) acquiring the data from the microphone at the same clock frequency; and (3) converting the signal from PDM at 2.4 MHz to the PCM format at 8 kHz with a 24-bit resolution.

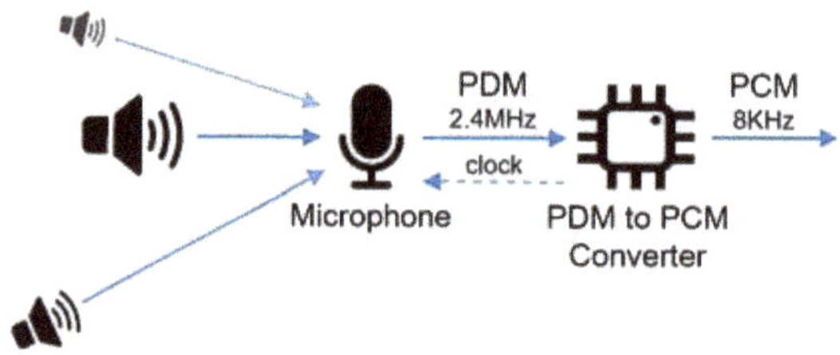

Figure 4. Data acquisition system with the PDM to PCM converter.

To convert the signal from PDM to PCM for each microphone [54], the processing steps depicted in Figure 5 were used. The process has three stages, starting with a low-pass filter that receives the PDM signal from the microphone and, after a quantization process, outputs a new signal to the next block. Since the microphones use a sampling frequency of 2.4 MHz, the next block performs a decimation by a factor of 300, which creates a new signal frequency of 8 kHz. These two stages are developed in the FPGA using a Cascade Integrator Comb (CIC) filter block. The last stage corresponds to a Finite Impulse Response (FIR) filter block, that performs a band-pass filter. This filter has the lower cut-off frequency at 70 Hz to remove the Direct Current (DC) component, and it has the upper cut-off frequency at 3 kHz, which reduces possible aliasing phenomena (audio and spatial aliasing). Although the bandwidth of the pass-band filter is between 70 Hz and 3 kHz, this range is enough for the human voice's fundamental frequency, which is commonly between 85 and 155 Hz for an adult male, and 165 to 255 Hz to an adult female [55].

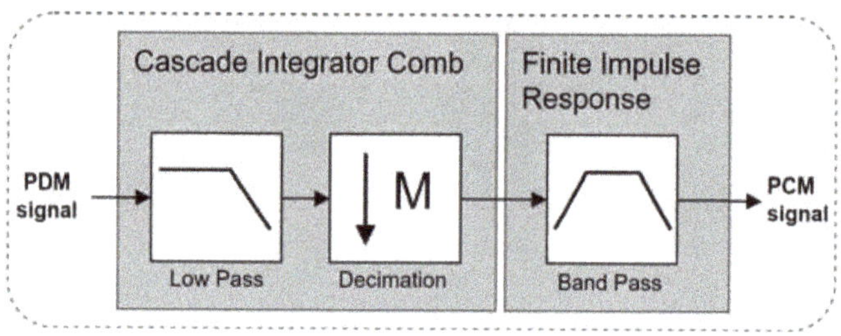

Figure 5. Block diagram of the conversion of PDM to PCM signal.

4.2. Sound Source Localization

The algorithm to localize the sound sources is presented in Figure 6. Since its working principle is based on the signals energy, it requires data from all microphones to calculate the DoA of the sound source. This task is performed in six sequential steps: (1) ***Absolute Value***; (2) ***Average Value***; (3) ***Noise Removal***; (4) ***Polar to Cartesian***; (5) ***DoA Calculation***; and (6) ***Get Angle***.

(1) ***Absolute Value:*** In this step, for each microphone, the input data (in PCM format) are received at the same frequency of the sampling frequency (8 kHz) to calculate its absolute value.
(2) ***Average Value:*** This step receives data from the previous block and calculates the moving average for each microphone signal.
(3) ***Noise Removal:*** This block receives the data from the previous step and a user-defined noise threshold signal (***Noise Threshold***). If the average value of the central microphone is less than the ***Noise Threshold***, then it is considered only background noise in the environment, and the new average value for each microphone is set to

zero. Otherwise, the average value of the central microphone is subtracted from the remaining microphone's data.

(4) ***Polar to Cartesian:*** This stage calculates, for each of the six UCA microphones, its cartesian position multiplied by its corresponding average value. The output from this block is the weight vector for each microphone according to the signal energy.

(5) ***DoA Calculation:*** In this step, the resultant of all vectors to output a cartesian vector with the DoA estimation is calculated.

(6) ***Get Angle:*** This stage calculates the DoA angle from the cartesian vector.

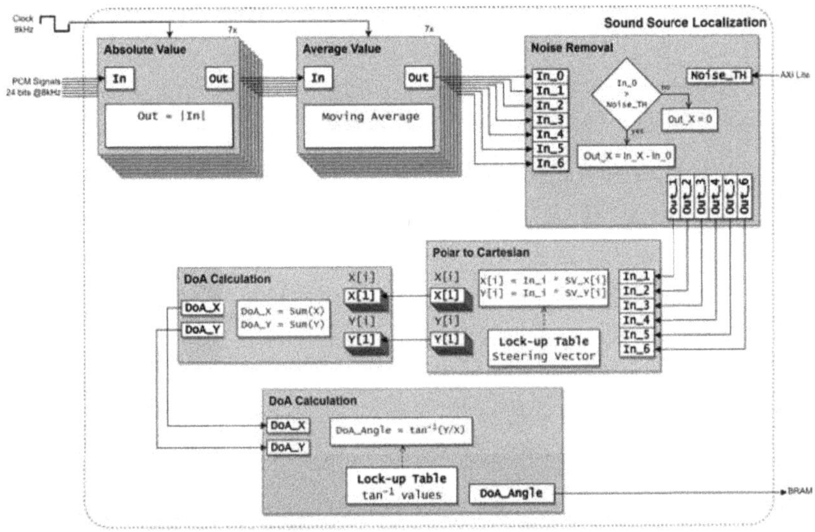

Figure 6. Block diagram of the energy sound source localization algorithm.

To optimize the conversion steps, the ***Polar to Cartesian*** and the ***Get Angle***, make use of look-up tables. Additionally, the output from the ***Polar to Cartesian*** stage provides information to control the LEDs in the microphone array. The output data corresponds to the weight of each microphone according to the location of the sound source. This data are processed to generate a Pulse-width Modulation (PWM) signal for each LED, which results in a brighter light on the LEDs closer to the sound source.

4.3. Interface between the PL and the PS

The communication between the PS and the PL is made through the AXI-Lite protocol over the AMBA bus. System data can be classified as: (1) data acquired by the microphones and processed in hardware that includes the DoA estimation and the respective PCM signal, transferred from the PL to the PS; and (2) control data transferred from the PS to the PL that is used to configure the acquisition and localization systems. In the first case, that is, data from the PL to the PS, the process is performed in two ways:

(1) the PCM signal data are written to the Block Random Access Memory (BRAM) directly through a BRAM controller that defines the writing position. This data are accessed by the PS through an AXI interface connected to the BRAM;

(2) the DoA data are directly set available to the PS via AXI-Lite interface.

In the second case, where the PL receives the control signals from PS, the AXI-Lite protocol is used, where the PS can access control registers to enable the ***PDM-to-PCM Converter*** and ***Signal Processing*** modules, and to configure the noise threshold inside the sound source localization module.

4.4. Software Stack

Figure 7 depicts the software stack that is supported by the PS. The ROS2 system, supported by a virtualized Linux, is used to ease the integration of the microphone array with the ROS network standard. As previously shown in Figure 2 from Section 2, it provides specific interfaces of nodes and topics, allowing both for system flexibility, scalability, and interoperability features. Moreover, to configure and capture the audio data packages from the microphone array within the OS, a device driver was developed and included in the custom Linux image. Finally, to guarantee data security and prevent unwanted accesses from third parties to the audio data that flows through the system, an additional virtualization layer was added through our in-house made static partitioning hypervisor Bao [56].

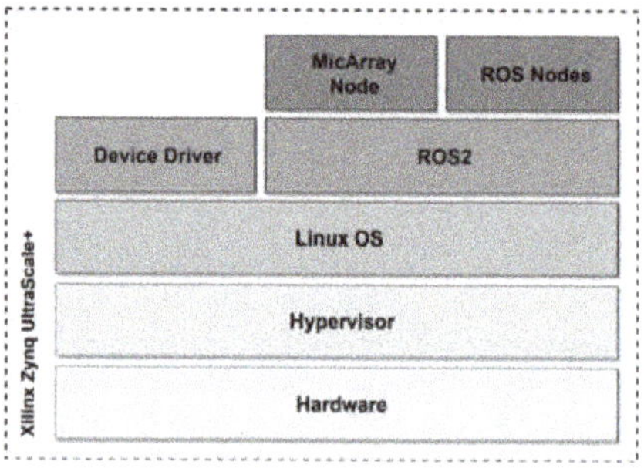

Figure 7. System software stack.

5. Evaluation

In order to test and evaluate the speaker localization and identification prototype, we have used the following experimental setup: (1) the sensor system prototype (Figure 8a); (2) an audio sound source; and (3) a laptop computer with a ROS interface that subscribes to the ROS topics. The laptop is connected to the processing platform through an Ethernet interface (in a wired ad-hoc network using an SSH session) to control the prototype system and store the acquired and processed data (signal data and DoA from each sound source). Regarding the sensor system, three different steps were executed to demonstrate its behavior with just one sound source:

(1) first step test evaluates and verifies the acquisition system, i.e., sampling, PDM to PCM format conversion, and data filtering;
(2) second step evaluates the accuracy and precision of the localization system;
(3) lastly, the third step evaluates the localization system in the presence of a moving sound source, checking the DoA and verifying the resulting angle with the actual position of the sound source.

The center microphone is used as the reference to calculate the DoA, and the angle measurements, with resolution of 1°, start at 0° on the positive x-axis of the unit circle graph (corresponding to the microphone number 5), and go counter-clockwise around the circle until they are back at 360° (Figure 8b). Since the sensor system is to be placed in the center of the vehicle's roof and to bring the sound sources closer to the conventional passenger locations, during the tests, the sound source was placed between 50 to 150 cm away from the microphone array with an elevation of 45°.

(**a**) (**b**)

Figure 8. Sensor system prototype to localize and identify sound sources in an SAV and location of each microphone in the PCB to calculate the DoA. (**a**) Sensor system prototype. (**b**) Microphone's location on a XY referential.

5.1. Data Acquisition

To test the acquisition modules, the system was adapted to bypass the sound source localization module and each microphone's data were directly sent from the PDM-to-PCM module to the device driver using the BRAM controller. During the tests, the sensor system was exposed to different sound sources, which allowed the analysis of the behavior of the PDM-to-PCM module at different frequencies and distances. Figure 9 depicts the results for the three types of microphones available in the microphone array. The sound source used to test the data acquisition was a controlled signal (sine wave) at different frequencies: 220 Hz (Figure 9a); 440 Hz (Figure 9b), which is currently used as the reference frequency for tuning musical instruments; 880 Hz (Figure 9c); and 1760 Hz (Figure 9d). All the sound sources have their fundamental frequency within the bandwidth defined for the band-pass filter.

The results show that the microphones and the conversion module can achieve a good performance in collecting sound within the frequency ranges defined by the band-pass filter since the main characteristics of the original signals are present on the acquired samples. However there are some differences in the received signal's amplitude, which are mainly associated with the receiving power of the collected signal, which can be affected by: (1) the location and position of the microphones in the array; (2) the aperture size of the microphone's sound port; (3) and the aperture size of the PCB hole for the sound port. For instance, the INMP621 and the ICS-51360 are located in the bottom layer of the microphone array, which result in signals with lower amplitude values. Moreover, Figure 9a presents some signal's distortion and lower amplitudes, which are mainly due to the speaker's ability to generate lower frequencies.

For testing the band-pass filter we have generated sound waves at frequencies above the filter's cutoff frequency which is 3 KHz. Figure 10 shows the results in each microphone type for two different frequencies, 3520 Hz and 7040 Hz. When the frequency is nearby the cutoff frequency (3 KHz), the sound is attenuated according to the low-pass component of the filter. However, at 3520 Hz, there are still some signal components since, by definition, at the cutoff frequency the output drops below 70.7% of its input. For the 7040 Hz frequency, there is only noise and the audio signal is nearly zero. These results show the correct operation of the band-pass filter which is used to avoid temporal and spatial aliasing.

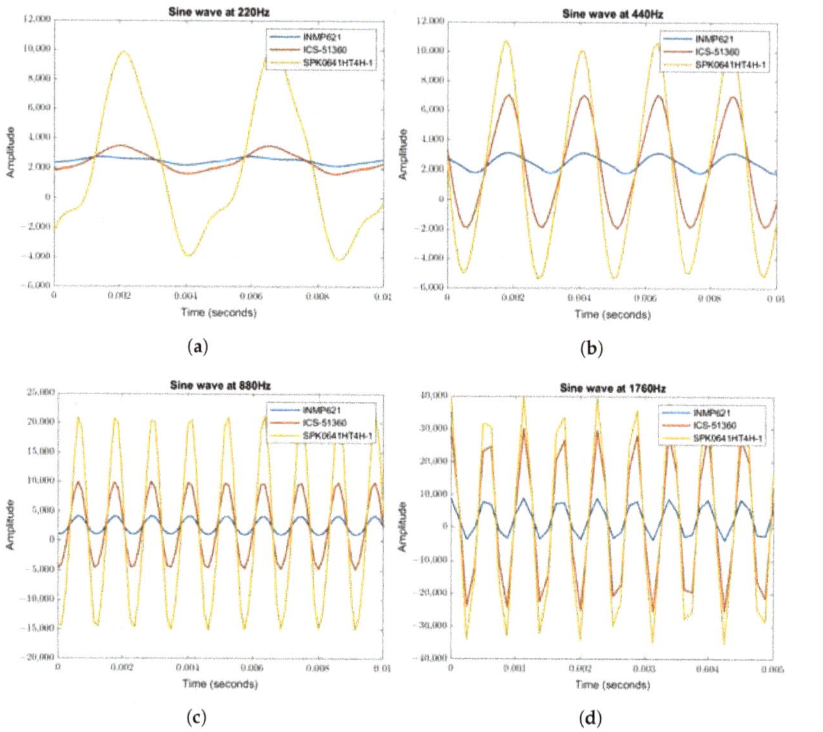

Figure 9. Data acquisition by the three types of microphones at different frequencies. (**a**) Sine wave at 220 Hz. (**b**) Sine wave at 440 Hz. (**c**) Sine wave at 880 Hz. (**d**) Sine wave at 1760 Hz.

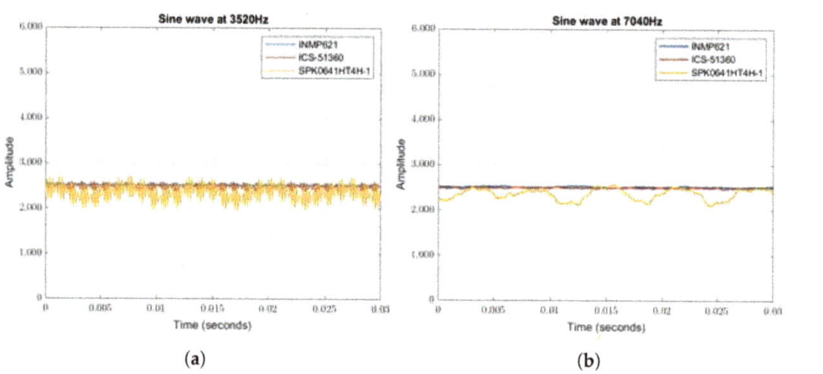

Figure 10. Sine wave acquisition for frequencies above the filter's passband. (**a**) Sine wave at 3520 Hz. (**b**) Sine wave at 7040 Hz.

5.2. Sound Source Localization

To evaluate the localization process, which is the most important goal of this project, two different tests were executed: (1) a set of measurements to evaluate the sensor system accuracy in terms of DoA, and (2) a tracking test. The accuracy test was executed with a sound source at the 180° position using the controlled signal (sine wave) at different frequencies: 220 Hz, 440 Hz, 880 Hz, and 1760 Hz. For each frequency, a set of measurements was executed at different distances: 50 cm, 75 cm, 100 cm, and 150 cm. A total of 200 DoA measurements were acquired for each experiment. Figure 11 presents the box plots of the experiments. The results present a similar and high accuracy for all the frequencies tested. However, it is also possible to conclude that, with increasing distance, the precision is affected.

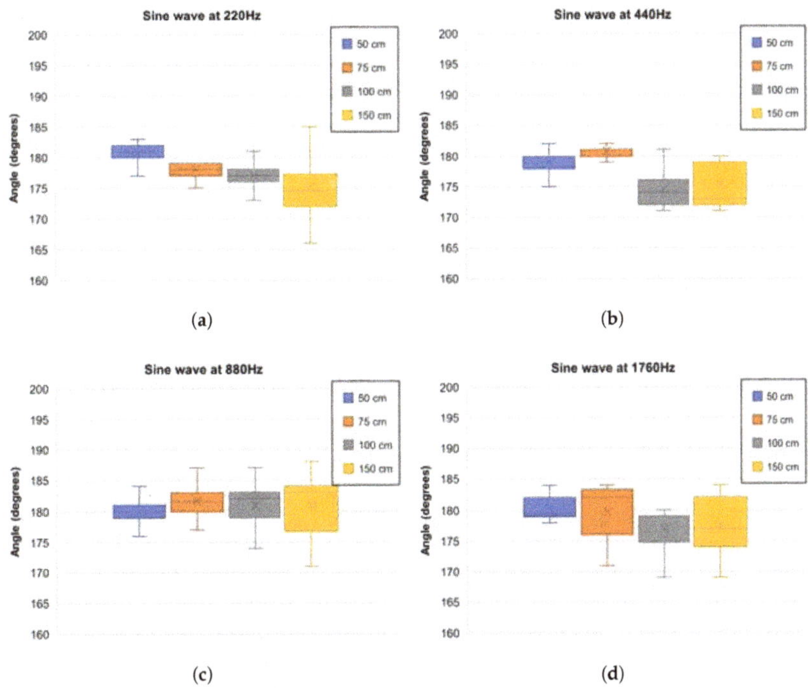

Figure 11. Box plot of the DoA measurements at different frequencies and distances. (**a**) Sine wave at 220 Hz. (**b**) Sine wave at 440 Hz. (**c**) Sine wave at 880 Hz. (**d**) Sine wave at 1760 Hz.

Table 1 presents the accuracy calculated in each experiment. Regarding the distance parameter, the accuracy is higher for smaller distances. When the test was performed at 50 cm, the accuracy reached at least 99.54%. At 880 Hz the system obtains the best overall results, as the accuracy reached values above 99.20% for all tested distances.

Table 1. DoA accuracy at different frequencies and distances.

		Sound Source Distance			
		50 cm	**75 cm**	**100 cm**	**150 cm**
Sine Wave Frequency	**220 Hz**	99.62 %	98.92 %	98.33 %	97.33 %
	440 Hz	99.54 %	99.56 %	96.94 %	97.45 %
	880 Hz	99.92 %	99.20 %	99.48 %	99.46 %
	1760 Hz	99.76 %	99.86 %	98.24 %	98.67 %

In the second evaluation, the system was tested with a moving sound source around the microphone array that followed the pattern shown in Figure 12a. The result presented in Figure 12b demonstrates with precision the location of the moving sound source. Because the algorithm needs 64 samples to localize the sound source and the sampling frequency is 8 kHz, after the first DoA calculation is performed, the following are always available with an 8 ms delay.

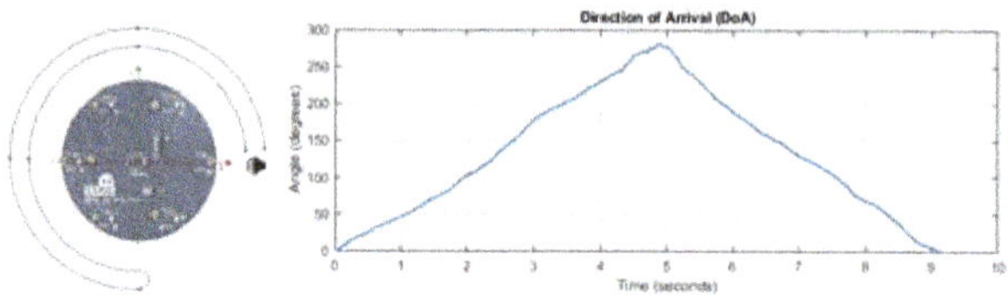

(a) (b)

Figure 12. Moving sound pattern and respective calculated DoA. (**a**) Sound source path. (**b**) Direction of arrival of a sound source over the time.

5.3. FPGA Hardware Resources

The FPGA implementation requires the resources described in Table 2. In terms of LookUp Table (LUT) and LUTRAM, the implementation uses 8.09% and 4.47% of the resources available in the platform, which corresponds to 18628 LUT and 4450 LUTRAM. From the available 21,725 Flip Flop (FF) units, the system uses a total of 21,725 FF, which corresponds to 4.71% of the available resources. The BRAM module is the most used resource, requiring a total of 191 BRAM units corresponding to 61.22% of the available BRAM in the platform. Due to the acquisition system that using the FIR and CIC blocks, the system requires 49 of the available 1728 Digital Signal Processor (DSP) units (2.84%). The hardware also requires 14 Input/Output (IO) pins, which corresponds to the seven microphone inputs, one output for the clock signal used to drive the microphones, and six outputs to control the LEDs PWM signal. Finally, to support the clock generator module, the system requires one of the eight available Mixed-Mode Clock Manager (MMCM) units, and five of the 544 existing Global Clock Buffer (BUFG) blocks.

Table 2. FPGA resources utilization.

Resource	Utilization	Available	Utilization (%)
LUT	18,628	230,400	8.09%
LUTRAM	4550	101,760	4.47%
FF	21,725	460,800	4.71%
BRAM	191	312	61.22%
DSP	49	1728	2.84%
IO	14	360	3.89%
BUFG	5	544	0.92%
MMCM	1	8	12.50%

6. Conclusions

This article presents a sensor system solution to monitor sound events in an SAV cabin. This solution is composed of a microphone array connected to a processing platform, which provides the localization of the sound sources to higher-level application through an ROS interface. Regarding the proposed solution and the tests performed, the implemented system is able to acquire data from all microphones, filter the collected signals, and calculate the DoA of one sound source with good accuracy results. Through individual ROS topics, the *MicArray Node* makes the acquired audio samples available to other high-level applications, in order to identify and classify the sound events. This way, it is possible to identify the type of event that occurs and act accordingly. Concerning the architecture, the system allows the deployment of independent hardware blocks for customization and acceleration purposes.

We believe that, with the growing interest in developing autonomous vehicles, passenger monitoring solutions like the one proposed in this article will surely contribute one step further towards the option of SAV. To the best of our knowledge, there are no audio-only solutions in the literature that are intended to monitor passengers inside an SAV. From a broader perspective, this solution, as a concept, can be integrated in other applications beyond SAV, where the localization of a sound source in a real-time approach is a major priority.

7. Future Work

Current work encompasses the exploration of more advanced and efficient algorithms to localize multiple sound sources, such as variable step-size least mean square. Since the individual data of each sound source are required, in a scenario where there are multiple and simultaneous sources, it becomes mandatory to use a sound source separation algorithm, such as independent component analyses with fast convergence. Moreover, an open issue related to the vehicle interior is noise, and no matter how acoustically isolated the vehicle is, the noise will always be present at different amplitudes. Thus, the next step is to use localization and separation algorithms with self-adapting resources that change the noise threshold, preserving the signal in the presence of noise. Furthermore, the sensor system also needs to be tested in a situation closed to the SAV reality, for example, inside a vehicle's roof.

Author Contributions: Conceptualization, T.G. and S.P. (Sandro Pinto); formal analysis, C.L.; investigation, T.G., I.M., J.S., B.S., D.C., P.S., S.P. (Samuel Pereira) and A.S.; writing—original draft preparation, T.G., I.M.; writing—review and editing, I.M. and T.G.; supervision, T.G. and S.P. (Sandro Pinto); project administration, N.H., T.G. and S.P. (Sandro Pinto). All authors have read and agreed to the published version of the manuscript.

Funding: This work is supported by: European Structural and Investment Funds in the FEDER component, through the Operational Competitiveness and Internationalization Programme (COMPETE 2020) [Project n° 039334; Funding Reference: POCI-01-0247-FEDER-039334].

Conflicts of Interest: The authors declare no conflict of interest.

References

1. Litman, T. *Autonomous Vehicle Implementation Predictions*; Victoria Transport Policy Institute: Victoria, BC, Canada, 2021.
2. Roriz, R.; Cabral, J.; Gomes, T. Automotive LiDAR Technology: A Survey. *IEEE Trans. Intell. Transp. Syst.* **2021**, 1–16. [CrossRef]
3. Liu, L.; Chen, C.; Pei, Q.; Maharjan, S.; Zhang, Y. Vehicular edge computing and networking: A survey. *Mob. Netw. Appl.* **2021**, *26*, 1145–1168. [CrossRef]
4. Daily, M.; Medasani, S.; Behringer, R.; Trivedi, M. Self-Driving Cars. *Computer* **2017**, *50*, 18–23. [CrossRef]
5. Badue, C.; Guidolini, R.; Carneiro, R.V.; Azevedo, P.; Cardoso, V.B.; Forechi, A.; Jesus, L.; Berriel, R.; Paixão, T.M.; Mutz, F.; et al. Self-driving cars: A survey. *Expert Syst. Appl.* **2021**, *165*, 113816. [CrossRef]
6. Rojas-Rueda, D.; Nieuwenhuijsen, M.J.; Khreis, H.; Frumkin, H. Autonomous vehicles and public health. *Annu. Rev. Public Health* **2020**, *41*, 329–345. [CrossRef]
7. Jones, E.C.; Leibowicz, B.D. Contributions of shared autonomous vehicles to climate change mitigation. *Transp. Res. Part D Transp. Environ.* **2019**, *72*, 279–298. [CrossRef]
8. Chaudhry, B.; El-Amine, S.; Shakshuki, E. Passenger safety in ride-sharing services. *Procedia Comput. Sci.* **2018**, *130*, 1044–1050. [CrossRef]
9. Cartenì, A. The acceptability value of autonomous vehicles: A quantitative analysis of the willingness to pay for shared autonomous vehicles (SAVs) mobility services. *Transp. Res. Interdiscip. Perspect.* **2020**, *8*, 100224. [CrossRef]
10. Paddeu, D.; Parkhurst, G.; Shergold, I. Passenger comfort and trust on first-time use of a shared autonomous shuttle vehicle. *Transp. Res. Part C Emerg. Technol.* **2020**, *115*, 102604. [CrossRef]
11. Fouad, R.M.; Onsy, A.; Omer, O.A. Improvement of Driverless Cars' Passengers on Board Health and Safety, using Low-Cost Real-Time Heart Rate Monitoring System. In Proceedings of the 2018 24th International Conference on Automation and Computing (ICAC), Newcastle upon Tyne, UK, 6–7 September 2018; pp. 1–6.
12. Koojo, I.; Machuve, D.; Mirau, S.; Miyingo, S.P. Design of a Passenger Security and Safety System for the Kayoola EVs Bus. In Proceedings of the 2021 IEEE AFRICON, Arusha, Tanzania, 13–15 September 2021; pp. 1–6.
13. Costa, M.; Oliveira, D.; Pinto, S.; Tavares, A. Detecting Driver's Fatigue, Distraction and Activity Using a Non-Intrusive Ai-Based Monitoring System. *J. Artif. Intell. Soft Comput. Res.* **2019**, *9*, 247–266. [CrossRef]
14. Chakravarty, P.; Mirzaei, S.; Tuytelaars, T.; Van hamme, H. Who's Speaking? Audio-Supervised Classification of Active Speakers in Video. In Proceedings of the 2015 ACM on International Conference on Multimodal Interaction, ICMI '15, Seattle, WA, USA, 9–13 November 2005; pp. 87–90. [CrossRef]
15. Qian, R.; Hu, D.; Dinkel, H.; Wu, M.; Xu, N.; Lin, W. Multiple sound sources localization from coarse to fine. In Proceedings of the European Conference on Computer Vision, Glasgow, UK, 23–28 August 2020; pp. 292–308.

16. Senocak, A.; Oh, T.H.; Kim, J.; Yang, M.H.; Kweon, I.S. Learning to localize sound source in visual scenes. In Proceedings of the IEEE Conference on Computer Vision and Pattern Recognition, Salt Lake City, UT, USA, 18–2 June 2018; pp. 4358–4366.
17. Stachurski, J.; Netsch, L.; Cole, R. Sound source localization for video surveillance camera. In Proceedings of the 2013 10th IEEE International Conference on Advanced Video and Signal Based Surveillance, Krakow, Poland, 27–30 August 2013; pp. 93–98.
18. Pieropan, A.; Salvi, G.; Pauwels, K.; Kjellström, H. Audio-visual classification and detection of human manipulation actions. In Proceedings of the 2014 IEEE/RSJ International Conference on Intelligent Robots and Systems, Chicago, IL, USA, 14–18 September 2014; pp. 3045–3052.
19. Tamai, Y.; Kagami, S.; Amemiya, Y.; Sasaki, Y.; Mizoguchi, H.; Takano, T. Circular microphone array for robot's audition. In Proceedings of the SENSORS, 2004 IEEE, Vienna, Austria, 24–27 October 2004; pp. 565–570.
20. Grondin, F.; Michaud, F. Lightweight and optimized sound source localization and tracking methods for open and closed microphone array configurations. *Robot. Auton. Syst.* **2019**, *113*, 63–80. [CrossRef]
21. Wakabayashi, M.; Okuno, H.G.; Kumon, M. Multiple sound source position estimation by drone audition based on data association between sound source localization and identification. *IEEE Robot. Autom. Lett.* **2020**, *5*, 782–789. [CrossRef]
22. Hulsebos, E.; Schuurmans, T.; de Vries, D.; Boone, R. Circular Microphone Array for Discrete Multichannel Audio Recording. Audio Engineering Society Convention 114. Audio Engineering Society. March 2003. Available online: http://www.aes.org/e-lib/browse.cfm?elib=12596 (accessed on 4 January 2022).
23. Subramanian, A.S.; Weng, C.; Watanabe, S.; Yu, M.; Yu, D. Deep learning based multi-source localization with source splitting and its effectiveness in multi-talker speech recognition. *Comput. Speech Lang.* **2022**, *10*, 101360. [CrossRef]
24. Danès, P.; Bonnal, J. Information-theoretic detection of broadband sources in a coherent beamspace MUSIC scheme. In Proceedings of the 2010 IEEE/RSJ International Conference on Intelligent Robots and Systems, Taipei, Taiwan, 18–22 October 2010; pp. 1976–1981. [CrossRef]
25. Pavlidi, D.; Puigt, M.; Griffin, A.; Mouchtaris, A. Real-time multiple sound source localization using a circular microphone array based on single-source confidence measures. In Proceedings of the 2012 IEEE International Conference on Acoustics, Speech and Signal Processing (ICASSP), Kyoto, Japan, 25–30 March 2012; pp. 2625–2628.
26. Rafaely, B.; Peled, Y.; Agmon, M.; Khaykin, D.; Fisher, E. *Spherical Microphone Array Beamforming*; Springer: Berlin, Germany, 2010; pp. 281–305.
27. Kurc, D.; Mach, V.; Orlovsky, K.; Khaddour, H. Sound source localization with DAS beamforming method using small number of microphones. In Proceedings of the 2013 36th International Conference on Telecommunications and Signal Processing (TSP), Rome, Italy, 2–4 July 2013; pp. 526–532. [CrossRef]
28. Dehghan Firoozabadi, A.; Irarrazaval, P.; Adasme, P.; Zabala-Blanco, D.; Játiva, P.P.; Azurdia-Meza, C. 3D Multiple Sound Source Localization by Proposed T-Shaped Circular Distributed Microphone Arrays in Combination with GEVD and Adaptive GCC-PHAT/ML Algorithms. *Sensors* **2022**, *22*, 1011. [CrossRef]
29. Busso, C.; Hernanz, S.; Chu, C.W.; Kwon, S.i.; Lee, S.; Georgiou, P.G.; Cohen, I.; Narayanan, S. Smart room: Participant and speaker localization and identification. In Proceedings of the (ICASSP'05), IEEE International Conference on Acoustics, Speech, and Signal Processing, Philadelphia, PA, USA, 23 March 2005; Volume 2, pp. ii/1117–ii/1120.
30. Chen, X.; Shi, Y.; Jiang, W. Speaker tracking and identifying based on indoor localization system and microphone array. In Proceedings of the 21st International Conference on Advanced Information Networking and Applications Workshops (AINAW'07), Niagara Falls, ON, Canada, 21–23 May 2007; Volume 2, pp. 347–352.
31. Murthi, M.; Rao, B. Minimum variance distortionless response (MVDR) modeling of voiced speech. In Proceedings of the 1997 IEEE International Conference on Acoustics, Speech, and Signal Processing, Munich, Germany, 21–24 April 1997; Volume 3, pp. 1687–1690. [CrossRef]
32. Gupta, P.; Kar, S. MUSIC and improved MUSIC algorithm to estimate direction of arrival. In Proceedings of the 2015 International Conference on Communications and Signal Processing (ICCSP), Melmaruvathur, India, 2–4 April 2015; pp. 757–761. [CrossRef]
33. Roy, R.; Kailath, T. ESPRIT-estimation of signal parameters via rotational invariance techniques. *IEEE Trans. Acoust. Speech Signal Process.* **1989**, *37*, 984–995. [CrossRef]
34. Das, A. Real-Valued Sparse Bayesian Learning for Off-Grid Direction-of-Arrival (DOA) Estimation in Ocean Acoustics. *IEEE J. Ocean. Eng.* **2021**, *46*, 172–182. [CrossRef]
35. He, D.; Chen, X.; Pei, L.; Zhu, F.; Jiang, L.; Yu, W. Multi-BS Spatial Spectrum Fusion for 2-D DOA Estimation and Localization Using UCA in Massive MIMO System. *IEEE Trans. Instrum. Meas.* **2021**, *70*, 1–13. [CrossRef]
36. Yun, W.; Xiukun, L.; Zhimin, C. DOA Estimation of Wideband LFM Sources based on Narrowband Methods Integration Using Random Forest Regression. In Proceedings of the 2021 OES China Ocean Acoustics (COA), Harbin, China, 14–17 July 2021; pp. 816–820. [CrossRef]
37. Jalal, B.; Yang, X.; Igambi, D.; Ul Hassan, T.; Ahmad, Z. Low complex direction of arrival estimation method based on adaptive filtering algorithm. *J. Eng.* **2019**, *2019*, 6214–6217. [CrossRef]
38. Tiete, J.; Domínguez, F.; Silva, B.D.; Segers, L.; Steenhaut, K.; Touhafi, A. SoundCompass: A Distributed MEMS Microphone Array-Based Sensor for Sound Source Localization. *Sensors* **2014**, *14*, 1918–1949. [CrossRef]
39. Hoshiba, K.; Washizaki, K.; Wakabayashi, M.; Ishiki, T.; Kumon, M.; Bando, Y.; Gabriel, D.; Nakadai, K.; Okuno, H.G. Design of UAV-Embedded Microphone Array System for Sound Source Localization in Outdoor Environments. *Sensors* **2017**, *17*, 2535. [CrossRef]

40. He, W.; Motlicek, P.; Odobez, J.M. Deep Neural Networks for Multiple Speaker Detection and Localization. In Proceedings of the 2018 IEEE International Conference on Robotics and Automation (ICRA), Brisbane, Australia, 21–25 May 2018; pp. 74–79. [CrossRef]
41. Purwins, H.; Li, B.; Virtanen, T.; Schlüter, J.; Chang, S.Y.; Sainath, T. Deep Learning for Audio Signal Processing. *IEEE J. Sel. Top. Signal Process.* **2019**, *13*, 206–219. [CrossRef]
42. Xu, W.; Jia, M.; Gao, S.; Li, L. Multiple Sound Source Separation by Using DOA Estimation and ICA. In Proceedings of the 2021 4th International Conference on Information Communication and Signal Processing (ICICSP), Shanghai, China, 24–26 September 2021; pp. 249–253. [CrossRef]
43. Li, H.; Chen, K.; Wang, L.; Liu, J.; Wan, B.; Zhou, B. Sound Source Separation Mechanisms of Different Deep Networks Explained from the Perspective of Auditory Perception. *Appl. Sci.* **2022**, *12*, 832. [CrossRef]
44. Butt, U.M.; Khan, S.A.; Ullah, A.; Khaliq, A.; Reviriego, P.; Zahir, A. Towards Low Latency and Resource-Efficient FPGA Implementations of the MUSIC Algorithm for Direction of Arrival Estimation. *IEEE Trans. Circuits Syst. I Regul. Pap.* **2021**, *68*, 3351–3362. [CrossRef]
45. Da Silva, B.; Braeken, A.; Touhafi, A. FPGA-based architectures for acoustic beamforming with microphone arrays: Trends, challenges and research opportunities. *Computers* **2018**, *7*, 41. [CrossRef]
46. Jung, Y.; Jeon, H.; Lee, S.; Jung, Y. Scalable ESPRIT Processor for Direction-of-Arrival Estimation of Frequency Modulated Continuous Wave Radar. *Electronics* **2021**, *10*, 695. [CrossRef]
47. Nsalo Kong, D.F.; Shen, C.; Tian, C.; Zhang, K. A New Low-Cost Acoustic Beamforming Architecture for Real-Time Marine Sensing: Evaluation and Design. *J. Mar. Sci. Eng.* **2021**, *9*, 868. [CrossRef]
48. Ribeiro, Â.; Rodrigues, C.; Marques, I.; Monteiro, J.; Cabral, J.; Gomes, T. Deploying a Real-Time Operating System on a Reconfigurable Internet of Things End-device. In Proceedings of the IECON 2019-45th Annual Conference of the IEEE Industrial Electronics Society, Lisbon, Portugal, 14–17 October 2019; Volume 1, pp. 2946–2951.
49. Marques, I.; Rodrigues, C.; Tavares, A.; Pinto, S.; Gomes, T. Lock-V: A heterogeneous fault tolerance architecture based on Arm and RISC-V. *Microelectron. Reliab.* **2021**, *120*, 114120. [CrossRef]
50. Brandstein, M.; Ward, D.; Lacroix, A.; Venetsanopoulos, A. (Eds.) *Microphone Arrays: Signal Processing Techniques and Applications*, 1st ed.; Digital Signal Processing; Springer: Berlin/Heidelberg, Germany, 2001.
51. InvenSense. Wide Dynamic Range Microphone with PDM Digital Output Data Sheet ADMP62. In *DS-INMP621-00 Datasheet Rev 1.3*; InvenSense Inc.: San Jose, CA, USA, 2016.
52. InvenSense. Bottom Port PDM Digital Output Multi-Mode Microphone. In *ICS-51360 Datasheet Rev 1.0*; InvenSense Inc.: San Jose, CA, USA, 2016.
53. Knowles. Digital SiSonic Microphone With Multiple Performance Modes. In *Datasheet SPK0641HT4H-1 Rev A*; Knowles Electronics, LLC: Itasca, IL, USA, 2016.
54. Hegde, N. Seamlessly interfacing MEMs microphones with blackfin processors. In *EE-350 Engineer-to-Engineer Note*; Analog Devices, Inc.: Norwood, MA, USA, 2010.
55. Re, D.E.; O'Connor, J.J.; Bennett, P.J.; Feinberg, D.R. Preferences for very low and very high voice pitch in humans. *PLoS ONE* **2012**, *7*, e32719. [CrossRef]
56. Martins, J.; Tavares, A.; Solieri, M.; Bertogna, M.; Pinto, S. Bao: A lightweight static partitioning hypervisor for modern multi-core embedded systems. In Proceedings of the Workshop on Next Generation Real-Time Embedded Systems (NG-RES 2020), Bologna, Italy, 21 January 2020.

electronics

Article

Performance Evaluation of VANET Routing Protocols in Madinah City

Mohammad A. R. Abdeen [1,*], Abdurrahman Beg [2,3], Saud Mohammad Mostafa [2], AbdulAziz AbdulGhaffar [2], Tarek R. Sheltami [2,3] and Ansar Yasar [4]

1 Department of Computers and Information Systems, The Islamic University of Madinah, Al-Madinah 42351, Saudi Arabia
2 Department of Computer Engineering, King Fahd University of Petroleum and Minerals, Dhahran 31261, Saudi Arabia; g201703830@kfupm.edu.sa (A.B.); g201706290@kfupm.edu.sa (S.M.M.); g201703470@kfupm.edu.sa (A.A.); tarek@kfupm.edu.sa (T.R.S.)
3 Interdisciplinary Research Center of Smart Mobility and Logistics, King Fahd University of Petroleum and Minerals, Dhahran 31261, Saudi Arabia
4 Transportation Research Institute (IMOB), Hasselt University, 3500 Hasselt, Belgium; ansar.yasar@uhasselt.be
* Correspondence: mabdeen@iu.edu.sa

Abstract: Traffic management challenges in peak seasons for popular destinations such as Madinah city have accelerated the need for and introduction of autonomous vehicles and Vehicular ad hoc networks (VANETs) to assist in communication and alleviation of traffic congestions. The primary goal of this study is to evaluate the performance of communication routing protocols in VANETs between autonomous and human-driven vehicles in Madinah city in varying traffic conditions. A simulation of assorted traffic distributions and densities were modeled in an extracted map of Madinah city and then tested in two application scenarios with three ad hoc routing protocols using a combination of traffic and network simulation tools working in tandem. The results measured for the average trip time show that opting for a fully autonomous vehicle scenario reduces the trip time of vehicles by approximately 7.1% in high traffic densities and that the reactive ad hoc routing protocols induce the least delay for network packets to reach neighboring VANET vehicles. From these observations, it can be asserted that autonomous vehicles provide a significant reduction in travel time and that either of the two reactive ad hoc routing protocols could be implemented for the VANET implementation in Madinah city. Furthermore, we perform an ANOVA test to examine the effects of the factors that are considered in our study on the variation of the results.

Keywords: VANET; V2X; autonomous vehicles; routing protocols; ad hoc protocols; wireless communication

Citation: Abdeen, M.A.R.; Beg, A.; Mostafa, S.M.; AbdulGhaffar, A.; Sheltami, T.R. and Yasar, A. Performance Evaluation of VANET Routing Protocols in Madinah City. *Electronics* **2022**, *11*, 777. https://doi.org/10.3390/electronics11050777

Academic Editors: Calin Iclodean, Bogdan Ovidiu Varga and Felix Pfister

Received: 27 January 2022
Accepted: 1 March 2022
Published: 2 March 2022

1. Introduction

Vehicle-to-Vehicle (V2V) and Vehicle-to-Infrastructure (V2I) communications play a vital role in building an Intelligent transportation system in Vehicular ad hoc networks (VANETs) [1,2]. The vehicles and infrastructure communicate and exchange important information about the traffic situations, road conditions, and many more [3,4]. This information helps autonomous vehicles to make decisions dynamically and avoid potentially dangerous conditions. Furthermore, utilizing V2V and V2I concepts in smart cities will enhance the safety of the vehicle on the roads as well as optimize the flow of traffic [5–7].

Figure 1 describes a VANET of a smart city environment with V2I and V2V wireless ad hoc communication. Vehicles transmit and relay packets across the network informing of incidents, such as accident scenarios, at specific locations. The ad hoc protocols employed by the literature and this work proactively or reactively send out packets to the nearest neighbors based on the communication range and antenna. The figure further elaborates on the V2I concept in the VANET environment, where a base station node with a larger communication range receives the relayed packet and forwards it to vehicles further away or via cable to other entities in the network.

The concurrent presence of human-driven and fully automated vehicles on a road network poses new research challenges. Especially in the case that this type of mixed fleet needs to cooperate under a real-life urban or extra-urban environment [8,9]. The key challenge in these environments is traffic congestion, a regular phenomenon, which causes dynamic changes in the environment and in the decisions the automated vehicles will make (lane blockage, delays, illegal parking, short-time cars' stops) to be self-navigated within the urban area [10,11]. To make things worse, the coexistence of heavy pedestrian traffic and adverse weather conditions, such as heavy rain, extreme heat, strong winds, and hail, can make the co-presence of conventional (human-driven) and automated vehicles even more challenging [12]. We need to find a way that will not lead to an increase in the traffic jam due to this necessary cooperation of the two different types of vehicle fleets. Traffic congestion induces serious infrastructure degradation in metropolitan areas. Considering that malfunction cooperation among conventional and automated vehicles can increase traffic jams and consequently the cost, it is clear that civil infrastructures should be equipped with novel sensors and software tools to enable cooperative functionality among the conventional and automated vehicles [13].

For the seamless coexistence of automated and conventional vehicles, it is clear that new signaling and traffic management methods are required. These methods should be dynamic and adapted according to the real-time traffic flow conditions, which will allow for increased efficiency. In this paper, we develop wireless communication between Vehicles and Infrastructure. Both human-driven and autonomous vehicles will need to wirelessly communicate with each other: Vehicle-to-Vehicle and with the surrounding road infrastructure (V2I). The transportation network utilizes traffic modeling and available data to simulate the future state. First, we use OpenStreetMap (OSM) [14,15] to extract the roads network map of Madinah city. Next, we use the Simulation of Urban MObility (SUMO) [16,17] traffic modeling tool to generate various traffic scenarios for our simulation. We select OMNeT++ [18,19] as the network simulation platform in our study and import the Madinah city map as well as SUMO-generated traffic scenarios to perform simulations and obtain the results. For this study, we select three routing protocols, including Ad hoc On-Demand Distance Vector (AODV) [20,21], Destination Sequenced Distance Vector (DSDV) [22,23], and Dynamic Manet on Demand (DYMO) [24–27], which are discussed later. This study takes into account the conditions of the city of Madinah, especially at peak times of Hajj and Umrah. To the best of our knowledge, very few works are conducted on autonomous vehicles in the Kingdom of Saudi Arabia.

The contribution of our work is as follows:

- We study the effect of different routing protocols in VANET communication systems. Two types of routing protocols (proactive and reactive) are considered, and a performance evaluation is conducted of the VANET system under varying traffic scenarios.
- Furthermore, we demonstrated the effect of various populations of autonomous to human-driven vehicles in the smart city by analyzing the impact of the introduction of autonomous vehicles on the trip times of the vehicles in the VANET scenario.
- Finally, we evaluated the main factors that influence the performance of the VANET system through the analysis of variance (ANOVA) test, and our study illustrated the contrasting influence of the factors captured on multiple metrics in a smart city simulation.

The organization of this paper is as follows: Section 2 provides a brief summary of recent studies, while Section 3 provides a detailed methodology of our work. Section 4 explains the simulation details and various test scenarios we considered in our study. Section 5 elaborates and analyzes the results and findings of our simulation. Lastly, the conclusion is presented in Section 6.

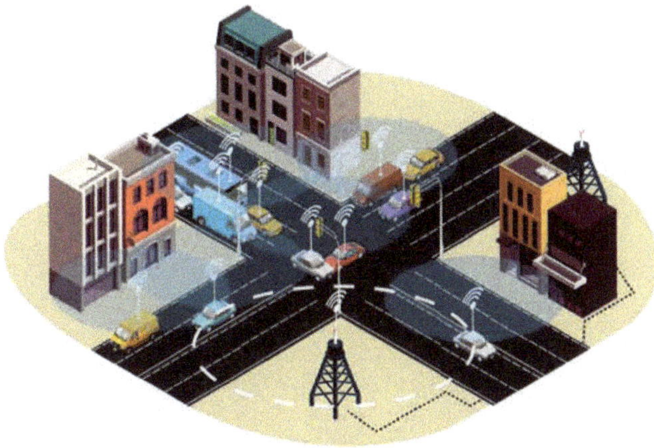

Figure 1. Vehicles in a VANET communicating when an accident occurs at the traffic light intersection.

2. Related Works

There are various research areas in the VANET domain [28]. We focus on the communication aspects of VANETs. Tremendous work regarding the communication routing protocols for VANETs is studied. The authors in [29] discuss the details of the protocol stack, application, and challenges of VANET. In [30], the authors classify the routing protocols based on the type of architecture and mode of operation. They discuss the features of the current known routing protocols and how bio-inspired protocols can improve the performance of the routing process. Several factors might affect the performance of the routing protocols. Hence, the authors conclude that depending on the VANET application, the routing protocol needs to be designed or tailored.

The authors in [31] evaluate the performance of AODV and Dynamic Source Routing (DSR) [32,33] routing protocols in VANET with dense and sparse car traffic density. The simulation was carried out in the OPNET Network Simulator [34] using IEEE 802.11b standard [35] to study the impact on VANET. The Packet Delivery Ratio (PDR), throughput, and end-to-end delay were the performance metrics used for the study, and the authors concluded that AODV was better in dense traffic density. However, the authors only considered reactive routing protocols in their analysis.

Another study [36] analyzed the performance of DSR, AODV, and DSDV routing protocols in terms of PDR, average throughput, delay, and total energy under high traffic density. The authors compared the performance of the routing protocols in a highly congested area of Khartoum to find out the most suitable routing protocol. Regardless, the authors did not consider different traffic scenarios or routes that might affect the performance. In [37], the authors studied the performance of AODV, DSR, and DSDV routing protocols in terms of PDR, throughput, and Normalized Routing Load (NRL) as performance metrics. The intended map of the city of Casablanca was generated using OSM, and the mobility model was created using SUMO. Simulations were run using the Network Simulator 2 (NS-2) [38] tool with a high traffic density. The authors concluded that the AODV protocol outperforms DSR and DSDV protocols under a heavy traffic load. However, the authors did not consider different traffic scenarios and the autonomy of vehicles in VANET systems.

Ghori et al. [39] studied several routing protocols to identify the most suitable protocol for video streaming in VANET. They classified and examined the routing protocols and discussed the pros and cons of each routing protocol. The authors evaluated the performance of AODV and DSR routing protocols in terms of throughput and delay using OPNET as the network simulation tool. Road-side Units (RSUs) [40] were used to simulate a complex traffic scenario, and the authors concluded that AODV is the best routing protocol for

VANET. However, the authors did not consider any proactive routing protocols in their study. The researchers in [41] analyzed AODV, Optimized Link State Routing Protocol (OLSR) [42], and DSDV routing protocols in terms of PDR, goodput, routing overhead, and end-to-end delay as performance metrics under different node densities and velocities. The mobility model was generated using the BonnMotion tool [43], and the simulation was carried out in NS-3 [44]. The authors concluded that the OLSR routing protocol performs best in their scenario. Nonetheless, the authors did not consider a more realistic scenario with traffic lights, etc. Additionally, they did not consider high node density in their study.

The authors in [45] present a detailed classification of the routing protocols in VANET with their benefits and drawbacks. They simulate a VANET environment of Oujda city using OSM, SUMO, and NS-3 tools to compare the topology-based and position-based routing protocols. The performance of DSDV, AODV, Greedy Perimeter Stateless Routing (GPSR) [46], OLSR, and Greedy Perimeter Coordinator Routing (GPCR) [47] routing protocols are evaluated in terms of PDR, end-to-end delay, throughput, and routing overhead by varying the node density. The authors observe that the OLSR protocol outperforms other protocols in terms of PDR and throughput. GPSR and GPCR protocols perform better concerning the routing overhead and end-to-end delay. Additionally, the authors propose a new greedy forwarding technique based on the angle direction, speed variation, density, and distance to the next-hop node to improve the GPSR and GPCR protocols. However, no simulation was conducted based on the proposed technique. Furthermore, the authors did not consider autonomous traffic distribution in their traffic scenarios.

Shi et al. [48] evaluate the performance of two communication technologies, including 802.11p and LTE-V, in the V2X scenario. The authors consider a scenario where the vehicles communicate an accident at an intersection. To perform the experiments, the authors deploy two vehicles in a real-world test field, which communicate with each other using 802.11p and LTE-V technologies. The Packet Delivery Ratio (PDR) and latency are selected as two performance metrics. From the results, the authors conclude that 802.11p provides much lower latency compared to LTE-V; however, the PDR of LTE-V is much better. Nevertheless, the study only provides a small-scale deployment and does not provide an extensive comparison between the two technologies. Furthermore, the implementation scenario is oversimplified in this study.

The authors of [49] conduct a performance evaluation of routing protocols for VANETs using a simulation system called Cellular Automaton-based VEhicular NETwork (CAVENET). CAVENET generates mobility behavior in one-dimensional cellular automata, and the network is simulated on NS-2. They implemented two reactive routing protocols, AODV and DYMO, and one proactive routing protocol, OSLR. Packet delivery ratio is used as the primary evaluation metric between the protocols. They claim that DYMO is the best routing protocol because of its route maintenance ability. Their simulation is limited as it only considers 30 nodes from a 10 to 90 s simulation time. Further, only two nodes can deliver the majority of the packets, while some cannot deliver any packet at all because of disappearing routes over multi-hop communication, and the PDR is dependent on the number of hops in this simulation, not the routing protocol.

García-Campos et al. [50] perform a comparison study of ad hoc reactive routing protocols for VANETs in urban settings. The routing protocols studied were AODV, DSR, DYMO, and Location-Aided Routing (LAR) [51], and the MAC protocol used was IEEE 802.11p. The performance evaluation was conducted on the ns-2 simulator for the network simulation and BonnMotion for the mobility generator. The number of nodes in the VANET ranged up to a maximum of 175 with increments of 25 vehicles. Performance metrics captured were throughput, end-to-end delay, max. route activity time, number of hops, jitter, and more. Their results show that Dymo and AODV perform best for the jitter metric, Dymo for route activity time, and LAR for the remaining metrics. Their study is limited in that only one distribution of vehicles are used, and they only scale the traffic to 175 nodes.

3. Methodology

3.1. OpenStreetMap

The map of Madinah city was exported with the OpenStreetMap tool—an open-source geographic database of the world. Figure 2 shows the extracted map on the Java OpenStreetMap Editor (JOSM), a java tool to inspect and edit OSM maps. From this tool, ID extraction is possible of the edges (roads) and their connections to create custom trip routes of vehicles traveling from points of interest to the Prophet's Mosque in the center of the exported map.

Figure 2. The imported map from OpenStreetMap of Madinah city.

The selection criteria comprised of factors such as the inclusion of main highway routes towards the center, the Prophet's Mosque. These highways connect to a primary Ring Road that encircles the point of interest in the center of the city. In addition to all the highway routes, the roads within the encircling Ring Road need to be included as passenger vehicle trips end near the hotels at this location or the drop-off points near the mosque.

Some of these highways connecting to the Ring Road are primary sources of external traffic towards the city. For example, the west and south-west highways are the primary routes for traffic entering the city for pilgrims coming from Makkah city via road and from the main hub airport for the Hijaz region and pilgrims traveling to the country, King Abdulaziz International Airport (KAIA). In addition, the eastern highway welcomes visitors from the Madinah train station, the final stop for the high-speed haramain railway project connecting Makkah, Jeddah (KAIA airport), and King Abdullah Economic City (KAEC).

3.2. Simulators and Frameworks

A combination of a traffic simulation tool and a network simulation tool was used to model the traffic behavior and communication of autonomous and human-driven vehicles in Madinah city. The network simulator imported the traffic data generated on the traffic simulator to perform VANET communication in each scenario tested. The traffic simulator created a variety of traffic situations to match the traffic patterns of the population densities in Madinah city.

3.2.1. Simulation of Urban MObility (SUMO)

SUMO is an open-source traffic modeling simulation tool developed by the German Aerospace Center, now maintained by the Eclipse Foundation. SUMO enables researchers to model road traffic and traffic management systems and to perform in-depth analysis prior to launching the solution in real-world scenarios. In addition to road traffic comprising autonomous and human-driven vehicles, SUMO is capable of simulating public transportation and pedestrians.

SUMO provides a variety of tools with the ability to generate, execute, and evaluate traffic simulations that involve the importing or creation of road networks, route calculations with given parameters and constraints, visualization of the traffic, and emissions cost of each trip for the vehicles.

SUMO has been used in prior research projects to conduct a variety of application studies with a diverse set of research questions and objectives. Examples include an evaluation of proposed algorithms for traffic light control systems to improve vehicle throughput and reduce waiting times, artificial intelligence (AI) training of traffic light schedules, traffic effects of autonomous vehicles, simulation of traffic parking scenarios, and so on.

3.2.2. OMNeT++

OMNeT++ is a discrete-event simulation library and platform developed to perform network simulations. These network simulations can include anything from wired to wireless communication to domain-specific networks, such as wireless sensor networks, ad hoc networks, and so on. These networks are modeled using external framework projects and imported into the OMNeT++ simulator. INET [52] is a popular framework that contains models for a diverse range of network protocols, such as IPv6, Border Gateway Protocol (BGP), and many Ad hoc routing protocols.

OMNeT++ employs a modular architecture for models, referred to as components. These components are written in C++ language and then integrated using higher-level language NEtwork Description (NED) into larger components or models. This modular architecture enables researchers to easily import and embed external models into the applications for simulation.

In this research, OMNeT++ is the core simulation platform, INET provides the communication libraries such as routing protocols and wireless technologies, while Vehicles in network simulation (Veins) [53,54] provides cars and road network libraries to create VANETs. All of this is performed within the specific traffic models generated by SUMO on the imported OSM map, as described by the system design in Figure 3.

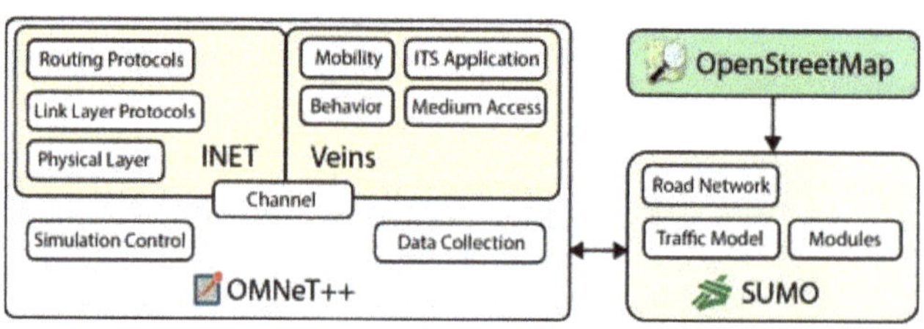

Figure 3. Simulationdesign of the experiments and a subset of the modules utilized from each component.

3.2.3. INET Framework

INET is an open-source framework model library for OMNeT++ network simulations designed for communication experiments. It consists of a variety of network protocols (e.g., Open Shortest Path First (OSPF), BGP), wired and wireless protocols for the data link layer (e.g., IEEE 802.11), mobility models, and numerous other models and components.

INET, such as OMNeT++, follows a modular design with the concept of modules communicating by passing messages using agents. These agents and protocols are the components that form the higher-level function of network devices, such as routers, switches, hosts, and other devices.

INET is used in this research to import and implement the link-layer protocol, IEEE 802.11p designed for VANETS, and the ad hoc routing protocols described later.

3.2.4. Veins

Veins is an open-source simulation framework for vehicular networking. The models provided by the framework libraries are imported into the OMNeT++ event-based simulation to interact with INET and SUMO to perform VANET communication simulation for a given road network and traffic characteristics.

The Veins framework provides a user-modifiable application-specific simulation code that enables researchers to tweak or adapt certain applications to their use-cases. In this research, Veins is used to simulate the VANET applications for the two test case scenarios performed in the simulation experiments described in detail later.

3.2.5. Simulation Process

Figure 4 illustrates in further detail the steps that were taken to perform the simulation experiments between the various simulation platforms. The map extracted from the OSM platform is refined by correcting disjoint roads, missing paths, and other miscellaneous modifications. The refined map or road network is then imported into the SUMO traffic modeling tool. Traffic is generated based on the predefined parameters for various experiment configurations. The resulting set of XML files for the specific scenario is configured with a set of modified routes to measure the trip time performance measure—explained in further detail later. The set of XML files and a configuration file are imported into an OMNeT++ network simulator project where the simulation start and end times are defined in the OMNeT++ configuration file for synchronizing the simultaneous simulations. The configuration file requires the path to the XML files to load the traffic into the road network.

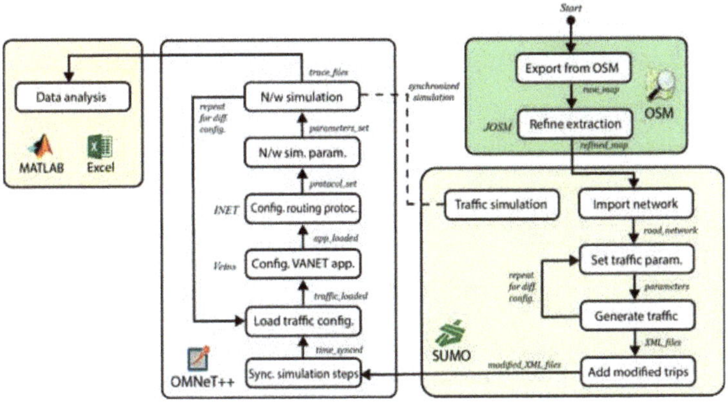

Figure 4. Flow diagram of the simulation processes.

Two tweaks are made on the Veins imported libraries to enable the Veins VANET car to perform routing operations on network packets. The Veins accident scenario application is also modified, enabling Veins cars to propagate the accident packet to their neighbors within the communication range specified. The receiving cars further relay the packet until all vehicles in the network are aware of the accident packet. These modified Veins applications in tandem with the routing protocols provided by INET are loaded in runtime for each vehicle simulated by the SUMO simulator. The simulations are repeated for the various configurations designed for this evaluation. At the end of the simulation time, the trace files are extracted and saved for later data analysis.

3.3. Ad Hoc Routing Protocols

Several routing protocols are implemented in the INET framework. The target of this research is to compare topology-based routing protocols [55] in certain traffic contexts. We chose the DSDV protocol as a proactive routing protocol. For the reactive routing protocol, we selected AODV and DYMO protocols for comparison.

Based on the Bellman–Ford algorithm [56,57], DSDV is a loop-free proactive routing protocol that maintains the routing table containing sequence numbers for each route in the network. A node will update its routing table when it receives a route update with a higher sequence number. If the route update has the same sequence number, a route with a better metric is selected. A periodic update of the routing tables is necessary for the DSDV protocol. Full dump or incremental update methods are used to perform the routing updates. In the full dump method, nodes transmit the whole routing table, whereas, in the incremental update, the node only transmits entries that have changed. Routing updates are broadcasted by either transmitting infrequent full dumps or frequent incremental updates. In a high dynamic topology, routing information needs to be updated more frequently, consuming more power.

Reactive protocols do not maintain routing information about all the nodes but rather only keep the information of the nodes that are present in the route. AODV is a loop-free reactive protocol where routes are generated based on demand from the source node. When a source node finds no route to the destination node, it initiates the route discovery process by flooding [58] the network with route request (RREQ) messages. After discovering the destination node, a unicast message is sent back to the source node in the form of a route reply (RREP) message. A route is established and kept in the routing table of the source node until the link is expired. If any node is unreachable due to broken links, a route error (RERR) message is broadcasted. Like AODV, DYMO is a reactive protocol in which routes are calculated on demand. However, DYMO does not support unnecessary HELLO messages and relies on the sequence numbers assigned to all the messages. Table 1 shows the characteristic comparison between Ad hoc routing protocols.

Table 1. Comparison of Ad-hoc routing protocols.

Characteristics	Routing Protocols		
	AODV	**DYMO**	**DSDV**
Protocol type	Reactive	Reactive	Proactive
Routing scheme	On-demand	On-demand	Table-driven
Routing loop	Loop-free	Loop-free	Loop-free
Control message overhead	Low	Low	High

3.4. Traffic Characteristics

The dataset used to perform experiments were generated using the SUMO traffic modeling simulator due to the lack of real-world datasets available for Madinah city. The synthetic dataset contained modeling of autonomous vehicles and human-driven vehicles with different traffic densities and distributions to measure the performance difference in each test case.

3.4.1. Distributions

Varying traffic distributions were considered when modeling the traffic to study the effect of a higher volume of autonomous vehicles to human-driven vehicles on the performance metrics captured. Five distribution types were modeled:

- t1—Autonomous 0: 100 Human-Driven;
- t2—Autonomous 25: 75 Human-Driven;
- t3—Autonomous 50: 50 Human-Driven;
- t4—Autonomous 75: 25 Human-Driven;
- t5—Autonomous 100: 0 Human-Driven.

3.4.2. Densities

Varying traffic densities were considered when modeling the traffic to study the effect of a higher volume of vehicles on the performance metrics captured from the two test case scenarios detailed later. Four densities of traffic were modeled:

- 500 vehicles (low);
- 1000; vehicles;
- 2500 vehicles;
- 5000 vehicles (high).

Both low- and high-density numbers of vehicles are labeled to differentiate between the performance metrics captured and discussed in the results and analysis section of this paper.

3.5. Performance Metrics

The performance metrics measured in the experiments reflected the objective of the two application scenarios:

$$Average\ trip\ time = \frac{1}{n}\sum_{i=1}^{n} departTime_i - arrivalTime_i \tag{1}$$

$$Average\ latency = \frac{1}{n}\sum_{i=1}^{n} packetRx_i - packetTx_i \tag{2}$$

The average trip time is measured through the SUMO simulator that tracks the vehicle trajectory over the road network and measures the departure time and arrival time, even the waiting time for each vehicle that completed its route. In the case of latency, the trace file of the experiment provides time stamps of the packet transmission time and packet receiving time for specified nodes within the communication range. We compute the latency from the time the first 'accident' packet is sent until the last node in the communication range receives the packet and acknowledges it.

4. Simulations

4.1. Simulation Parameters

Table 2 shows the simulation parameters used in our OMNet++ simulation. For radio medium, we select IEEE 802.11 Dimensional Radio with a radio band of 5.9 GHz. The number of channels for each node (vehicle) is three, which represents the number of channels within the band. The transmission power for the vehicles is set to 20 mW, while the bandwidth is 10 MHz. As mentioned earlier, we selected AODV, DYMO, and DSDV as the three routing protocols in our simulations. The simulation can be visualized on OMNeT++, as seen in Figure 5.

Table 2. Simulation Parameters—OMNeT++.

Parameters	Value
Radio Type	IEEE 802.11 Dimensional Radio
Radio Band	5.9 GHz
Number of Channels	3
Transmitter Power	20 mW
Bandwidth	10 MHz
Routing Protocols	AODV, DYMO, DSDV

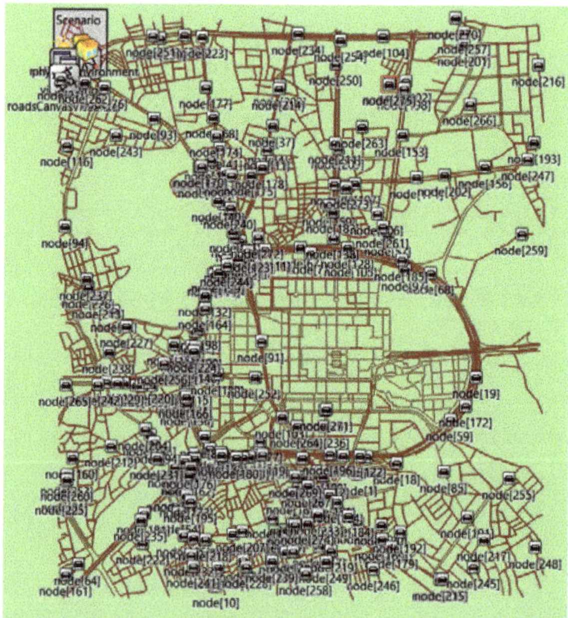

Figure 5. OMNeT++ simulation of the traffic model generated by SUMO.

Table 3 details the list of parameters and their respective values for the traffic model simulations. Four traffic densities were considered with their respective arrival rates departing from a simulation time equal to 5 s. The fringe factor refers to the probability of the routes generated starting or ending at the edges of the network given to the SUMO route generator. All vehicles were of passenger class with human-driven vehicles implementing the IDM car-following model. Each traffic model was generated three times per distribution with fixed seeds, respectively, and repeated for each of the five distributions.

Table 3. Simulation Parameters—SUMO.

Parameters	Value
Traffic Densities	500, 1000, 2500, 5000
Traffic Distributions	0:100, 25:75, 50:50, 75:25, 100:0
Fringe Factor	100
Vehicle Class	Passenger
Car-Following Model	IDM
Custom Trips	R1, R2, R3
Custom Trip Lengths	1.8 km, 1.8 km, 1.9 km
Departure Time	5 s
Simulation Time	500 s
Repetitions & Seeds	R = 3, S = 1, 2, 3

Car-Following Model

Drivers in SUMO may react to environmental changes at specific intervals measured in simulation time, for example, by changing the velocity of the vehicle or changing lanes. The reaction time of the driver depends on the car-following model prescribed to the vehicles in the traffic distribution configuration. The frequency of the drivers' decision-making can be decoupled from the simulation time using the *actionStepLength* parameter, as presented in Table 4.

Table 4. Car-following model parameters (human-driven model).

Parameters	Value
model	IDM (Intelligent Driver Model)
minGap	2.5 m
accel, decel, emergencyDecel	$2.6\ \mathrm{m/s^2}, 4.5\ \mathrm{m/s^2}, 9.0\ \mathrm{m/s^2}$
tau	1 s
delta	4
stepping	0.25 s
actionStepLength	1 s

By assigning 1 s, the drivers evaluate their surroundings and make a decision every simulation time of 1 s. The parameter *minGap* defines the distance maintained by the driver at a full stop or 'standing' situations, such as congestions and traffic light stops. The table further describes the way the IDM car-following model works through the use of parameters such as the acceleration, deceleration, and emergency deceleration rates, predefined for passenger vehicles in this instance. *Tau* and *delta* are tuning parameters. The former is the minimum gap measured in the time unit that the driver will try to maintain while behind another vehicle, and the latter is the recommended value of the acceleration exponent. The acceleration exponent delta is part of the IDM differential equations to calculate the approaching rate for a given vehicle. Finally, the *stepping* parameter defines the internal step length in seconds for the calculation of the following speed.

4.2. Simulation Environment Specifications

Tables 5 and 6 describe the physical host and software specifications, respectively, utilized for the simulation experiments and analysis of the captured metrics.

Table 5. Physical host specification.

Item	Specification
CPU	Intel Core i5-8250U CPU @ 1.60GHz
Operating System	Windows 10 Build 18363 (64-bit)
Main Memory	20 GB

Table 6. Software specification.

Item	Specification
Network Simulation	OMNeT++ v5.6.2
Traffic Simulation	SUMO v1.11.0
Data Analysis Tool 1	MATLAB R2021b
Data Analysis Tool 2	Microsoft Excel

4.3. Test Scenarios

4.3.1. Custom Trips

In this scenario, three custom trips were created for each of the various traffic distributions and densities to measure the trip time for a vehicle from a point of interest moving towards the Prophet's Mosque. This is an example case of a taxi vehicle providing a service to the pilgrims arriving in the city. The three routes use a primary highway and move towards the major Ring Road encircling the Prophet's Mosque and nearby accommodations, then move towards the local roads to drop off the pilgrim at the nearest location to the mosque.

This test case scenario, executed and averaged over repetitions, provides the researchers with an estimate of the impact of traffic distributions and densities on the single trip journey of pilgrims from various points of interest. Three custom vehicle trips were considered moving towards the Prophet's Mosque:

- West Highway to Prophet's Mosque (Figure 6);
- South Highway to Prophet's Mosque (Figure 7);
- North Highway to Prophet's Mosque (Figure 8).

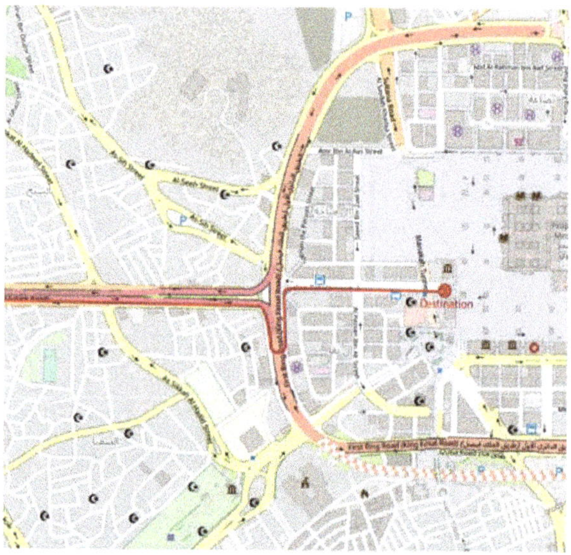

Figure 6. Route 1—West Highway to Prophet's Mosque.

Figure 7. Route 2—South Highway to Prophet's Mosque.

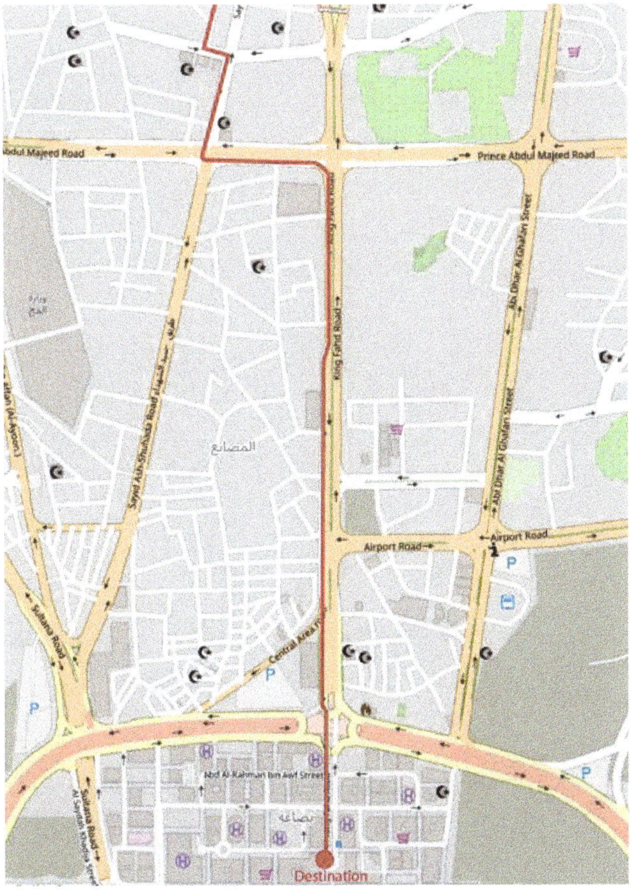

Figure 8. Route 3—North Highway to Prophet's Mosque.

4.3.2. Communication Latency

This test scenario aims at collecting information on the packet transmission delays when a VANET node sends a packet to all neighboring nodes within a specified communication range, determined by the communication power of the radio antenna. In this case, an accident is simulated on a vehicle in the network, blocking the road. Then, a packet alert is broadcasted to the nearest VANET vehicles, warning them of an accident. These vehicles relay the packet to their neighbors until all the vehicles in the environment have received the accident packet. Once the alert is received, the alternative route to the destination is taken by vehicles to avoid the congestion caused by the accident.

5. Results and Analysis

The results for our experiments are based on the traffic characteristics discussed earlier. We consider studying the average trip time and average latency with different traffic distributions under varying traffic densities. Furthermore, we performed an Analysis Of Variance (ANOVA) test to study the impact of various factors on the system performance in terms of average trip time and latency.

5.1. Average Trip Time

The average trip time is calculated using Equation (1) for different routes with varying traffic densities and distributions. The graphs are plotted with confidence intervals with a confidence level of 90% to measure the uncertainty and establish if the results are statistically significant.

The average trip time for Route 1 with varying traffic distributions can be seen in Figure 9. The difference in the average trip time for low and high traffic densities is about 10 s. The average trip time is reduced by up to 7.1% for a fully autonomous traffic distribution under high traffic density. Traffic distributions t1, t2, and t3 have similar average trip times for low or high traffic densities. However, traffic distributions t4 and t5 achieved shorter average trip times for both types of traffic density. We consider a 90% confidence level attributing to the confidence internals in the graphs. The confidence intervals are barely non-overlapping for each group, denoting they are statistically significant and not random.

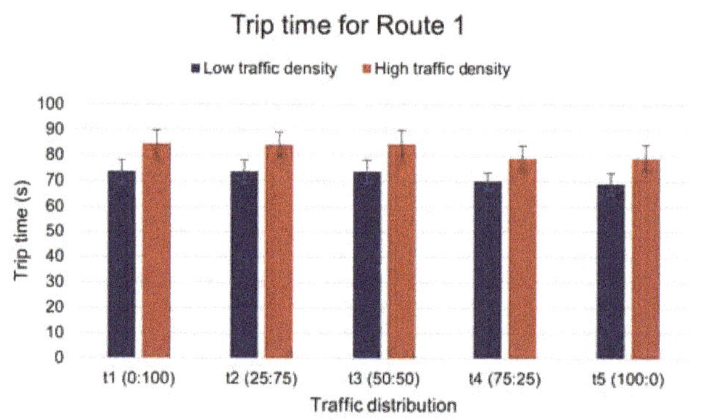

Figure 9. Average trip time for Route 1 in low and high traffic densities.

In Figure 10, the average trip time for Route 2 against different traffic distributions is shown. At a low traffic density, the average trip time decreases with an increase in autonomous level. The average trip times for t1, t2, and t3 traffic distributions are similar at a high traffic density. The average trip time further decreases for t4 and t5 traffic distributions. A fully autonomous traffic distribution incurs the least average trip time at a high density. The confidence intervals overlap, signifying that the results are not significant. Hence, we tend to perform ANOVA tests to investigate the results and study the effects of the factors on the results.

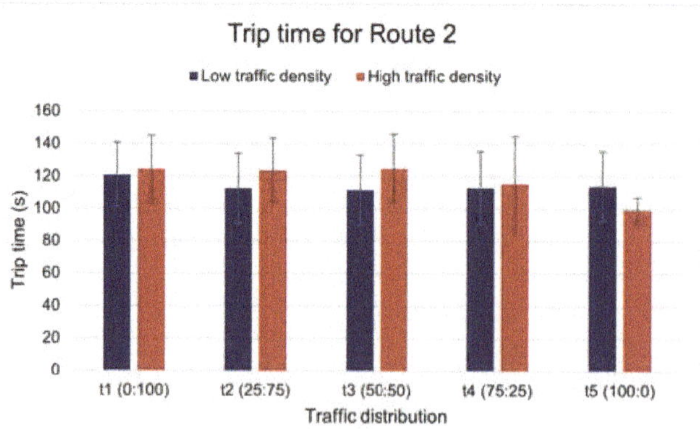

Figure 10. Average trip time for Route 2 in low and high traffic densities.

Figure 11 shows the average trip time for Route 3 against varying traffic distributions under low and high traffic densities. The average trip time would be higher for any traffic distribution with a high traffic density. The difference in average trip time at low and high traffic densities is highest for semi-autonomous traffic distribution. A fully autonomous traffic distribution incurs the shortest average trip time at any traffic density. Although the confidence intervals overlap, the mean does not fall in the confidence intervals for traffic distributions t2, t4, and t5. Hence, an ANOVA test would predict the effect of the factors on the variation of the results.

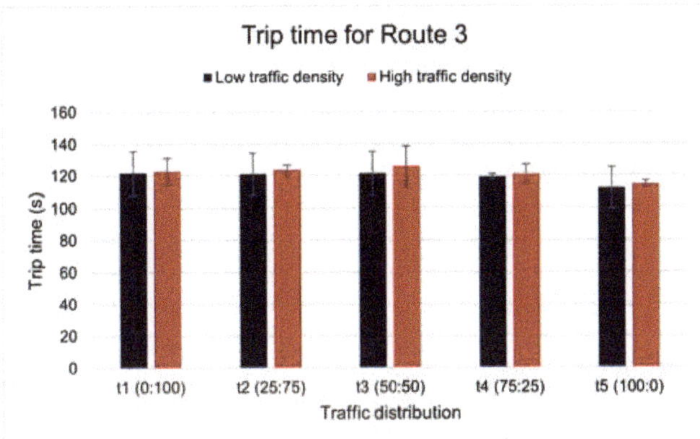

Figure 11. Average trip time for Route 3 in low and high traffic densities.

5.2. Average Latency

The latency graph for fully human-driven traffic distribution with varying traffic densities is shown in Figure 12. It can be seen that latency increases with a higher traffic density. Looking at this graph, it can be observed that the difference between reactive and proactive protocols is not very significant. At a lower traffic density, i.e., less than 3300 vehicles, DSDV outperforms both reactive protocols (AODV and DYMO). Once the vehicle density increases beyond 3300, the reactive protocols provide less latency. The reason behind the increased latency in a higher density is because more nodes will send the acknowledgments to the sender, compared to the lower traffic density, which will take more time to wait and send in order to avoid collisions.

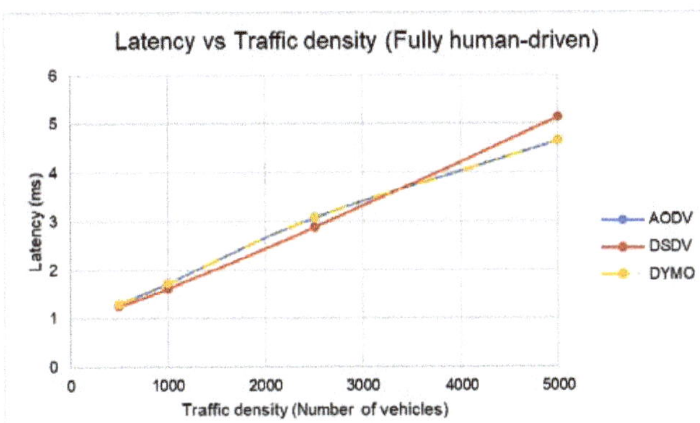

Figure 12. Latency graph for fully human-driven vehicles with varying traffic densities.

For a fully autonomous traffic distribution, as seen in Figure 13, the latency is slightly higher for AODV and DYMO protocols when the traffic density is less than 1900 vehicles.

With increasing traffic density, the DSDV protocol is outperformed by the AODV and DYMO protocols. The difference in latency is about 2.38 milliseconds, with DSDV having significantly higher latency. Therefore, as the traffic density increases, reactive protocols tend to perform better for a fully autonomous traffic scenario. The reason behind the increased latency in the DSDV protocol is the control packet overhead caused by the periodic exchange of packets. As mentioned earlier, the DSDV protocol exchanges control packets regularly to update the routing tables. Since we observe the latency from the time the first 'accident' packet is sent until the last node in the communication range receives the packet and acknowledges it. With high traffic density, there is an increased chance that a DSDV control packet transmission will start while the nodes are reporting the accident. This exact scenario happens in this case. When there is a transmission of control (HELLO) packets, the nodes will wait until the channel is free to transmit the accident or acknowledgment packets, which results in higher latency.

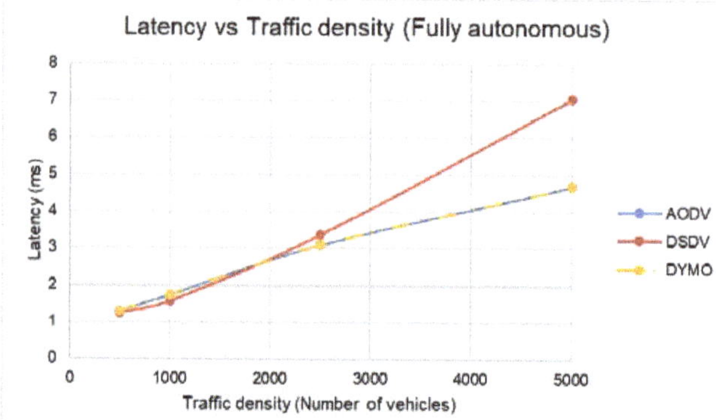

Figure 13. Latency graph for fully autonomous vehicles with varying traffic densities.

In Figure 14, we plot the latency graph for semi-autonomous traffic distribution where 50% of vehicles are human-driven, and the rest are autonomous vehicles. For increasing traffic density, latency for reactive and proactive protocols increases linearly. It is clear that there is no significant difference between the reactive and proactive protocols in this scenario as the density of traffic increases. The analysis of variance presented later further confirms this observation.

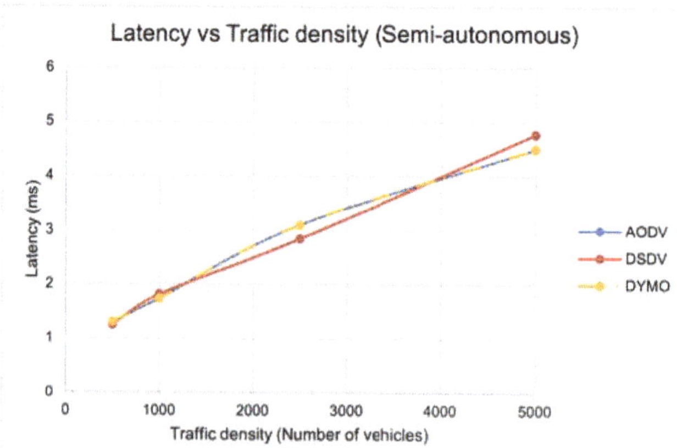

Figure 14. Latency graph for semi-autonomous vehicles with varying traffic densities.

Figure 15 shows the latency graph with varying traffic distributions under low traffic density. For any traffic distribution, DSDV performs slightly better than reactive protocols at a low traffic density. The difference in latency is about 0.04 milliseconds between reactive and proactive protocols.

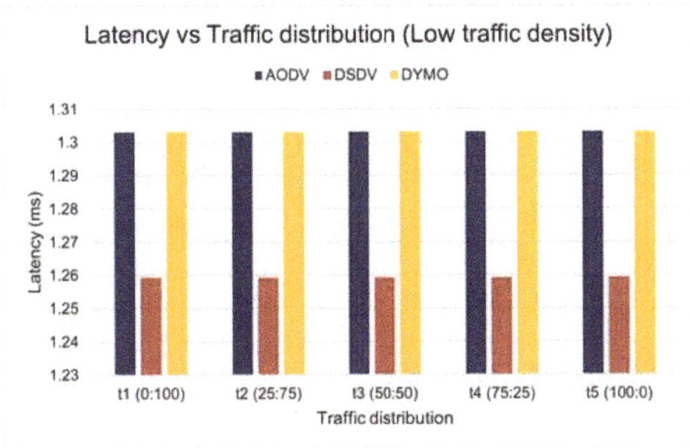

Figure 15. Latency graph for low traffic density with varying traffic distribution.

A latency graph with varying traffic distributions under a high traffic density can be seen in Figure 16. For a fully human-driven traffic distribution, t1, DSDV provides higher latency compared to reactive protocols. For semi-autonomous (t3) or 25% autonomous vehicle (t2) traffic distribution, reactive protocols perform slightly better than DSDV. In 75% autonomous traffic distribution (t4), the DSDV protocol is outperformed by the AODV and DYMO protocols. For a fully autonomous traffic distribution, AODV and DYMO outperform DSDV in high traffic density. The higher latency in the DSDV protocol is due to the control packet overhead. When the control packets (HELLO packets) are sent by any node, the neighboring nodes will not send any data packets as they sense the channel is busy. The latency is observed from the time the first 'accident' packet is sent until the last node in the communication range receives the packet and acknowledges it. Thus, in the case of DSDV, if there is a transmission of control (HELLO) packets, the nodes will wait until the channel is free to transmit the accident or acknowledgment packets, which results in higher delays.

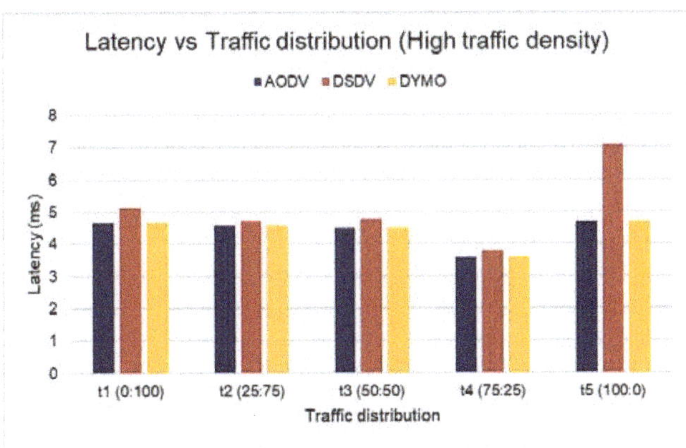

Figure 16. Latency graph for high traffic density with varying traffic distribution.

5.3. Analysis of Variance (ANOVA)

For the analysis of variation study, we used a Full Factorial Design with K = 3 factors and five (5) levels [59,60] for trip time and latency metrics. The formulation of the ANOVA is listed below for each performance metric captured. Factors A, B, and C are different for the trip time and latency factorial designs:

y_{ijkl} = observation in the Ith replication of experiment with factors A, B, and C at levels i, j, and k, respectively;
μ = mean observation;
α_i = effect of factor A at level i;
β_j = effect of factor B at level j;
ξ_k = effect of factor C at level k;
$\gamma_{AB_{ij}}$ = interaction between A and B at levels i and j;
$\gamma_{ABC_{ijk}}$ = interaction between A, B, C at levels i, j, k.

5.3.1. ANOVA of Trip Time

Table 7 shows the factors and the levels considered for the trip time variance analysis. The factors selected for the trip time were the routes, traffic distribution, and traffic density. The route factor had three levels, traffic distribution had five (5) levels, and traffic density had two (2) levels, as denoted in the table.

Table 7. Factors and their respective levels for trip time.

Symbol	Factor	Levels				
		1	2	3	4	5
R	Routes	R1	R2	R3		
T	Traffic Distribution	0:100	25:75	50:50	75:25	100:0
D	Traffic Density	500	5000			

We quantify the impact of the main factors on the trip time by performing ANOVA tests. The ANOVA for the trip time is given in Table 8. We can observe that the main factors contribute about 97% to the variation of the trip time results, and the first and second-order interactions between the factors contribute around 2% and 1%, respectively. The route (R) has the most significant effect on the results with a 91.8% variation. Traffic distribution (T) shows a variation of 3.2% on the trip time results. However, traffic density (D) has the least significant effect on the results adding up to 1.6%. Our assumption that the residuals are normally distributed is proven by the Quantile-Quantile (Q-Q) for all three factors, as shown in Figure 17. The visual tests show that the residuals fall evenly around the least-squares line forming a heavy-tailed linear line.

Table 8. ANOVA of trip time.

Component	Sum of Squares	Perc. Variation	DF	Mean Square
y	339,533.0005		30	
$\bar{y}$	327,189.6338		1	
$y - \bar{y}$	12,343.3667	100.00%	29	
Main Effects	11,928.2186	96.6%	7	1704.0
D	201.0703	1.6%	1	
T	400.6816	3.2%	4	
R	11,326. 4667	91.8%	2	
First-order Interactions	264.6889	2.1%	14	18.9
D-T	92.6519		4	
D-R	92.5407		2	
T-R	79.4963		8	
Second-order Interactions	150.4593	1.2%	8	18.8
D-T-R	150.4593		8	

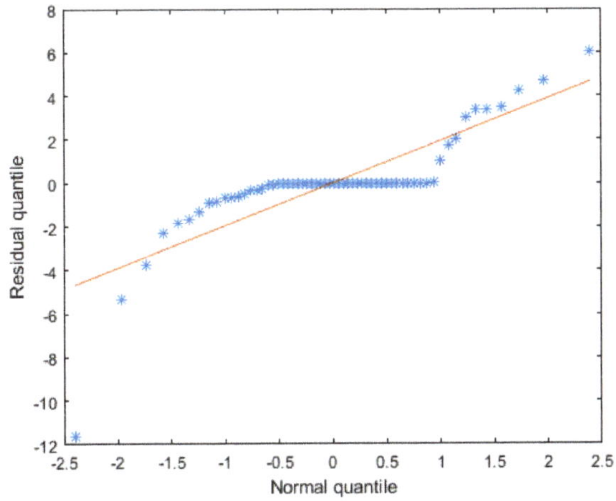

Figure 17. Quantile—Quantile Plot of Trip Time Residual.

5.3.2. ANOVA of Latency

The factors and levels considered for the latency variance analysis are shown in Table 9. Routing protocols had three (3) levels, traffic distribution had five (5) levels, and traffic density had four (4) levels, as listed in the table.

Table 9. Factors and their respective levels for latency.

Symbol	Factor	Levels				
		1	2	3	4	5
RP	Routing Protocols	AODV	DSDV	DYMO		
T	Traffic Distribution	0:100	25:75	50:50	75:25	100:0
D	Traffic Density	500	1000	2500	5000	

For the latency results, the ANOVA is shown in Table 10. From the table, we can observe that the main effects contribute 92.7% of the variation in the latency results. The traffic density (D) has the highest impact, with 91.1% of the variation. While the routing protocol (RP) has the least contribution with 0.2%. Traffic distribution (T) contributes 1.4% of latency variation. Additionally, the highest interaction is between the traffic density and distribution that has the most significant effect with 3.94% variation on the latency results compared to any other interaction combination. This indicates that the above factors interact to influence the latency result. The interaction between all three factors accounts for only 1.9% of the variation. The Q-Q plot of latency residuals is shown in Figure 18. We can observe that the residuals fall evenly around the least-squares line forming a trend with a heavy tail due to outliers. Hence, our simulations are concrete, and the variations in the results are not due to randomness.

Table 10. ANOVA of latency.

Component	Sum of Squares	Perc. Variation	DF	Mean Square
y	538.0728		60	
$\bar{y}$	427.9197		1	
$y - \bar{y}$	110.1531	100.00%	59	
Main Effects	102.0573	92.7%	9	11.3
D	100.3077	91.1%	3	
T	1.5098	1.4%	4	
RP	0.2398	0.2%	2	
First-order Interactions	6.0242	5.5%	26	0.2
D-T	4.3165		12	
D-RP	0.9441		6	
T-RP	0.7635		8	
Second-order Interactions	2.0716	1.9%	24	0.1
D-T-RP	2.0716		24	

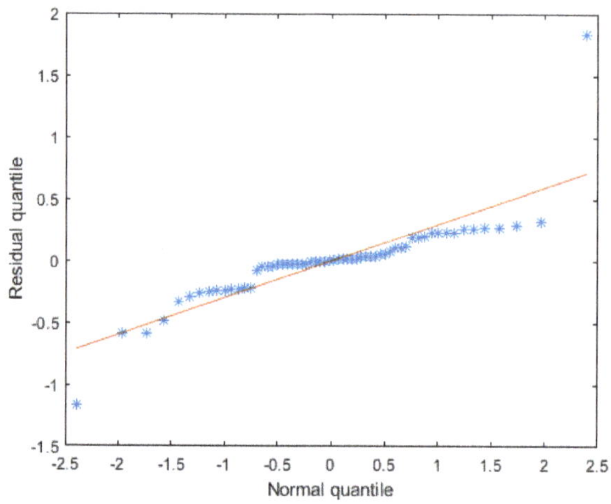

Figure 18. Quantile—Quantile Plot of Latency Residual.

6. Conclusions

In this work, we evaluated the performance of various VANET routing protocols in Madinah city under different traffic scenarios. Two reactive routing protocols, AODV and DYMO, are selected, while one proactive protocol, DSDV, is considered. We considered both fully autonomous and human-driven vehicles in our study to replicate the state of future traffic. We further model multiple traffic distributions with the varying volume of autonomous and human-driven vehicles, along with several traffic densities to represent different traffic conditions. We performed simulations to analyze the average trip time and average communication latency between vehicles. For an average trip time, we found that fully autonomous traffic distribution achieved the shortest trip time, with a reduction of around 7.1% compared to other distributions with human-driven vehicles. For average latency, we observed that the DSDV protocol performs better in fully human-driven scenarios. However, in the case of fully autonomous traffic, both reactive protocols outperform the DSDV protocol, reducing the latency by 2.38 milliseconds (33.7% improvement). From our experiments, we recommend the use of reactive routing protocols compared to the proactive DSDV protocol, as the control message overhead of DSDV is much higher. Furthermore, we performed an ANOVA test to examine the effect of factors on the trip time and latency results. For the average trip time, we observed that the routes compared to traffic density and distribution had the most significant effect (91.8%) on the variation of the results. On the other hand, traffic density had the highest impact on the latency result (91.1%) compared to other factors. The ANOVA table also highlighted the effects of the interactions between the factors on the results. The visual tests observed from the Q-Q plots exhibit that the residuals fall evenly around the least-squares line, forming a heavy-tailed linear line. This confirms our assumption that the residuals are normally distributed. Hence, the variation in the results is not random but due to factors considered in our study.

6.1. Limitations

Our focus was studying and evaluating the routing protocols under different traffic scenarios. We considered 500 vehicles as a low traffic density and 5000 vehicles as a high traffic density for our study. It might not be the case in a real-world scenario for a highly congested area. Furthermore, we did not assess dynamic vehicle velocity changes based on conditions in our simulation. Although we considered a reasonable size of map area for the city of Madinah, we might need to take into account a larger area to study the overall performance. Other environmental factors that might affect the performance are out of scope for this particular study.

6.2. Future Works

This research can be extended by conducting further studies on the various issues considered in this article in the future. For the traffic model designed and adopted in this work, multiple factors can be expanded to outline the effect of those factors. For example, the car-following model for the human-driven vehicle can be another factor to consider by comparing against models, such as Krauss, EIDM, Wiedemann, and so on. Further, the parameters for each respective car-following model can be studied, such as minimum gaps kept from the vehicle in front, acceleration, and deceleration rates and other parameters. Another future study could outline the effect of even higher traffic densities in Madinah city with further distributions of not just passenger vehicles but also buses, motorcycles, etc. Routing protocols in this study could be expanded to include other ad hoc protocols, such as OLSR and DSR. Performance metrics can be furthered by considering the impact of the packet delivery ratio and the network throughput of the accident notification and relaying packets on traffic congestion. Other metrics could also be studied such as the effect of the network behavior on the carbon dioxide emissions and waiting time of passenger vehicles in the traffic model.

Author Contributions: Conceptualization, M.A.R.A., A.B., S.M.M., A.A., T.R.S. and A.Y.; methodology, M.A.R.A., A.B., S.M.M., A.A., T.R.S. and A.Y.; software, A.B., S.M.M. and A.A.; writing—original draft, A.B., S.M.M. and A.A.; writing—review and editing, M.A.R.A., A.B., S.M.M., A.A., T.R.S. and A.Y. All authors have read and agreed to the published version of the manuscript.

Funding: This research was funded by the Islamic University of Madinah project number 580, Tamayoz-2 program.

Institutional Review Board Statement: Not applicable.

Informed Consent Statement: Not applicable.

Acknowledgments: The authors would like to acknowledge the research deanship of the Islamic University of Madinah and the King Fahd University of Petroleum and Minerals for the support of this research.

Conflicts of Interest: The authors declare no conflict of interest.

References

1. Miller, J. Vehicle-to-vehicle-to-infrastructure (V2V2I) intelligent transportation system architecture. In Proceedings of the 2008 IEEE Intelligent Vehicles Symposium, Eindhoven, The Netherlands, 4–6 June 2008; IEEE: Piscataway, NJ, USA, 2008; pp. 715–720.
2. Arena, F.; Pau, G.; Severino, A. A review on IEEE 802.11 p for intelligent transportation systems. *J. Sens. Actuator Networks* **2020**, *9*, 22. [CrossRef]

3. Chen, S.; Hu, J.; Shi, Y.; Peng, Y.; Fang, J.; Zhao, R.; Zhao, L. Vehicle-to-everything (V2X) services supported by LTE-based systems and 5G. *IEEE Commun. Stand. Mag.* **2017**, *1*, 70–76. [CrossRef]
4. Yuan, Y.; Zheng, G.; Wong, K.K.; Letaief, K.B. Meta-reinforcement learning based resource allocation for dynamic V2X communications. *IEEE Trans. Veh. Technol.* **2021**, *70*, 8964–8977. [CrossRef]
5. Abboud, K.; Omar, H.A.; Zhuang, W. Interworking of DSRC and cellular network technologies for V2X communications: A survey. *IEEE Trans. Veh. Technol.* **2016**, *65*, 9457–9470. [CrossRef]
6. Biswas, S.; Tatchikou, R.; Dion, F. Vehicle-to-vehicle wireless communication protocols for enhancing highway traffic safety. *IEEE Commun. Mag.* **2006**, *44*, 74–82. [CrossRef]
7. Arena, F.; Pau, G.; Severino, A. V2X communications applied to safety of pedestrians and vehicles. *J. Sens. Actuator Networks* **2020**, *9*, 3. [CrossRef]
8. Wang, J.; Shao, Y.; Ge, Y.; Yu, R. A survey of vehicle to everything (V2X) testing. *Sensors* **2019**, *19*, 334. [CrossRef]
9. Ihsan, A.; Chen, W.; Zhang, S.; Xu, S. Energy-efficient NOMA multicasting system for beyond 5G cellular V2X communications with imperfect CSI. *IEEE Trans. Intell. Transp. Syst.* **2021**, 1–15. [CrossRef]
10. Elhoseny, M.; Shankar, K. Energy efficient optimal routing for communication in VANETs via clustering model. In *Emerging Technologies for Connected Internet of Vehicles and Intelligent Transportation System Networks*; Springer: Berlin, Germany, 2020; pp. 1–14.
11. Ali, I.; Li, F. An efficient conditional privacy-preserving authentication scheme for Vehicle-To-Infrastructure communication in VANETs. *Veh. Commun.* **2020**, *22*, 100228. [CrossRef]
12. MacHardy, Z.; Khan, A.; Obana, K.; Iwashina, S. V2X access technologies: Regulation, research, and remaining challenges. *IEEE Commun. Surv. Tutorials* **2018**, *20*, 1858–1877. [CrossRef]
13. Zeadally, S.; Hunt, R.; Chen, Y.S.; Irwin, A.; Hassan, A. Vehicular ad hoc networks (VANETS): Status, results, and challenges. *Telecommun. Syst.* **2012**, *50*, 217–241. [CrossRef]
14. OpenStreetMap. Available online: https://www.openstreetmap.org/ (accessed on 27 January 2022).
15. Haklay, M.; Weber, P. Openstreetmap: User-generated street maps. *IEEE Pervasive Comput.* **2008**, *7*, 12–18. [CrossRef]
16. Eclipse SUMO—Simulation of Urban Mobility. Available online: https://www.eclipse.org/sumo/ (accessed on 27 January 2022).
17. Behrisch, M.; Bieker, L.; Erdmann, J.; Krajzewicz, D. SUMO–simulation of urban mobility: An overview. In Proceedings of the SIMUL 2011, The Third International Conference on Advances in System Simulation, Barcelona, Spain, 23–28 October 2011; ThinkMind: Barcelona, Spain, 2011.
18. OMNeT++ Discrete Event Simulator. Available online: https://omnetpp.org/ (accessed on 27 January 2022).
19. Varga, A.; Hornig, R. An overview of the OMNeT++ simulation environment. In Proceedings of the 1st International Conference on Simulation Tools and Techniques for Communications, Networks and Systems & Workshops, Marseille, France, 3–7 March 2008; ICST: Brussels, Belgium, 2008; pp. 1–10.
20. Perkins, C.; Belding-Royer, E.; Das, S. *RFC3561: Ad Hoc On-Demand Distance Vector (AODV) Routing*; IETF: Santa Barbara, CA, USA, 2003.
21. Chakeres, I.D.; Belding-Royer, E.M. AODV routing protocol implementation design. In Proceedings of the 24th International Conference on Distributed Computing Systems Workshops, Tokyo, Japan, 23–24 March 2004; IEEE: Piscataway, NJ, USA, 2004; pp. 698–703.
22. Perkins, C.E.; Bhagwat, P. Highly dynamic destination-sequenced distance-vector routing (DSDV) for mobile computers. *ACM SIGCOMM Comput. Commun. Rev.* **1994**, *24*, 234–244. [CrossRef]
23. He, G. Destination-sequenced distance vector (DSDV) protocol. *Netw. Lab. Hels. Univ. Technol.* **2002**, *135*.
24. Chakeres, I. Dynamic MANET On-Demand (DYMO) Routing. 2008. Available online: http://tools.ietf.org/html/draft-ietf-manet-dymo-14 (accessed on 21 January 2022).
25. Sommer, C.; Dressler, F. The DYMO routing protocol in VANET scenarios. In Proceedings of the 2007 IEEE 66th Vehicular Technology Conference, Baltimore, MD, USA, 30 September–3 October 2007; IEEE: Piscataway, NJ, USA, 2007; pp. 16–20.
26. Billington, J.; Yuan, C. On modelling and analysing the dynamic MANET on-demand (DYMO) routing protocol. In *Transactions on Petri Nets and Other Models of Concurrency III*; Springer: Berlin, Germany, 2009; pp. 98–126.
27. Gupta, A.K.; Sadawarti, H.; Verma, A.K. Implementation of DYMO routing protocol. *arXiv* **2013**, arXiv:1306.1338.
28. Kanellopoulos, D.; Cuomo, F. Recent Developments on Mobile Ad-Hoc Networks and Vehicular Ad-Hoc Networks. *Electronics* **2021**, *10*, 364. [CrossRef]
29. Cunha, F.; Villas, L.; Boukerche, A.; Maia, G.; Viana, A.; Mini, R.A.; Loureiro, A.A. Data communication in VANETs: Protocols, applications and challenges. *Ad Hoc Netw.* **2016**, *44*, 90–103. [CrossRef]
30. Alves Junior, J.; Wille, E.C. Routing in vehicular ad hoc networks: Main characteristics and tendencies. *J. Comput. Netw. Commun.* **2018**, *2018*, 1302123. [CrossRef]
31. Shaheen, A.; Gaamel, A.; Bahaj, A. Comparison and Analysis Study between AODV DSR Routing Protocols in Vanet with IEEE 802.11. *J. Ubiquitous Syst. Pervasive Netw.* **2016**, *7*, 7–12.
32. Johnson, D.B.; Maltz, D.A.; Broch, J. DSR: The dynamic source routing protocol for multi-hop wireless ad hoc networks. *Ad Hoc Netw.* **2001**, *5*, 139–172.
33. Johnson, D.; Hu, Y.C.; Maltz, D. *RFC 4728: The Dynamic Source Routing Protocol (DSR) for Mobile Ad Hoc Networks for IPv4*; Technical Report; IETF: Fremont, CA, USA, 2007.

34. Chang, X. Network simulations with OPNET. In *WSC'99. 1999 Winter Simulation Conference Proceedings. 'Simulation-A Bridge to the Future' (Cat. No. 99CH37038)*; IEEE: Piscataway, NJ, USA, 1999; Volume 1, pp. 307–314.
35. Hiertz, G.R.; Denteneer, D.; Stibor, L.; Zang, Y.; Costa, X.P.; Walke, B. The IEEE 802.11 universe. *IEEE Commun. Mag.* **2010**, *48*, 62–70. [CrossRef]
36. Abdelgadir, M.; Saeed, R.A.; Babiker, A. Mobility routing model for vehicular ad-hoc networks (VANETs), smart city scenarios. *Veh. Commun.* **2017**, *9*, 154–161. [CrossRef]
37. Kandali, K.; Bennis, H. Performance Assessment of AODV, DSR and DSDV in an Urban VANET Scenario. In *International Conference on Advanced Intelligent Systems for Sustainable Development*; Springer: Berlin, Germany, 2018; pp. 98–109.
38. Issariyakul, T.; Hossain, E. Introduction to network simulator 2 (NS2). In *Introduction to Network Simulator NS2*; Springer: Berlin, Germany, 2009; pp. 1–18.
39. Ghori, M.R.; Sadiq, A.S.; Ghani, A. VANET routing protocols: Review, implementation and analysis. *J. Physics Conf. Ser.* **2018**, *1049*, 012064. [CrossRef]
40. Ali, F.; Shaikh, F.K.; Ansari, A.Q.; Mahoto, N.A.; Felemban, E. Comparative analysis of VANET routing protocols: On road side unit placement strategies. *Wirel. Pers. Commun.* **2015**, *85*, 393–406. [CrossRef]
41. Sallum, E.E.A.; dos Santos, G.; Alves, M.; Santos, M.M. Performance analysis and comparison of the DSDV, AODV and OLSR routing protocols under VANETs. In Proceedings of the 2018 16th International Conference on Intelligent Transportation Systems Telecommunications (ITST), Lisboa, Portugal, 15–17 October 2018; IEEE: Piscataway, NJ, USA, 2018; pp. 1–7.
42. Clausen, T.; Jacquet, P.; Adjih, C.; Laouiti, A.; Minet, P.; Muhlethaler, P.; Qayyum, A.; Viennot, L. Optimized Link State Routing Protocol (OLSR). 2003. Available online: https://www.ietf.org/rfc/rfc3626.txt (accessed on 21 January 2022).
43. Aschenbruck, N.; Ernst, R.; Gerhards-Padilla, E.; Schwamborn, M. Bonnmotion: A mobility scenario generation and analysis tool. In Proceedings of the 3rd International ICST Conference on Simulation Tools and Techniques, Malaga, Spain, 15–19 March 2010; ICST: Brussels, Belgium, 2010; pp. 1–10.
44. Riley, G.F.; Henderson, T.R. The ns-3 network simulator. In *Modeling and Tools for Network Simulation*; Springer: Berlin, Germany, 2010; pp. 15–34.
45. Bengag, A.; Bengag, A.; Elboukhari, M. Routing Protocols for VANETs: A Taxonomy, Evaluation and Analysis. *Adv. Sci. Technol. Eng. Syst. J.* **2020**, *5*, 77–85. [CrossRef]
46. Karp, B.; Kung, H.T. GPSR: Greedy perimeter stateless routing for wireless networks. In Proceedings of the 6th Annual International Conference on Mobile Computing and Networking, Boston, MA, USA, 6–11 August 2000; Association for Computing Machinery: New York, NY, USA, 2020; pp. 243–254.
47. Lochert, C.; Mauve, M.; Füßler, H.; Hartenstein, H. Geographic routing in city scenarios. *ACM Sigmobile Mob. Comput. Commun. Rev.* **2005**, *9*, 69–72. [CrossRef]
48. Shi, M.; Lu, C.; Zhang, Y.; Yao, D. DSRC and LTE-V communication performance evaluation and improvement based on typical V2X application at intersection. In Proceedings of the 2017 Chinese Automation Congress (CAC), Jinan, China, 20–22 October 2017; IEEE: Piscataway, NJ, USA, 2017; pp. 556–561.
49. Spaho, E.; Barolli, L.; Mino, G.; Xhafa, F. Simulation Evaluation of AODV, OLSR and DYMO Protocols for Vehicular Networks using CAVENET. In Proceedings of the 2011 International Conference on Complex, Intelligent, and Software Intensive Systems, Seoul, Korea, 30 June–2 July 2011; IEEE: Piscataway, NJ, USA, 2011; pp. 152–159.
50. García-Campos, J.; Reina, D.; Toral, S.; Bessis, N.; Barrero, F.; Asimakopoulou, E.; Hill, R. Performance evaluation of reactive routing protocols for VANETs in urban scenarios following good simulation practices. In Proceedings of the 2015 9th International Conference on Innovative Mobile and Internet Services in Ubiquitous Computing, Santa Cantarina, Brazil, 8–10 July 2015; IEEE: Piscataway, NJ, USA, 2015; pp. 1–8.
51. Ko, Y.B.; Vaidya, N.H. Location-Aided Routing (LAR) in mobile ad hoc networks. *Wirel. Netw.* **2000**, *6*, 307–321. [CrossRef]
52. INET Framework. Available online: https://inet.omnetpp.org/ (accessed on 27 January 2022).
53. Veins. Available online: https://veins.car2x.org/ (accessed on 27 January 2022).
54. Sommer, C.; Eckhoff, D.; Brummer, A.; Buse, D.S.; Hagenauer, F.; Joerer, S.; Segata, M. Veins: The open source vehicular network simulation framework. In *Recent Advances in Network Simulation*; Springer: Berlin, Germany, 2019; pp. 215–252.
55. Qureshi, K.N.; Abdullah, H. Topology based routing protocols for VANET and their comparison with MANET. *J. Theor. Appl. Inf. Technol.* **2013**, *58*, 707–715.
56. Cheng, C.; Riley, R.; Kumar, S.P.; Garcia-Luna-Aceves, J.J. A loop-free extended Bellman-Ford routing protocol without bouncing effect. *ACM Sigcomm Comput. Commun. Rev.* **1989**, *19*, 224–236. [CrossRef]
57. Goldberg, A.; Radzik, T. *A Heuristic Improvement of the Bellman-Ford Algorithm*; Technical Report; Stanford University, Computer Science Department: Stanford, CA, USA, 1993.
58. Lim, H.; Kim, C. Flooding in wireless ad hoc networks. *Comput. Commun.* **2001**, *24*, 353–363. [CrossRef]
59. Jain, R. *The Art of Computer Systems Performance Analysis: Techniques for Experimental Design, Measurement, Simulation, and Modeling*; John Wiley & Sons: Hoboken, NJ, USA, 1990.
60. Girden, E.R. *ANOVA: Repeated Measures*; Sage Publications, Inc.: Thousand Oaks, CA, USA, 1992; Number 84.

electronics

MDPI

Article

Efficient Deep Reinforcement Learning for Optimal Path Planning

Jing Ren [1,*], Xishi Huang [2] and Raymond N. Huang [3]

1 Department of Electrical, Computer, and Software Engineering, Ontario Tech University, Oshawa, ON L1G 0C5, Canada
2 RS OPTO Tech Ltd., Suzhou 215100, China
3 Department of Mechanical and Industrial Engineering, University of Toronto, Toronto, ON M5S 3G8, Canada
* Correspondence: jing.ren@ontariotechu.ca

Abstract: In this paper, we propose a novel deep reinforcement learning (DRL) method for optimal path planning for mobile robots using dynamic programming (DP)-based data collection. The proposed method can overcome the slow learning process and improve training data quality inherently in DRL algorithms. The main idea of our approach is as follows. First, we mapped the dynamic programming method to typical optimal path planning problems for mobile robots, and created a new efficient DP-based method to find an exact, analytical, optimal solution for the path planning problem. Then, we used high-quality training data gathered using the DP method for DRL, which greatly improves training data quality and learning efficiency. Next, we established a two-stage reinforcement learning method where, prior to the DRL, we employed extreme learning machines (ELM) to initialize the parameters of actor and critic neural networks to a near-optimal solution in order to significantly improve the learning performance. Finally, we illustrated our method using some typical path planning tasks. The experimental results show that our DRL method can converge much easier and faster than other methods. The resulting action neural network is able to successfully guide robots from any start position in the environment to the goal position while following the optimal path and avoiding collision with obstacles.

Keywords: deep reinforcement learning; global optimal path planning; dynamic programming; mobile robots; shortest path; continuous state space; collision avoidance

Citation: Ren, J.; Huang, X.; Huang, R.N. Efficient Deep Reinforcement Learning for Optimal Path Planning. *Electronics* **2022**, *11*, 3628. https://doi.org/10.3390/electronics11213628

Academic Editors: Calin Iclodean, Bogdan Ovidiu Varga and Felix Pfister

Received: 4 September 2022
Accepted: 28 October 2022
Published: 7 November 2022

1. Introduction

Deep reinforcement learning (DRL) has been a powerful tool in many applications, including optimal path planning for mobile robots [1–10]. However, traditional deep reinforcement learning approaches use the trial-and-error method, which is extremely time-consuming and often does not converge [11]. The learning efficiency has become the bottleneck in applying DRL to more real-time path planning problems. One key factor that determines the efficiency of the learning process is the quality of the training data. Traditionally, DRL neural networks learn from randomly generated experience training data. Consequently, training is often a lengthy process and it is easy to get trapped in a local minimum and fail. In order to overcome this fundamental challenge of DRL, in this paper, we propose an improvement to the training data quality by using dynamic programming (DP)-based optimal data collection. This new DP-based data collection method is an excellent match for the path planning problems because the shortest distance path planning problems can be mapped to be a DP problem that can therefore achieve the global optimal solution. This DP-based data collection method can provide an abundant optimal training dataset for DRL.

Path planning is a popular research topic with many recently published studies [12–15]. However, it remains an active research area with much to be explored. In path planning, researchers strive to achieve the global optimal solution and one such method that can guarantee this is DP. There are many DP algorithms used to solve various problems that can be applied in navigating the global optimal path for mobile robots, which commonly use grid discretization to approximate the global optimal solution [16–19]. Different from most works, in this study, we first mapped the dynamic programming method to the typical, shortest traveling distance path planning for mobile robots and then used DP to find the exact, analytical, optimal solution for the continuous path planning problems. Specifically, we first employed DP to find the optimal solution for the center of each grid cell and then we created a novel method to compute the optimal solution for any continuous start position in the continuous workspace using only local information of neighbor cells. As a result, our method can achieve the optimal solution for any start position during robot continuous navigation compared to previous works using DP-based path planning methods.

Although DP-based data collection can provide optimal experience training data for DRL, it is likely that the learning process will still be too long or even get trapped in a local minimum and fail to converge if the data are fed to the DRL algorithm directly. To further improve the convergence performance and speed up the learning process, we propose the use of extreme learning machines (ELM) prior to the DRL in order to start DRL at a near-optimal starting point. This staged learning method allows DRL to fully focus on the most challenging part of the path planning, such as the areas around the obstacles or the ridges along which multiple optimal paths diverge. This initialization can be very fast due to the time efficiency feature of ELM.

By using DRL in this study, we aimed to learn the optimal action policy, i.e., closed-loop feedback for the real-time navigation of mobile robots in a 2D environment. Our optimal closed-loop action policy, which applies to any start position and covers all trajectories in the free workspace, is better than the optimal open-loop action sequence that is restricted to a single start position and therefore one trajectory. The optimal action policy allows for a real-time, fast, optimal response as the robot moves around in a complex environment. The closed-loop action policy eliminates the shortcomings of open-loop action sequences. It is also robust to disturbances or noise and reduces the deviations caused by disturbances since the effects of the disturbances are automatically compensated for. For example, in the case of the robot's state deviating from the optimal path caused by disturbance, the closed-loop action policy can mitigate the deviation from the original trajectory without accumulative errors. The optimal action policy can guarantee that the robot can still move along a new optimal path starting from the new disturbed state or position, which is close to the original optimal trajectory while the open-loop action sequence can cause the robot to move far away from the original optimal trajectory due to accumulative errors. Therefore, the closed-loop optimal policy is crucial for the real-time optimal navigation for mobile robots.

The contributions of this work are multifold: First, we mapped the dynamic programming method to typical optimal path planning problems for mobile robots and created an efficient dynamic programming-based method to find an exact, analytical, optimal solution for the continuous path planning problem. We then used high-quality training data gathered from the dynamic programing method for DRL, which greatly improves data quality and learning efficiency. We established a new two-stage learning method where we employed ELM to learn from initial optimal experience data in order to initialize the actor and critic neural networks to a near-optimal solution prior to DRL. Finally, we illustrated our proposed method using typical path planning tasks.

We organized the paper as follows. In Section 2, we provide an overview of the most relevant and recent papers in the fields of path planning, dynamic programming, optimal path planning, and DRL for path planning. In Section 3, we detail the new DRL approach implemented for this study. Experimental results are presented in Section 4. We conclude the paper and offer a few pointers for future research in Section 5.

2. Related Work

Path planning is a research field studying the moving strategies of robots or vehicles. In path planning, the goal is to safely move an agent from a start position to its corresponding final destination while evading any obstacles or other agents [12–15]. With numerous applications, such as improving robot movement efficiency and safety, increasingly efficient and powerful algorithms have been developed. To this end, research has been focused on optimal path planning, which can be defined with one or more of the following constraints: shortest time [20], shortest distance [21], and optimal energy consumption [22].

Conventional path planning methods, such as RRT, A*, and Bug approaches [13,14,23–26], aim to find one path from only one start position to the goal position, i.e., an open-loop sequence of points along the trajectory. When the start position changes or the mobile robot deviates from the optimal path caused by disturbances, path planning is reinitialized from the new start position. In contrast, our proposed method produces a closed-loop optimal action and control policy. With this policy, once the continuous optimal policy is learned via DRL, the robot can always move along the optimal path starting from any new position even when disturbed in the free workspace, based on the optimal policy and without replanning. Therefore, the purpose of our proposed approach is different from RRT, A*, and Bug approaches; our approach is better.

Deep learning (DL) is a new learning paradigm that performs nonlinear transformation using a multi-layer network structure [11,27–33] and has been gaining popularity in the last decade. In [28], Yann LeCun et al. presented a landmark work that employed the back-propagation method to acquire kernel coefficients directly from imaging representations of human written numbers, resulting in automated learning. We can categorize DL into three main groups: supervised learning [29], unsupervised learning [30], and reinforcement learning [31]. Unsupervised learning aims to find the hidden structures in unlabeled data but is ineffective in path planning problems due to the nature of this type of learning. On the other hand, supervised learning is much better with path planning problems due to its ability to easily converge and its lack of need to specify how the task should be performed. However, it is not without any flaws; in some applications, it can be hard to collect enough labelled data. In addition, the performance of this method is also limited: the robot cannot outperform the "supervisor", i.e., the training data, in supervised learning. In comparison, the last main method, DRL, does not need labelled data for training and can adequately generalize to new scenarios. However, DRL suffers the drawback of low sample efficiency [32].

In the last decade, DRL has been successfully applied to more and more applications. DRL performs better than human players in various fields by trying different strategies. In particular, DRL has been successfully used in mobile robotics or automated vehicle motion planning [1–11]. For example, in [10], the authors investigated the performance of a DRL-based deterministic policy gradient method for the dynamic environment. The algorithm was applied to multiple vehicle path planning. In [2], the authors studied a complex, big environment in which conventional algorithms often result in failures. They used a DRL-based algorithm to map vehicle's control actions to sensory inputs [2]. In [3]. The authors focused on a mixed environment and used DRL for vehicle motion planning. This algorithm was used for automated, multi-vehicle scenarios. In addition, by understanding and predicting the motions of moving obstacles using the equipped sensors, DRL can be programmed to avoid dynamic obstacles [4]. The elements of social interaction were introduced to DRL [5], where social rules were used to guide DRL. When we consider non-communicating multi-agent path planning problems, traditionally, the computation time can be prohibitively long due to unobservable agents' goals. However, DRL can be effectively used to reduce online computation time [6] and enable real-time navigation. While it is more challenging for DRL to avoid collision when the dynamic obstacles or agents do not follow any behavior rules, we can mitigate this issue by adding an LSTM segment for flexible observation size. This allows the DRL algorithm to achieve a better performance as the number of mobile obstacles increases [7]. A proposed method with low computation time and low energy consumption while achieving optimal or close to optimal path formulates a clustered IoT network as a combinatorial optimization problem. The authors in [8] used a seq2seq network and DRL to train this network using information of the clusters and the UAV's start or end as the input. In [9], the authors introduced a new combination of elite-duplication genetic-assisted path planning with DRL. Using this method, they can optimally generate sparse waypoints in a constrained space. In practical path planning applications, such as for warehouse robots, there are multiple agents. To address this problem, in [10], the authors formulate the problem as a decentralized, partially observable Markov decision process and use a DRL approach to solve the problem by feeding global and local map representations into convolutional layers.

In the literature, many works have been presented in the research area of dynamic programming and DP-based optimal path planning. However, we found that most of these works cannot offer an exact continuous solution to the global optimal problem. They tend to use a distance transformation algorithm. This approach aims to find the paths from the goal position back to the start position. In order to use this method, we generate a distance wave front, which is propagated to cover all free space beginning from the destination [16–19]. The authors in [13] presented a constrained traveling distance transformation algorithm, which can search for the shortest distance between any two points with the presence of static obstacles. This method calculated the optimal traveling distance by discretizing the workspace into image pixels and approximating the traveling distance to the closest pixel of reference for every grid point. The authors in [17] extended the idea of the distance transform method for 2D path planning. They defined the propagated cost as a weighted sum of the traveling distance to the destination and the total cost of obstacles moving closer. The authors in [18,19] presented a real-time obstacle avoidance path planning for robots, which is applicable to scenarios with dynamic targets and obstacles. In this particular path planning scenario, the authors aimed to minimize both the cumulative local penalty functions along the path and the sum of the current known distance to a target. In all of the above works, the environment was discretized; researchers could not find an exact, analytical, continuous optimal solution for the path planning problem.

In [34], the authors proposed an actor-critic deep reinforcement learning method with experience replay. The sampling method described in this paper is very efficient. In [35], the authors proposed a Q algorithm-based ELM (Q-ELM) to tackle a slow convergence problem. In this ELM, the input was the mobile robot's perception of the external environment information and the output was the corresponding reward and punishment for each action decision, which was the Q value.

The essence of conventional extreme learning machines (ELM) [36–44] is to use a single hidden layer feedforward neural network for training and learning. Later, ELM has been extended to use neural networks with multi-layer hidden nodes for different applications [36]. Since hidden nodes do not need to be iteratively tuned, ELM learns much faster and yields more promising performance compared to multiple layer perceptron (MLP)-based algorithms [37]. ELM can also provide a unified learning platform with a widespread type of feature mappings and can be applied in regression and multiclass classification applications [38]. However, this method is not flawless. Compared with MLP, in order to achieve comparable accuracy, ELM often needs much more hidden nodes [39]. In [40], the authors directly used ELM in a path planning method. They use a multi-layer ELM to calculate the cost function of the A* algorithm and determine the accurate search direction by evaluating the impact of obstacles. The authors in [41] designed an adaptive fuzzy neural network planning method based on ELM. The ELM is used to solve classification and regression problems and is applied to quickly and accurately reduce the computational complexity of the traditional adaptive neural network. In this work, we used ELM to initialize DRL actor and critic neural networks, which were able to quickly produce the solution to the vicinity of the global optimum due to its short computation time and high-quality optimal initial training data.

3. Methodology

In this study, we evaluated our approach in a multi-obstacle environment, a typical layout for path planning applications, such as warehouse fulfillment. In this scenario, we assume that the robot can accurately measure the distance between itself and its surrounding environment with common sensors, such as Lidars and cameras.

Our main aim was to enhance the performance of learning algorithms and training data quality associated with DRL. The two basic steps in DRL are the data collection and the training process. At the data collection step, the traditional DRL usually collects random samples in the form of (s_t, a_t, r_t, s_{t+1}). These random samples are generally of low quality and result in a much longer training process in the learning stage. To solve this problem, we propose using DP that can find the global optimal trajectories from any start position to generate high-quality training data. In the following Section 3.1, we show how the path planning problem was formulated into a DP problem and solved for optimal paths, which in turn provides best quality data for DRL training. Conventionally, during the training step, the robot uses DRL algorithms directly to learn the optimal paths. In this Section 3.2 of this paper, we propose and discuss a detailed, novel two-stage deep reinforcement learning algorithm for fast learning.

Specifically, DP-based training data contain the global optimal information. The derived optimal experience data include not only local information but also key global useful information, which can effectively guide the reinforcement learning process. Therefore, the learning process can efficiently learn from these best experience data and converge fast. For example, global optimal moving directions (i.e., actions) are the highest quality information from the environment and are much more useful for learning than low-quality randomly collected experience data.

The major challenge of complex continuous action policy representation using the neural network is the tendency to fall in a local minimum. ELM can effectively deal with this problem in our application scenarios of this study. ELM is one type of supervised learning and is much more efficient at learning than general reinforcement learning algorithms since the target information is generally not available in conventional deep reinforcement learning approaches. The reasons are as follows:

- DP-based optimal training data provide global optimal moving directions or actions and can be used as the optimal learning target.
- Given the input-to-hidden parameters of the actor and critic neural networks in DRL, ELM formulates the learning problem as a quadratic optimization problem, which has the closed-form solution, resulting in rapid non-iterative learning. More importantly, ELM can achieve the global optimal solution, which effectively reduce or overcome the local minimum problem that is inherent in DNN learning. Therefore, ELM provides an excellent starting point close to the global optimal solution for DRL algorithms.

3.1. *Global Optimal Solution for Discrete Grid Cell Centers Using the DP Method*

In this section, we present a two-step DP-based path planning method. This method uses a distance propagation method to find the shortest distance from any position in the workplace to its corresponding destination. Specifically, we separated the task into two steps: a path planning step and a robot navigation step. During the first step, we divided the workspace into grid cells and use a DP-inspired algorithm to find the shortest distance from each cell center to the goal position. Next, during the navigation step, we generated continuous optimal trajectories from any start position in the workspace by using only the local information of neighboring cells.

3.1.1. DP-Inspired Algorithm for Global Optimal Path Planning for Discrete Grid Cells

In this study, we began the planning step by first partitioning the environment into grid cells and setting the goal with zero distance. We then calculated the distances from all the neighbor cells of the goal, eight in total, to the final goal position. Specifically, the shortest distance from each cell center to the destination was calculated, as shown in

Figure 1. Next, this distance generation method propagated through all the remaining cells, finally obtaining the shortest distance for each cell. A map was built for the whole workspace.

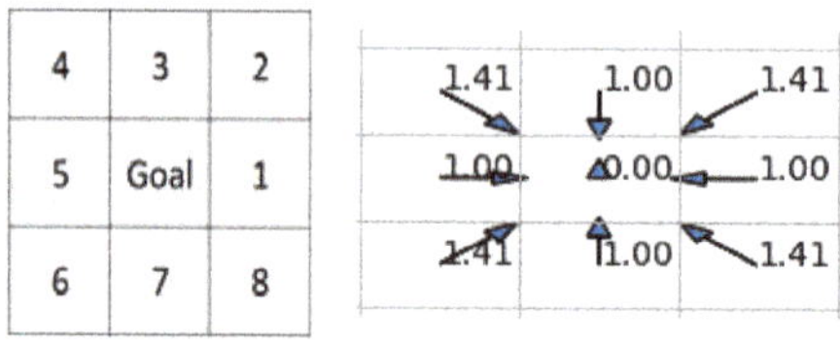

Figure 1. Goal and its eight neighbors (**left**) and shortest distance of the neighbor cells to goal (**right**).

We store the information for each cell, such as the center of the cell, the shortest length from the center of the cell to the goal, the optimal moving direction for the cell, and flags, to show the information, such as whether the cell is (1) in an obstacle, (2) already visited, or (3) in the priority queue (PQ).

The state space in our method is continuous in the navigation step. Any point in the free workspace is associated with the shortest distance from the final goal calculated by using only local information. In contrast, in this planning step, we used a discretized map of cells for planning. Based on the idea of dynamic programming, we illustrate in Figure 2 the cells that are organized by layers in our proposed method. Each cell is associated with the shortest length from the center of the cell to the goal. Each cell also contains key information, such as the optimal moving direction, which is used to guide the robot to the destination along the optimal path. In our method, only local information provided by its valid, available neighbors is required to calculate the shortest distance of a cell. We define valid neighbor cells as those that are not in any obstacle and define the available cells as the ones with the shortest lengths, which are already calculated from the previous steps.

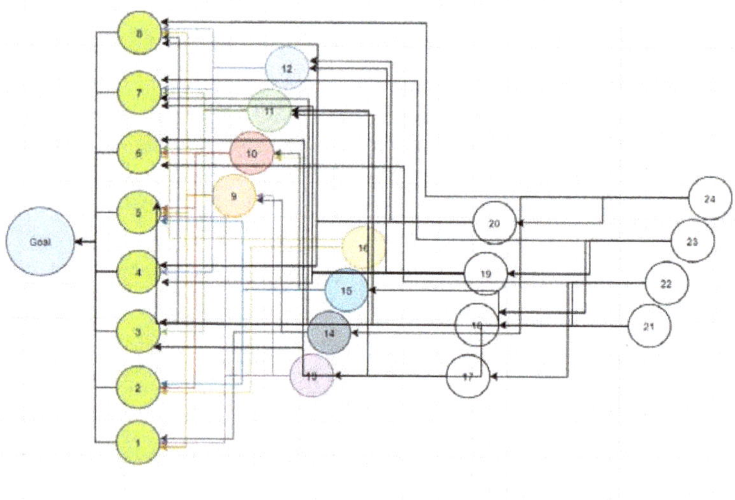

Figure 2. The illustration of the DP-like, layered-structure representation of the cells.

The DP-based formula for calculating the shortest traveling distance is as follows:

$$V_k(X_k) = \min_{X_{nbr} \in D_{nbr}(X_k)} \{ d(X_k, X_{wpm}) + V_{k+1}(X_{wpm}) \} \tag{1}$$

$$X_{wpm} = X_{nbr} + V_d(X_{nbr}) \tag{2}$$

where $d(X_k, X_{wpm})$ stands for the Euclidean distance from X_k to X_{wpm} and $V_{k+1}(X_{wpm})$ is the shortest distance to the goal. X_{nbr} is the center of X_k's neighbor cell, X_{wpm} is the corresponding waypoint of X_{nbr}, and $V_d(X_{nbr})$ is the optimal moving direction at X_{nbr}. As for a valid neighbor cell X_{nbr}, notice that there is no known obstacle between X_k and $X_{wp}(X_{nbr})$, i.e., the line segment of X_k and X_{wp}, does not pass any obstacle.

The global optimal direction V_d from the cell center X_k can be calculated as follows:

$$V_d = X_{wpm*} - X_k \tag{3}$$

where X_{wpm*} is the optimal waypoint that produces the shortest traveling distance $V_k(X_k)$, which is obtained using Equation (1).

Figure 3 shows a general case of how the shortest distance is determined for each cell. For example, there are three valid available neighbor cells for cell 326, which are cell 315, cell 322, and cell 312. We can compute the shortest distance and the optimal moving direction for cell 326 based on only the local information of these three surrounding cells.

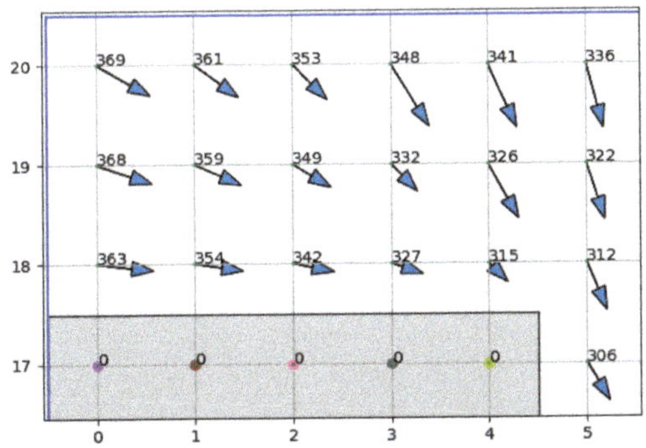

Figure 3. Calculation of the shortest distance for the cell with three valid neighbors available.

Note that not all surrounding cells have a valid shortest distance due to the presence of the obstacles. The center of a cell may be inside an obstacle and may have no valid value. As long as we can find one or two neighbors of the cell close to the obstacle with valid shortest distance values, we can still calculate the shortest distance for that cell. Lastly, it is also important to check whether the line connecting the cell and the waypoint passes through any obstacle.

In Figure 4, we show how to employ only local information to obtain the global shortest distance of a cell to the goal. In this example, the shortest distance of cell 37 is calculated using only the information of three neighbors: cell 13, cell 21, and cell 29. We also calculate the corresponding optimal waypoint P_{wp} using the centers and optimal moving directions of the neighboring cells using Equations (1) and (2). The shortest distance $V(P_{c37})$ is then calculated:

$$V(P_{c37}) = d(P_{c37}, P_{wp}) + V_{k+1}(P_{wp}) \tag{4a}$$

where $V_{k+1}(P_{wp})$ stands for the shortest distance from P_{wp} to the goal. $d(P_{37}, P_{wp})$ represents the distance between the center P_{c37} of cell 37 and the waypoint P_{wp}. In contrast with other methods [3], the optimal traveling distance from cell 37 does not need to pass through the center of neighbors 13 or 21; instead, it can find the optimal, accurate moving direction.

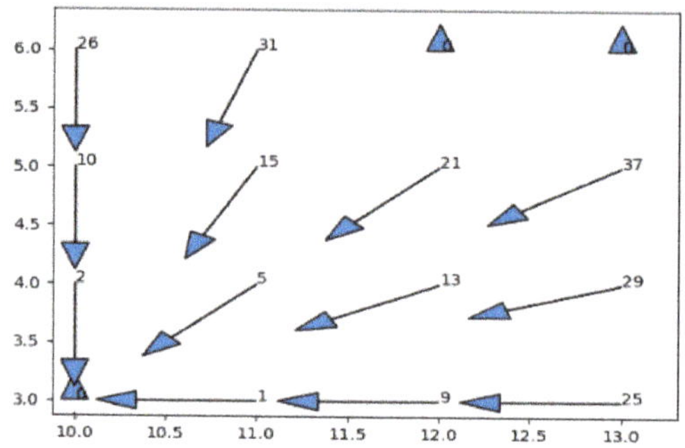

Figure 4. Computing global shortest distance of cell 37 from local neighbors: cells 13, 21, and 29.

In the implementation of our algorithm, we used a priority queue (PQ) to choose the next cell to consider and find the shortest distance from the cell to the goal. This process mimics the wave front propagation from the goal position. The cell priority $P_{dist}(C)$ was computed as follows:

$$P_{dist}(C) = \min_{C_j \in D_{nbr}(C)} \{d(C, C_j) + V_{k+1}(C_j)\} \tag{4b}$$

where $d(C, C_j)$ represents the Euclidean distance from the center of cell C to the center of cell, C_j, C_j is the valid available neighbor, $D_{nbr}(C)$ is the set of all the valid available neighbor cells of current cell C, and $V_{k+1}(C_j)$ is the shortest distance of C_j to the goal position. In Figure 5, we show the progress of distance propagation. In Figure 5a, we show the result of the propagation after 44 cells. In Figure 5b, we show the result of the propagation after 371 cells.

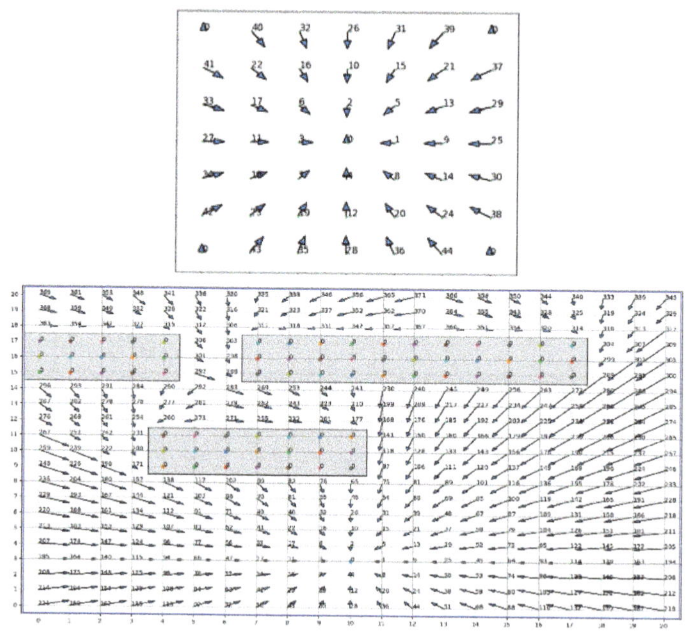

(**a**)

(**b**)

Figure 5. Progress of shortest distance propagation: (**a**) visited 44 cells, (**b**) visited 371 cells.

In this study, we developed a DP-based shortest traveling length of path planning algorithm, as shown in Algorithm 1.

Algorithm 1. DP-Inspired Shortest Distance Path Planning Algorithm

---

Step 1. Initialize the environment:
Step 1.1. Partition the whole map into $N \times M$ grid cells;
Step 1.2. Set cell properties for obstacles such as obstacle flag;
Step 1.3. Set the goal cell with zero distance.
Step 2. Process eight neighbors of the goal cell. For each neighbor, do the following operations:
Step 2.1. Calculate the shortest distance from the goal to the center of the cell, set cell properties such as the shortest distance, optimal moving direction, visit flags;
Step 2.2. Check each of eight neighbors of the current cell, add it to priority queue (PQ) if (not in PQ) AND (not visited yet) AND (not obstacle).
Step 3. Cell propagation for the whole free workspace:
Step 3.1. Take the top cell in PQ;
Step 3.2. Compute for the cell the shortest distance and optimal moving direction using only the local information of available, valid neighbor cells using Equations (1) and (3);
Step 3.3. Set cell properties such as the shortest distance, optimal direction, visit flags;
Step 3.4. Check each of eight neighbors of the cell, add it to priority queue (PQ) if (not in PQ) AND (not visited yet) AND (not obstacle);
Step 3.5. Back to step 3.1 until PQ is empty.
Step 4. Check and mark the cells close to ridge boundary based on local information of neighbor cells.

3.1.2. Navigation Step: Exact Shortest Distance Calculation for Any Point in the Map

This navigation step is a real-time process in which the robot travels along the optimal traveling path from any point in the continuous map. The trajectory does not need to go through grid cell centers. We calculated the optimal direction for any point using only local information of the neighbor cells. We discuss two cases in detail here.

In the first case, as shown in Figure 6a, the optimal moving directions of the valid neighbor cells are all pointing towards the same waypoint. We calculated the waypoint P_{wp} based on the information of any valid neighbor cell, the center position, and optimal moving direction of the cell. Then, we calculated the optimal traveling direction V_p and the shortest traveling distance $d(P)$ from the current point $P(x,y)$ as follows.

$$V_p = P_{wp} - P \tag{5}$$

$$d(P) = d(P, P_{wp}) + V_{k+1}(P_{wp}) \tag{6}$$

where $V_{k+1}(P_{wp})$ is the shortest distance from P_{wp} to the goal and $d(P, P_{wp})$ is the distance between points P and P_{wp}.

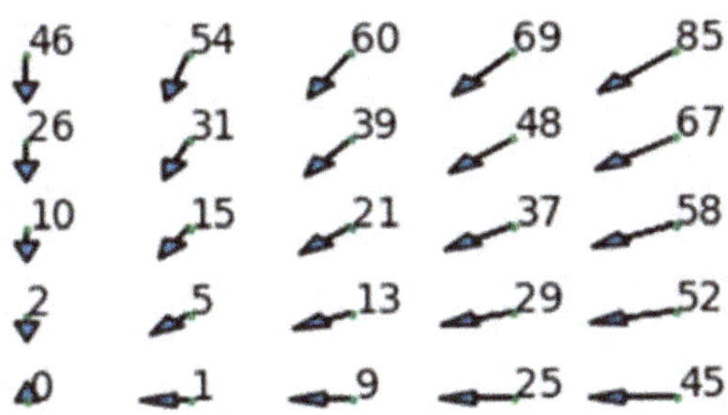

(a)

Figure 6. *Cont.*

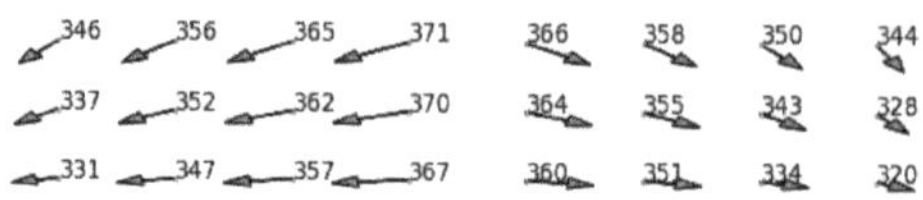

(**b**)

Figure 6. Calculation of Exact Shortest Distance from Any Point in the Map. (**a**) Case I, (**b**) Case II.

In Case II, we considered the robot position P, which is close to the ridge boundary as shown in Figure 6b, where there are several waypoints in the neighborhood. Some potential, optimal directions guide the waypoint on the left while others guide the waypoint on the right. Similar to Case I, we first calculated the waypoint set $D_{wp}(P)$ from all valid neighbors. Next, we minimized the distance from the point to the goal in order to obtain the best waypoint as follows.

$$d(P) = \min_{P_{wp} \in D_{wp}(P)} \{d(P, P_{wp}) + V_{k+1}(P_{wp})\} \tag{7}$$

Then, we calculated the shortest distance $d(P)$ from the point P to the goal using the best waypoint P_{wpm} and we also computed the corresponding optimal moving direction.

In this navigation step, for any arbitrary position in the environment, there are up to 9 valid cells in the neighborhood, including the cell containing the current point plus 8 neighbor cells. However, if the cell is not in the free space, some neighbor cell properties may not be available.

3.2. Two-Stage DRL

Although DRL has the potential to achieve superhuman performance in theory, in practice, it is very challenging to efficiently learn the parameters of the optimal continuous action policy, i.e., the continuous actor network for specific applications using DRL. Without an adequate initialization, DRL often converges too slowly or it may not even converge at all. In this paper, we propose a novel two-stage DRL algorithm based on the modified DDPG algorithm to efficiently learn the optimal policy for the mobile robot navigation. At the first stage, we employed the DP-based technique to generate many global optimal trajectories or experiences in the workspace. Based on these optimal trajectories, ELM was then used to calculate the favorable initial parameter values of the actor and critic networks (DNN) due to its non-iterative high computation speed and high-quality initial optimal training data. In the second stage, once the favorable initial values of the parameters are in the vicinity of the global optimum, DRL was used to fine-tune and ensure the high accuracy. More specifically, we employed both local experience data and global optimal experiences to guide the learning process to make the learning process converge to the optimal solution while avoiding the collision with obstacles using more local experiences in complicated regions, such as the regions close to obstacles and ridges. In the free workspace far away from obstacles or ridge regions, the optimal policy is relatively smoother and requires the smaller amount of training samples. By using this two-stage DRL method, we can significantly accelerate the learning process, reduce the probability of the robot getting trapped into a local minimum, and achieve the high-quality navigation policy.

3.2.1. Initial and Online Data Collection

In this study, the training data were partially collected prior to DRL training and partially during the training process. Before training, (I) the DP-based navigation approach was employed to generate multiple optimal trajectories by specifying random start positions and (II) for each grid cell, we produced experience data with optimal moving directions, obtained as described in Section 3.1. Experience data for the critic $Q(s,a)$ were then generated as follows. In the initial data collection, given a position or state s, there are two methods to produce the experience for $Q(s,a)$: (a) using the corresponding optimal moving direction a to advance one step to obtain the next state s_{t+1} and compute the optimal $Q(s,a) = -d(s,s_{t+1}) + V^*(s_{t+1})$ or (b) randomly generate a moving direction and advance one step to obtain the next state s_{t+1} and compute $Q(s,a) = -d(s,s_{t+1}) + V^*(s_{t+1})$, where $V^*(s_{t+1})$ is the negative shortest distance from state s_{t+1} to the goal. These trajectories were sampled to obtain initial training data. During training, further samples were collected and used in the later training process as the robot moves around in the environment and moves towards to the goal.

The data collection around the obstacles is particularly challenging as the robot (1) cannot collide with the obstacles and (2) needs to follow the optimal path, which often requires the robot to move closely around the obstacles. In order to avoid collision with obstacles, more samples were collected from the regions close to obstacles. Some samples are generated by advancing one step along the optimal moving directions while other samples are generated by advancing one step along random moving directions without collision with obstacles.

3.2.2. Using ELM for Near-Optimal Initialization

ELM learning was employed to rapidly initialize the actor and critic networks in deep reinforcement learning due to its short learning computation time and high-quality solution. Due to the fast computation, we can run ELM multiple times and select the best solution among multiple runs as the initial values of the DRL network parameters. ELM is able to achieve the global optimal solution for the quadratic programming problem for the given weights of input-to-hidden layers if the networks is with one hidden layer. Therefore, ELM initializes the parameters of the networks to a near-optimal solution when the training data are produced along the optimal moving directions generated in Section 3.1. By using ELM and global optimal initial training data, we also significantly reduced the probability of getting trapped in a local minimum.

Given a set of N distinct training samples $\{(x_i, t_i) \mid x_i \in R^d, t_i \in R^m, i = 1,2,\ldots, N\}$ where x_i is the training input data vector, t_i represents the target of each sample, and L denotes the number of hidden nodes, one single hidden layer neural network with L hidden neurons can be written as [39]:

$$y_i = \sum_{j=1}^{L} \beta_j g(w_j x_i + b_j) = \sum_{j=1}^{L} \beta_j h_j(x_i) = t_i + \varepsilon_i,\ i = 1,2,\ldots,N \tag{8}$$

where g is the activation function, w_j and b_j are random weights and biases, and ε is noise or error.

The matrix form of ELM is presented here:

$$\min_{\beta} \|H\beta - T\|^2 \tag{9}$$

$$\beta = \begin{bmatrix} \beta_1^T \\ \cdots \\ \beta_L^T \end{bmatrix} = \begin{bmatrix} \beta_{11} & \cdots & \beta_{1m} \\ \vdots & \ddots & \vdots \\ \beta_{L1} & \cdots & \beta_{Lm} \end{bmatrix}$$

$$T = \begin{bmatrix} t_1^T \\ \cdots \\ t_N^T \end{bmatrix} = \begin{bmatrix} t_{11} & \cdots & t_{1m} \\ \vdots & \ddots & \vdots \\ t_{N1} & \cdots & t_{Nm} \end{bmatrix}$$

$$H = \begin{bmatrix} h(x_1) \\ \cdots \\ h(x_N) \end{bmatrix} = \begin{bmatrix} h_1(x_1) & \cdots & h_L(x_1) \\ \vdots & \ddots & \vdots \\ h_1(x_N) & \cdots & h_L(x_N) \end{bmatrix} \tag{10}$$

where H is the hidden layer output matrix and T is the training data target matrix. The above quadratic optimization problem can be solved in a closed form: $\beta = H^+T$, H^+ is the Moore–Penrose generalized inverse of matrix H. Therefore, ELM is a non-iterative learning algorithm and is extremely fast.

During the ELM training phase, only the output weights β are adjusted according to the algorithm. The ELM training algorithm can be summarized in Algorithm 2.

Algorithm 2. Fast ELM Learning Algorithm

Step 1. Randomly assign the hidden node parameters, i.e., the input weights w_j and biases b_j for additional hidden nodes $j = 1, 2, \ldots, L$ in Equation (8).
Step 2. Calculate the hidden layer output matrix H using Equation (10).
Step 3. Compute the output weight vector β as follows:

$$\beta = H^{+}T \tag{11}$$

where H^{+} is the Moore–Penrose generalized inverse of matrix H.

Note that this algorithm is used in Algorithm 3 to initialize the actor and critic neural networks, where the parameters θ^{Q} or θ^{μ} each includes $\beta, w_1, w_2, \ldots, w_L, b_1, b_2, \ldots, b_L$.

3.2.3. Actor and Critic Neural Networks in DRL

The actor network and critic network consist of three layers each, using the sigmoid activation function in the hidden layer. In the actor network, the input is the two-dimensional state vector, i.e., the position $s = (x,y)$ in the workspace, and the output is a two-dimensional action vector, i.e., the moving direction $a = (vx,vy)$. Meanwhile, for the critic network, the input is a four-dimensional vector $(s,a) = (x,y,vx,vy)$ and the output is the negative distance $Q(s,a)$ from the current state s to the goal, taking the current action a. Both the actor and critic networks can be efficiently initialized using the rapid ELM algorithm.

3.2.4. Modified DDPG (MDDPG) for Fine-Tuning DRL Actor and Critic Networks

In this section, we modify the deep deterministic policy gradient method (DDPG) for planning the optimal path for the robot. The robot is assumed to interact with the environment E in discrete timesteps. At timestep t, the robot accomplishes three things: it takes an action a_t, moves one step from the state s_t, and receives a reward r_t. Both the action and state spaces are continuous in this section.

The action-value function depicts the expected return in state s_t after taking an action a_t. We detail the Bellman equation as follows:

$$Q^{\mu}(s_t,\ a_t) = E_{r_t, s_{t+1} \sim E}\left[r(s_t,\ a_t) + \gamma Q^{\mu}(s_{t+1},\ \mu(s_{s+1}))\right] \tag{12}$$

The loss function is defined as:

$$L\left(\theta^{Q}\right) = E_{s_{t \sim \rho^{\beta}},\ a_t \sim \beta, r^t \sim E}\left[\left(Q\left(\theta^{Q}\right) - y_t\right)^2\right] \tag{13}$$

where the reward signal $r(s_t,\ a_t)$ is the negative travel distance for each time step and $y_t = r(s_t,\ a_t) + \gamma Q(s_{t+1},\ \mu(s_{s+1}) | \theta^{Q})$.

The critic $Q(s,a)$ is learned using the Bellman equation. The actor is calculated by following the chain rule to the expected return from the derivative of J with respect to the actor parameters:

$$\nabla_{\theta^{\mu}} J \approx E_{s_{t \sim \rho^{\beta}}},\ \left[\nabla_{\theta^{\mu}} Q(s, a | \theta^{Q})\Big|_{s=s_t,\ a=\mu(s_t|\theta^{\mu})}\right] = E_{s_{t \sim \rho^{\beta}}},\ \left[\nabla_{a} Q(s, a | \theta^{Q})\Big|_{s=s_t,\ a=\mu(s_t)} \nabla_{\theta^{\mu}} \mu(s | \theta^{\mu})\Big|_{s=s_t}\right] \tag{14}$$

3.2.5. Two-Stage DRL Algorithm

Next, the DDPG algorithm (MDDPG) is modified to include DP-based data collection before and during the training time. ELM is employed to initialize the parameters of the actor and critic networks as well.

Algorithm 3. Two-Stage DRL Algorithm

Step 1. Prior to MDDPG training, DP-based data collection is performed to obtain the initial optimal actor data and optimal critic data.
Step 2. Randomly generate initial weights of θ^Q and θ^μ of actor and critic networks, respectively. Note that the parameters θ^Q and θ^μ include β, $w_1, w_2, \ldots, w_L,$, $b, b_2, \ldots, b_L$ in Algorithm 2.
Step 3. Based on initial training data, ELM is employed to rapidly compute hidden-to-output weights β^Q and β^μ for the actor network and critic network, respectively. We then replace the corresponding parts with β^Q and β^μ in the above randomly generated initial weights θ^Q and θ^μ.
Step 4. Initialization for MDDPG training:
Step 4.1. Initialize critic network $Q(\theta^Q)$ and actor network $\mu(\theta^\mu))$ with weights θ^Q and θ^μ, respectively;
Step 4.2. Initialize target network Q' and μ' with weights $\theta^{Q'} = \theta^Q$, $\theta^{\mu'} = \theta^\mu$;
Step 4.3. Initialize replay buffer R with N_0 collection steps with DP-based optimal experiences;
Step 4.4. Receive observation of start state s_1.
Step 5. Using MDDPG to fine-tune actor and critic neural networks
Step 5.1. Collect more training data for action exploration and stored in R;
Step 5.2. Select action $a_t = \mu(\theta^\mu) + N_t$ according to the current policy and exploration noise N_t;
Step 5.3. Execute action a_t and observe reward r_t and new state s_{t+1};
Step 5.4. Store transition (s_t, a_t, r_t, s_{t+1}) in R;
Step 5.5. Sample a random mini-batch of N transitions (s_t, a_t, r_t, s_{t+1}) from R;
Step 5.6. Compute.

$$y_i = r_i + \gamma Q'\left(s_{i+1}, \mu'\left(\theta^{\mu'}\right)|\theta^{Q'}\right) \tag{15}$$

Step 5.7. Update critic by minimizing the loss:

$$L = \frac{1}{N}\sum_{i=1}^{N}\left(y_i - Q(s_i, a_i|\theta^Q)\right)^2 \tag{16}$$

Step 5.8. Update the actor policy using the sampled policy gradient:

$$\nabla_{\theta^\mu} J \approx \frac{1}{N}\sum_i \nabla_a Q\left(\theta^Q\right)|_{s=s_i,a=\mu(s_i)}\,\mu(s|\theta^\mu)|_{s_i} \tag{17}$$

Step 5.9. Update the target networks:

$$\theta^{Q'} = \tau\theta^Q + (1-\tau)\,\theta^{Q'} \tag{18}$$

$$\theta^{\mu'} = \tau\theta^\mu + (1-\tau)\,\theta^{\mu'} \tag{19}$$

Step 5.10. Repeat Step 5.1 until reaching the maximum number of training or stop criteria.

4. Experimental Results

4.1. Environment

To evaluate the effectiveness of our method, we used our method in typical motion planning scenarios with multiple obstacles. The vehicle would be able to travel from any beginning position in the workspace and reach the goal position without colliding with any obstacles. One typical scenario is shown in Figure 7. The rectangle obstacles in the figure depict the barriers. The beginning position of the vehicle and the goal position are marked out in the figure.

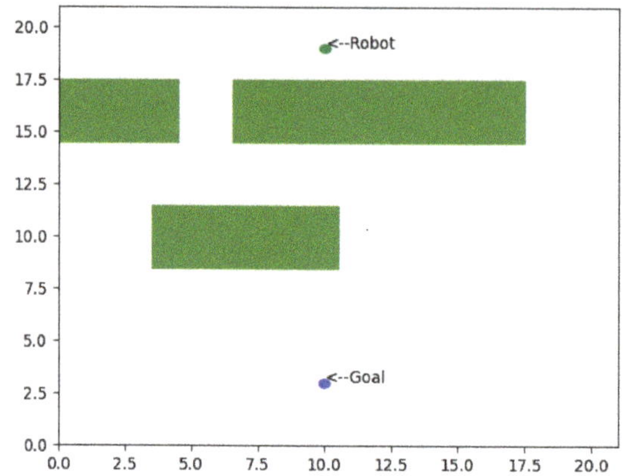

Figure 7. Robot working environment.

4.2. Vector Fields and Shortest Travelling Distance of the Optimal Path Map

To demonstrate the efficacy of the proposed method, we carried out the experiments for two typical scenarios with multiple obstacles. In Figure 8, we show the calculation order for the cell distance propagation. Every number before each arrow shows the order by which the cell is travelled. In Figure 9, we show the shortest travelling distance (i.e., the number before each arrow) from the goal position to each cell. In these two figures, every arrow represents the optimal moving direction for the center of the cell to the goal. The goal position is represented by a triangle with zero distance. Obstacles are represented by rectangular shapes. For any point in the free workspace, based only on the shortest distances and optimal moving directions of local neighbor cells as shown in Figure 9, we can efficiently calculate the corresponding optimal moving direction (i.e., action) and the shortest distance from the point to the goal position in order to navigate the robot to advance along the optimal path to the goal.

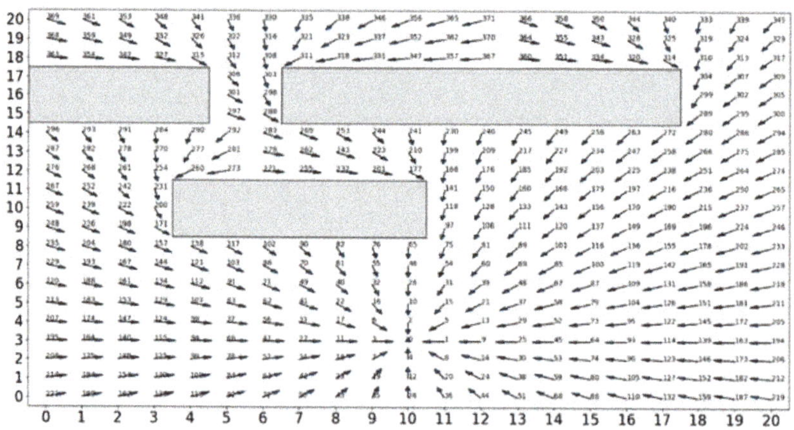

Figure 8. The computation order (each number before arrow) for the cell propagation.

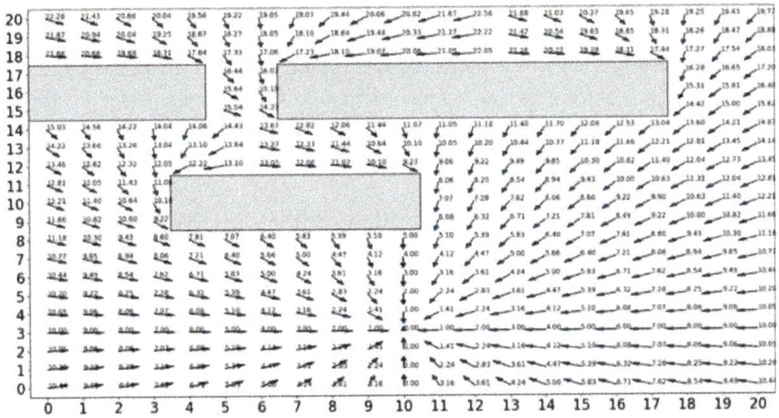

Figure 9. Shortest distance (the number before each arrow) and optimal moving vector (each arrow) fields of the map (x, y axes: indices of cells).

In Figure 10, we consider an environment with multiple obstacles with different shapes. We show the visitation order and the optimal moving vector fields of the entire environment.

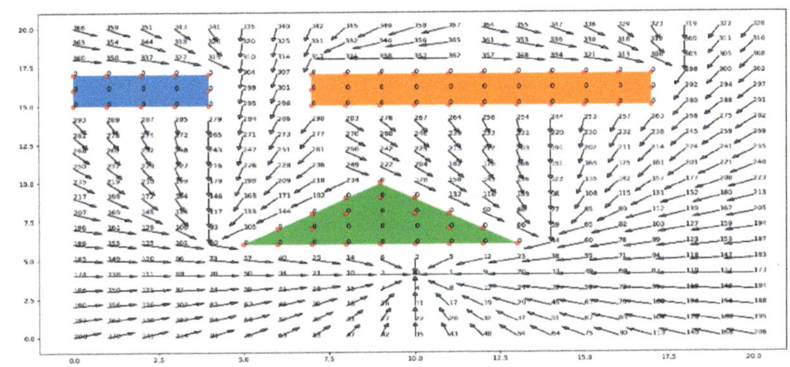

Figure 10. The visitation order and optimal moving vector fields of the map.

The Ridge Boundary

In Figure 11, we focus on the ridge boundary that separates two sets of moving directions. The starting points on the left of the ridge boundary take the paths to the far left and the starting points on the right of the ridge boundary take the paths to the far right. As a result, these starting points travel through different waypoints.

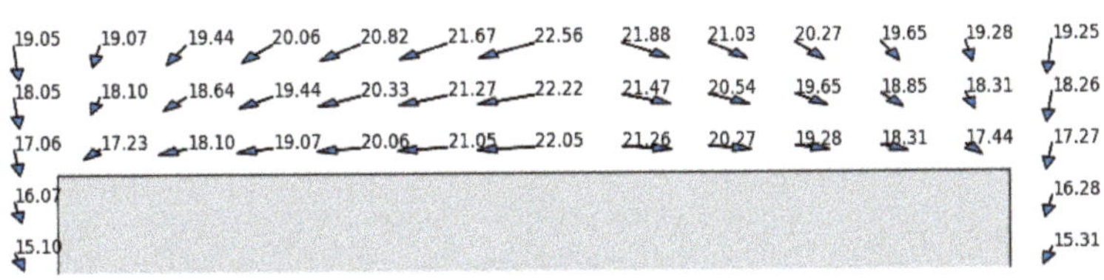

Figure 11. Ridge boundary in the middle of this figure to separate cells into left and right parts with different moving directions.

4.3. *Sample Optimal Trajectories for Optimal Data Collection*

By using the navigation algorithm described in Section 3.1.2, we randomly selected multiple starting points and computed their optimal trajectories, as shown in Figures 12 and 13. From any beginning point, the vehicle can move along the shortest traveling path all the time. These optimal trajectories can be sampled and fed into the two-stage DRL for training.

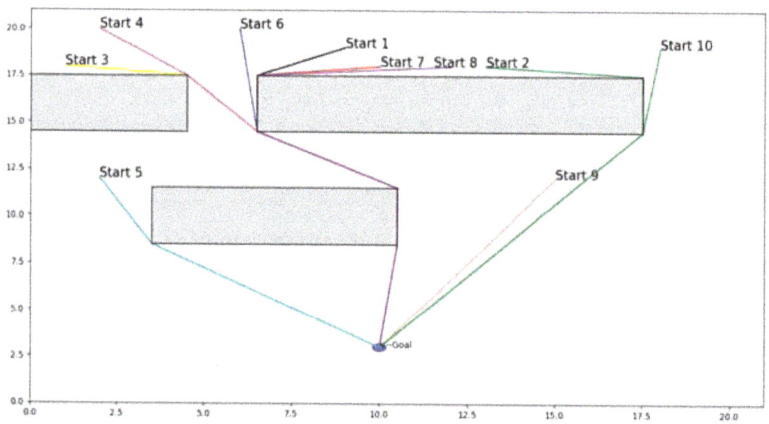

Figure 12. Typical sample optimal trajectories for data collection.

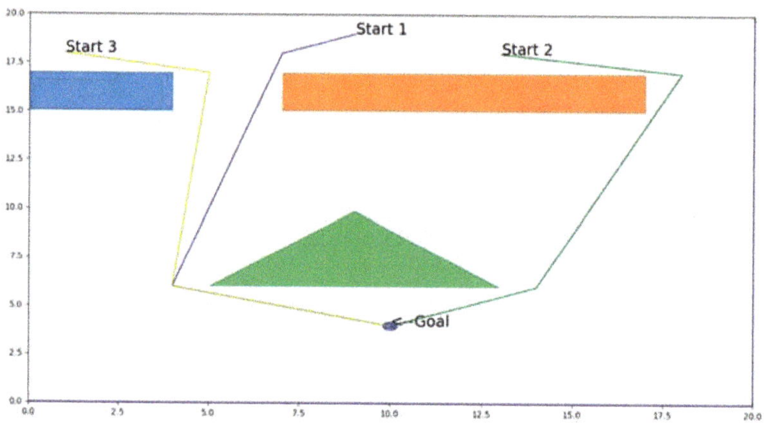

Figure 13. Sample trajectories for data collection in another scenario.

4.4. *Sample Optimal Trajectories Generated by the Two-Stage DRL Algorithm*

Once we collected the initial data of optimal trajectories using the DP-based navigation technique, we used our two-stage DRL to train the actor and critic neural networks. Instead of using random data for trial-and error-iterations, we used these high-quality data for initial training with ELM.

4.4.1. After First Stage: ELM Learning

The sample trajectories after ELM learning are shown in Figure 14. ELM can provide a near-optimal starting point for the neural networks in a short time. However, some path segments near obstacles may not be accurate. This problem with ELM learning is shown in Figure 16. Thus, in the second stage, we can use the modified DDPG algorithm to further improve the training accuracy.

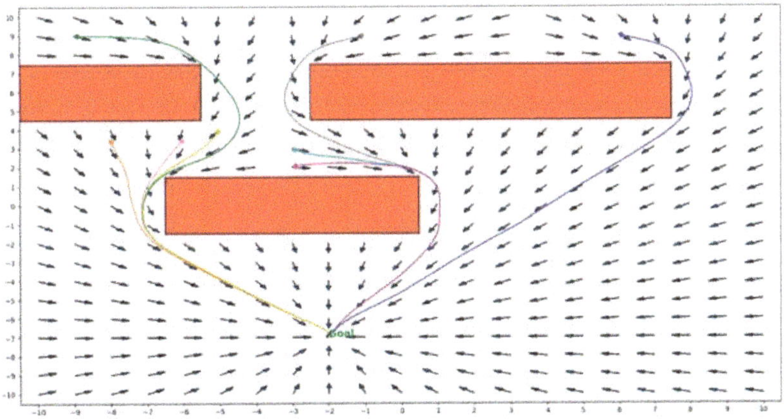

Figure 14. Sample trajectories after ELM initialization.

4.4.2. After Second Stage: DRL

The sample trajectories for one scenario after DLR learning are shown in Figure 15. In this stage, DRL focuses on the more challenging parts of the paths and regions. In Figure 15, we show that with DRL, we can improve the accuracy of the paths.

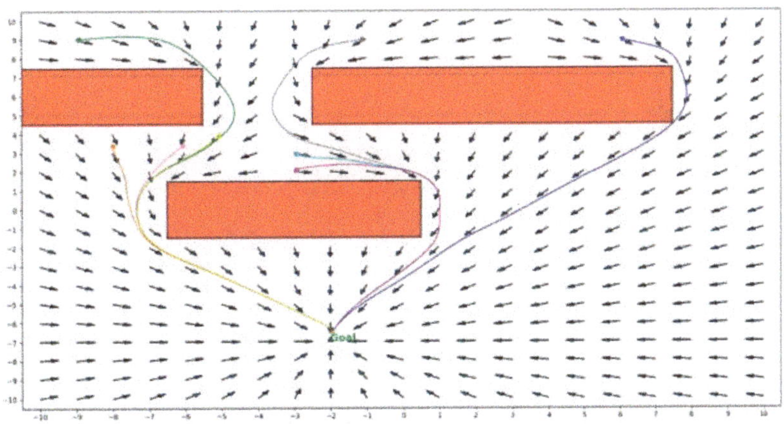

Figure 15. Sample trajectories after DRL fine-tuning.

In the regions close to obstacles, it is much more challenging to train the actor and critic networks. For example, regular experience data collection can lead to collision with the obstacle, which is shown by the green trajectory in Figure 16. After we added more experience data to the neighbor of the obstacle, the learned policy network was able to produce an optimal trajectory (blue line in Figure 16) while avoiding the collision with the obstacle.

In order to illustrate the advantages of our proposed DRL method over the original DDPG algorithm, we conducted the following experiment. In the same scenario as that depicted in Figure 15, we employed the DDPG algorithm to train the actor and critic neural networks with the same training experience samples. The only difference in this experiment using the original DDPG algorithm is that we employed random weights to initialize the actor and critic networks instead of using the weights from ELM learning results. We repeated the experiment for three times and they all failed to find the feasible paths. In Figure 17, we show a typical trajectory resulting from the original DDPG algorithm.

The robot starts from a position (−9.0, 9.0) and collides with the obstacles, even after 250,000 training iterations. The main reason for this failure is that, compared to supervised learning, deep reinforcement learning is much harder to converge, less efficient in learning, and easier to get stuck in local minima. These implementation issues associated with the conventional DRL algorithm result from many factors including weak guidance signals, obstacles, and long episodes in the complex environment. In contrast, our fast ELM learning with DP-based optimal data collection is able to initialize the corresponding weights to a near-optimal solution, which significantly improve the learning effectiveness and efficiency. In Figure 18, we show the trajectory of using our proposed method for the same scenario using only 70,000 training iterations. The robot can follow the optimal path, avoid obstacles, and reach the target successfully. This experiment demonstrated that our proposed method is much more efficient than the original DDPG algorithm.

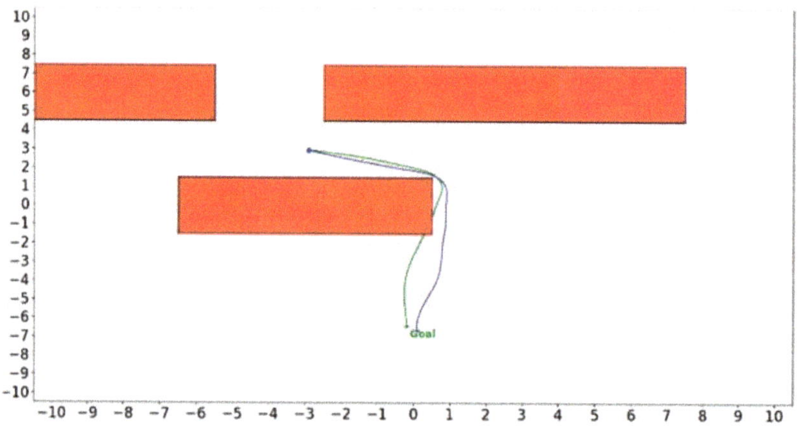

Figure 16. Collision with obstacles (**green**), collision avoidance by sampling more experiences close to obstacles (**blue**).

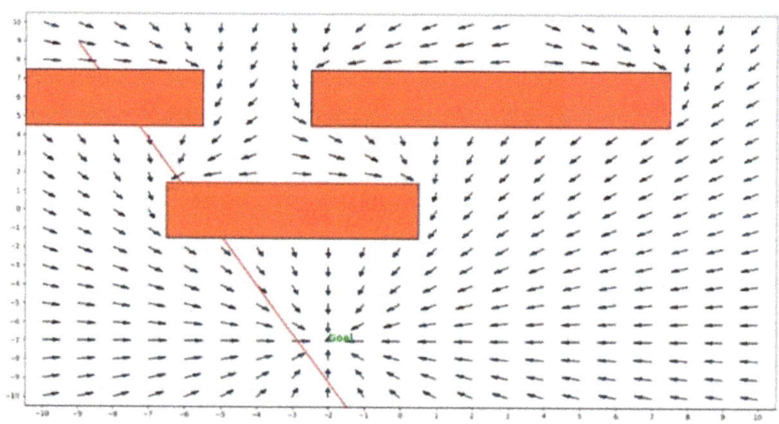

Figure 17. Collision with obstacles (red trajectory) with the conventional DDPG learning algorithm, even with the same training experience samples.

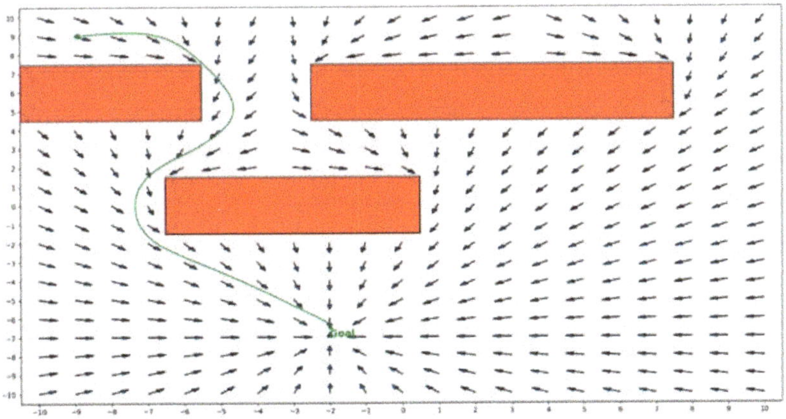

Figure 18. Trajectory produced by our proposed DRL method.

5. Conclusions and Future Work

In this paper, we presented a novel optimal path planning algorithm based on DRL with application to mobile robots. In order to improve training data quality and generate optimal training data for DRL, we mapped the DP method to typical optimal path planning problems and established an efficient DP-based method to find the exact, analytical, optimal solution. In order to accelerate the reinforcement learning process and improve the learning performance, we also created a two-stage DRL method, in which ELM was employed to initialize the weight parameters of the actor and critic networks. This algorithm is able to move the robot along an optimal path from any starting point in the continuous workspace to a specified goal location. For our next steps, we plan to conduct a comprehensive study to compare our proposed method with other existing techniques. We also plan on extending the capability of our algorithm to handle 3D environments and environments with obstacles of arbitrary shapes, moving obstacles, and multiple agents.

Author Contributions: Conceptualization, J.R. and X.H.; methodology, X.H., J.R. and R.N.H.; software, X.H. and R.N.H.; validation, J.R., X.H. and R.N.H.; formal analysis, X.H.; investigation, J.R. and X.H.; resources, X.H.; data curation, X.H.; writing—original draft preparation, J.R. and R.N.H.; writing—review and editing, X.H. and R.N.H.; visualization, R.N.H.; supervision, J.R.; project administration, J.R.; funding acquisition, J.R. All authors have read and agreed to the published version of the manuscript.

Funding: The authors would like to acknowledge the support of the Natural Sciences and Engineering Research Council of Canada (NSERC) [funding reference number 210471].

Data Availability Statement: Public datasets were not used in this study.

Conflicts of Interest: The authors declare no conflict of interest.

References

1. Qie, H.; Shi, D.; Shen, T.; Xu, X.; Li, Y.; Wang, L. Joint Optimization of Multi-UAV Target Assignment and Path Planning Based on Multi-Agent Reinforcement Learning. *IEEE Access* **2019**, *7*, 146264–146272. [CrossRef]
2. Wang, C.; Wang, J.; Shen, Y.; Zhang, X. Autonomous Navigation of UAVs in Large-Scale Complex Environments: A Deep Reinforcement Learning Approach. *IEEE Trans. Veh. Technol.* **2019**, *68*, 2124–2136. [CrossRef]
3. Makantasis, K.; Kontorinaki, M.; Nikolos, I. Deep reinforcement-learning-based driving policy for autonomous road vehicles. *IET Intell. Transp. Syst.* **2020**, *14*, 13–24. [CrossRef]
4. Garg, A.; Chiang, H.-T.L.; Sugaya, S.; Faust, A.; Tapia, L. Comparison of Deep Reinforcement Learning Policies to Formal Method for Moving Obstacle Avoidance. In Proceedings of the 2019 IEEE/RSJ International Conference on Intelligent Robots and Systems (IROS), Macao, China, 3–8 November 2019; pp. 3534–3541. [CrossRef]
5. Tung, T.X.; Ngo, D. Socially Aware Robot Navigation Using Deep Reinforcement Learning. In Proceedings of the 2018 IEEE Canadian Conference on Electrical & Computer Engineering, Quebec, QC, Canada, 13–16 May 2018.

6. Chen, Y.F.; Liu, M.; Everett, M.; How, J.P. Decentralized non-communicating multi-agent collision avoidance with deep reinforcement learning. In Proceedings of the IEEE International Conference on Robotics and Automation, Singapore, 29 May–3 June 2017; Volume 201, pp. 285–292. [CrossRef]
7. Everett, M.; Chen, Y.F.; How, J.P. Motion Planning among Dynamic, Decision-Making Agents with Deep Reinforcement Learning. In Proceedings of the 2018 IEEE/RSJ Intelligent Conference on Intelligent Robots and Systems (IROS), Madrid, Spain, 1–5 October 2018; pp. 3052–3059.
8. Zhu, B.; Bedeer, E.; Nguyen, H.H.; Barton, R.; Henry, J. Joint Cluster Head Selection and Trajectory Planning in UAV-Aided IoT Networks by Reinforcement Learning with Sequential Model. *IEEE Internet Things J.* **2022**, *9*, 12071–12084. [CrossRef]
9. Wang, N.; Zhang, Y.; Ahn, C.K.; Xu, Q. Autonomous Pilot of Unmanned Surface Vehicles: Bridging Path Planning and Tracking. *IEEE Trans. Veh. Technol.* **2021**, *71*, 2358–2374. [CrossRef]
10. Bayerlein, H.; Theile, M.; Caccamo, M.; Gesbert, D. Multi-UAV Path Planning for Wireless Data Harvesting with Deep Reinforcement Learning. *IEEE Open J. Commun. Soc.* **2021**, *2*, 1171–1187. [CrossRef]
11. Bengio, Y.; Lecun, Y.; Hinton, G. Deep learning for AI. *Turing Lect.* **2021**, *64*, 58–65. [CrossRef]
12. Souissi, O. Path planning: A 2013 survey. In Proceedings of the 2013 International Conference on Industrial Engineering and Systems Management, Rabat, Morocco, 28–30 October 2013; pp. 849–856.
13. Hart, P.E.; Nilsson, N.J.; Raphael, B. A Formal Basis for the Heuristic Determination of Minimum Cost Paths. *IEEE Trans. Syst. Sci. Cybern.* **1968**, *4*, 100–107. [CrossRef]
14. Lavalle, S.M. *Rapidly-Exploring Random Trees: A New Tool for Path Planning*; Computer Science Dept. Iowa State University: Ames, IA, USA, 1998.
15. Khatib, O. Real-time obstacle avoidance for manipulators and mobile robots. *Int. J. Rob. Res.* **1986**, *5*, 90–98. [CrossRef]
16. Pei, S.-C.; Horng, J.-H. Finding the optimal driving path of a car using the modified constrained distance transformation. *IEEE Trans. Robot. Autom.* **1998**, *14*, 663–670. [CrossRef]
17. Zelinsky, A. Using Path Transforms to Guide the Search for Findpath in 2D. *Int. J. Robot. Res.* **1994**, *13*, 315–325. [CrossRef]
18. Willms, A.; Yang, S. An efficient dynamic system for real-time robot-path planning. *IEEE Trans. Syst. Man Cybern. Part B Cybern.* **2006**, *36*, 755–766. [CrossRef]
19. Willms, A.R.; Yang, S.X. Real-Time Robot Path Planning via a Distance-Propagating Dynamic System with Obstacle Clearance. *IEEE Trans. Syst. Man Cybern. Part B Cybern.* **2008**, *38*, 884–893. [CrossRef]
20. Chen, M.; Zhu, D. Optimal Time-Consuming Path Planning for Autonomous Underwater Vehicles Based on a Dynamic Neural Network Model in Ocean Current Environments. *IEEE Trans. Veh. Technol.* **2020**, *69*, 14401–14412. [CrossRef]
21. Alexander, J.C.; Maddocks, J.H.; Michalowski, B.A. Shortest distance paths for wheeled mobile robots. *IEEE Trans. Robot. Autom.* **1998**, *14*, 657–662. [CrossRef]
22. Liu, S.; Sun, D. Minimizing Energy Consumption of Wheeled Mobile Robots via Optimal Motion Planning. *IEEE/ASME Trans. Mechatron.* **2013**, *19*, 401–411. [CrossRef]
23. Xu, Q.-L.; Yu, T.; Bai, J. The mobile robot path planning with motion constraints based on Bug algorithm. In Proceedings of the Chinese Automation Congress (CAC), Jinan, China, 20–22 October 2017; pp. 2348–2352. [CrossRef]
24. Lumelsky, V.; Stepanov, A. Dynamic path planning for a mobile automaton with limited information on the environment. *IEEE Trans. Autom. Control* **1986**, *31*, 1058–1063. [CrossRef]
25. Qi, J.; Yang, H.; Sun, H. MOD-RRT*: A Sampling-Based Algorithm for Robot Path Planning in Dynamic Environment. *IEEE Trans. Ind. Electron.* **2020**, *68*, 7244–7251. [CrossRef]
26. Tang, G.; Tang, C.; Claramunt, C.; Hu, X.; Zhou, P. Geometric A-Star Algorithm: An Improved A-Star Algorithm for AGV Path Planning in a Port Environment. *IEEE Access* **2021**, *9*, 59196–59210. [CrossRef]
27. Bengio, Y.; Courville, A.; Vincent, P. Representation Learning: A Review and New Perspectives. *IEEE Trans. Pattern Anal. Mach. Intell.* **2013**, *35*, 1798–1828. [CrossRef]
28. LeCun, Y.; Boser, B.; Denker, J.S.; Henderson, D.; Howard, R.E.; Hubbard, W.; Jackel, L.D. Backpropagation Applied to Handwritten Zip Code Recognition. *Neural Comput.* **1989**, *1*, 541–551. [CrossRef]
29. Hinton, G.; Deng, L.; Yu, D.; Dahl, G.E.; Mohamed, A.R.; Jaitly, N.; Senior, A.; Vanhoucke, V.; Nguyen, P.; Sainath, T.N.; et al. Deep Neural Networks for Acoustic Modeling in Speech Recognition. *IEEE Signal Process. Mag.* **2012**, *29*, 82–97. [CrossRef]
30. Szegedy, C.; Liu, W.; Jia, Y.; Sermanet, P.; Reed, S.; Anguelov, D.; Erhan, D.; Vanhoucke, V.; Rabinovich, A. Going Deeper with Convolutions. In Proceedings of the IEEE Conference on Computer Vision and Pattern Recognition (CVPR), Boston, MA, USA, 7–12 June 2015; pp. 1–9.
31. Kuutti, S.; Bowden, R.; Jin, Y.; Barber, P.; Fallah, S. A Survey of Deep Learning Applications to Autonomous Vehicle Control. *IEEE Trans. Intell. Transp. Syst.* **2020**, *22*, 712–733. [CrossRef]
32. Usama, M.; Qadir, J.; Raza, A.; Arif, H.; Yau, K.L.A.; Elkhatib, Y.; Hussain, A.; Al-Fuqaha, A. Unsupervised Machine Learning for Networking: Techniques, Applications and Research Challenges. *IEEE Access* **2019**, *7*, 65579–65615. [CrossRef]
33. Aradi, S. Survey of Deep Reinforcement Learning for Motion Planning of Autonomous Vehicles. *IEEE Trans. Intell. Transp. Syst.* **2020**, *23*, 740–759. [CrossRef]
34. Wang, Z.; Bapst, V.; Heess, N.; Mnih, V.; Munos, R.; Kavukcuoglu, K.; de Freitas, N. Sample Efficient Actor-Critic with Experience Replay. *arXiv* **2016**, arXiv:1611.01224. Available online: https://arxiv.org/abs/1611.01224 (accessed on 6 February 2022).

35. Ren, H.; Yin, R.; Li, F.; Wang, W.; Huo, M. Research on Q-ELM algorithm in robot path planning. In Proceedings of the 2016 Chinese Control and Decision Conference (CCDC), Yinchuan, China, 8–30 May 2016; pp. 5975–5979. [CrossRef]
36. Wang, J.; Lu, S.; Wang, S.-H.; Zhang, Y.-D. A review on extreme learning machine. *Multimed. Tools Appl.* **2021**, 1–50. [CrossRef]
37. Huang, G.-B. What are Extreme Learning Machines? Filling the Gap Between Frank Rosenblatt's Dream and John von Neumann's Puzzle. *Cogn. Comput.* **2015**, *7*, 263–278. [CrossRef]
38. Huang, G.-B.; Zhou, H.; Ding, X.; Zhang, R. Extreme Learning Machine for Regression and Multiclass Classification. *IEEE Trans. Syst. Man Cybern. Part B Cybern.* **2012**, *42*, 513–529. [CrossRef]
39. Zhang, R.; Lan, Y.; Huang, G.-B.; Xu, Z.-B. Universal Approximation of Extreme Learning Machine with Adaptive Growth of Hidden Nodes. *IEEE Trans. Neural Netw. Learn. Syst.* **2012**, *23*, 365–371. [CrossRef]
40. Yin, C.; Xia, Y.; Yang, R.; Yuan, Z.; Kuang, F.; Li, L. Path Planning Method Based on Multi-Layer ELM Optimized A. In Proceedings of the 2021 IEEE 21st International Conference on Communication Technology (ICCT), Tianjin, China, 13–16 October 2021; pp. 1423–1426. [CrossRef]
41. Yi, C.; Shan, C.; Hui, C.; Yuan, H.S. Motion planning of autonomous mobile robot based on ELANFIS. In Proceedings of the 2021 China Automation Congress (CAC), Beijing, China, 22–24 October 2021; pp. 4607–4612. [CrossRef]
42. Huang, G.-B.; Bai, Z.; Kasun, L.L.C.; Vong, C.M. Local Receptive Fields Based Extreme Learning Machine. *IEEE Comput. Intell. Mag.* **2015**, *10*, 18–29. [CrossRef]
43. Castellani, A.; Cornell, S.; Falaschetti, L.; Turchetti, C. tfelm: A TensorFlow Toolbox for the Investigation of ELMs and MLPs Performance. In Proceedings of the International Conference on Artificial Intelligence, Stockholm, Sweden, 13–19 July 2018; pp. 3–8.
44. Huang, G.B. An Insight into Extreme Learning Machines: Random Neurons, Random Features and Kernels. *Cogn. Comput.* **2014**, *6*, 376–390. [CrossRef]

electronics

MDPI

Article

PG-Based Vehicle-In-the-Loop Simulation for System Development and Consistency Validation

Weonil Son [1], Yunchul Ha [1], Taeyoung Oh [1], Seunghoon Woo [2], Sungwoo Cho [3] and Jinwoo Yoo [4,*]

1 Graduate School of Automotive Engineering, Kookmin University, Seoul 02707, Republic of Korea
2 Department of Automotive Engineering, Kookmin University, Seoul 02707, Republic of Korea
3 Evaluation Research Center, Korea Automobile Testing and Research Institute, Hwaseong 18247, Republic of Korea
4 Department of Automobile and IT Convergence, Kookmin University, Seoul 02707, Republic of Korea
* Correspondence: jwyoo@kookmin.ac.kr

Abstract: The concern over safety features in autonomous vehicles is increasing due to the rapid development and increasing use of autonomous driving technology. The safety evaluations performed for an autonomous driving system cannot depend only on existing safety verification methods, due to the lack of scenario reproducibility and the dynamic characteristics of the vehicle. Vehicle-In-the-Loop Simulation (VILS) utilizes both real vehicles and virtual simulations for the driving environment to overcome these drawbacks and is a suitable candidate for ensuring reproducibility. However, there may be differences between the behavior of the vehicle in the VILS and vehicle tests due to the implementation level of the virtual environment. This study proposes a novel VILS system that displays consistency with the vehicle tests. The proposed VILS system comprises virtual road generation, synchronization, virtual traffic manager generation, and perception sensor modeling, and implements a virtual driving environment similar to the vehicle test environment. Additionally, the effectiveness of the proposed VILS system and its consistency with the vehicle test is demonstrated using various verification methods. The proposed VILS system can be applied to various speeds, road types, and surrounding environments.

Keywords: Vehicle-In-the-Loop Simulation (VILS); autonomous driving; vehicle test; consistency validation; correlation validation; proving ground; path tracking; Adaptive Cruise Control (ACC)

Citation: Son, W.; Ha, Y.; Oh, T.; Woo, S.; Cho, S.; Yoo, J. PG-Based Vehicle-In-the-Loop Simulation for System Development and Consistency Validation. *Electronics* **2022**, *11*, 4073. https://doi.org/10.3390/electronics11244073

Academic Editors: Calin Iclodean, Bogdan Ovidiu Varga and Felix Pfister

Received: 11 November 2022
Accepted: 5 December 2022
Published: 7 December 2022

1. Introduction

Autonomous driving systems have been developed rapidly in recent years to achieve the goal of full autonomy. As technology advances, various systems are complexly linked to each other, and verifying the safety of autonomous driving systems will become increasingly important [1]. Consequently, various research institutes and governments are conducting extensive research on the development of related technologies and the verification of the safety of autonomous driving systems. Unlike the existing SAE Lv.2 level Advanced Driver Assistance System (ADAS), an autonomous driving system must respond to various situations on the road independently, and any accidents that occur are attributed to the vehicle instead of the driver [2]. To ensure the safety of the drivers in such systems, an effective safety verification method must be developed by testing the systems over millions or even billions of kilometers [3–5]. The safety verification testing methods are largely divided into simulation-based testing and vehicle testing. Currently, various simulation-based testing methods are being used, which include Model-In-the-Loop (MIL), Software-In-the-Loop (SIL), and Hardware-In-the-Loop (HIL) [6]. MIL and SIL are useful for testing algorithms quickly because the vehicle, sensor, actuator, and controller all consist of virtual models. Additionally, SIL can be implemented to test the software of an algorithm mounted on a vehicle, and verification can be performed at the ECU level without using an actual ECU. HIL can be implemented at the level of actual ECUs and hardware, and real-time verification, including verification of the characteristics of actual hardware, can be performed by including actual hardware such as sensors and actuators in the simulation. However, in the simulation-based testing method, there may be errors between the behavior of the actual vehicle and its dynamic characteristics due to the application of the virtual vehicle dynamics model and to the occupants being unable to feel the vehicle's behavior. Vehicle testing is the final step in verifying the effectiveness of the autonomous driving algorithm because it uses a real vehicle, along with real sensors and a real surrounding environment [7]. The effectiveness of the existing body control system and chassis control system can be verified through vehicle testing in the stage just before commercialization after the development process. However, verifying the effectiveness of the autonomous driving system, which responds to various dangerous traffic situations during the vehicle test, is difficult due to the expansion of Operational Design Domain (ODD). Vehicle testing requires various external factors such as the surrounding environment and objects, due to which it cannot be performed by the vehicle alone. Even if a test bed and several dummy obstacles such as vehicles and pedestrians are prepared at a high cost, it is not possible to replicate a specific scenario during the vehicle test; there is also a risk of collision with the surrounding objects [8].

Therefore, Vehicle-In-the-Loop Simulation (VILS), which is a verification method based on virtual environments and real vehicles, has gained considerable interest as a means to overcome the limitations of the simulation and vehicle testing methods. VILS can ensure repeatability and reproducibility and can considerably reduce the risk of collisions when compared to vehicle tests because the surrounding objects and environments are represented as virtual objects, as shown in Table 1. This method enables the verification of the effectiveness of an autonomous driving system and reflects the dynamic characteristics of the actual vehicle, unlike the existing simulation test methods, such as MIL, SIL, and HIL. Furthermore, the cost required to prepare the test environment is also reduced, thereby improving the economic efficiency.

Table 1. Comparison of various testing methods for a vehicle.

Testing Method	Vehicle Dynamics	Cost	Reproducibility	Danger
MIL	Model	Low	High	Low
SIL	Model	Low	High	Low
HIL	Partial Model	Mid	High	Low
Vehicle Test	Real	High	Low	High
VIL	Real	Mid	High	Low

VILS is divided into dynamo-VILS, which test vehicles on a fixed chassis dynamo, and PG-VILS, which are based on vehicles driving on a Proving Ground (PG) [9]. Related studies can be categorized according to the two different types of VILS. First, the related studies for dynamo-VILS are as follows.

Gletelink [10] proposed a method to verify the effectiveness of the Adaptive Cruise Control (ACC) function by attaching a vehicle to a dynamo and integrating a real robot target vehicle with a virtual surrounding vehicle within a limited indoor space (200 m × 40 m). Rossi [11] analyzed the energy efficiency of a verification vehicle using a dynamo-based ADAS test system called SERBER. Additionally, an Over-The-Air (OTA) sensor emulation method that uses a screen to project an image was employed to ensure that the camera sensor can directly capture an image of the virtual environment. Sieg [12] and Diewald [13] also tested the ADAS and autonomous driving functions using the dynamo and OTA-based radar emulation. Diewald developed a radar OTA emulation method for multiple objects and conducted a VILS test using a radar virtual model capable of replicating real conditions. Gao [14] developed an indoor dynamo-VILS environment and evaluated the performance of autonomous vehicle via Autonomous Emergency Braking (AEB) functions through the developed VILS environment. Zhang [15] also evaluated and analyzed the Energy Management Strategy (EMS) of Hybrid Electric Vehicles (HEV) through dynamo-VILS.

Unlike in the vehicle test, the dynamo-VILS used in the above study was tested in an indoor space, which enables flexible and repeatable testing based on a virtual environment. However, building the dynamo and emulator equipment for each sensor is extremely expensive and requires a large amount of space. Furthermore, the steering angle is limited due to the mechanical link of the dynamo, which requires additional equipment for testing [9].

In the case of the PG-VILS, the vehicle runs on the actual PG, and the vehicle dynamic characteristics are identical to those in the vehicle test, as shown in Figure 1. The acceptance of an autonomous driving system can be evaluated by the passengers in the vehicle. Additionally, the cost of building an environment is significantly lower when compared to dynamo-VILS. Therefore, an efficient verification environment can be configured and unexpected situations on the road, such as traffic congestion, can be virtually realized to verify the effectiveness of the autonomous driving function. The studies related to PG-VILS are as follows.

Figure 1. Concept of PG-based VILS.

Park [16] confirmed the validity and safety of the VILS method by establishing a VILS verification environment and performing an ADAS evaluation based on the EuroNCAP test scenarios. Kim [17] conducted tests using VILS in situations that were difficult to verify due to the high risk of collision, such as crossing vehicles and pedestrians on the road. He employed a PG with sufficient width to eliminate the risk of an accident during the test. Tettamanti [18] implemented various traffic scenarios by integrating the VILS environment, which is a virtual realization of an actual test bed, with SUMO, a traffic simulator. Lee [19] verified the effectiveness of Cooperative Adaptive Cruise Control (CACC) using a real vehicle and a virtual vehicle implemented using VILS. Kim [20] and Solmaz [21] constructed a virtual road environment and verified the effectiveness of the lane change algorithm through PG-VILS.

As above, various studies related to dynamo-VILS and PG-VILS have been conducted. The previous studies were primarily focused on the methodology related to VILS implementation or the verification of the target system results using VILS. As a result, there is insufficient research on securing the driving environment at the time of vehicle test and analyzing the consistency of VILS. Therefore, this study aims to verify the effectiveness of an autonomous driving system by establishing a test system using a PG-VILS. A VILS system that can reproduce the driving conditions during the vehicle test was developed, and the validity of the proposed VILS system is analyzed based on the consistency between the vehicle tests and VILS tests. The remainder of this paper is organized as follows. Section 2 describes the overall configuration of the proposed VILS system and presents the details of the system components used to improve the consistency of the vehicle and VILS tests. Section 3 describes the test vehicle used for the VILS applications, and Section 4 describes the procedure and verification method used to verify the consistency of the test results. In Section 5, the performance of the proposed VILS system is verified by comparing the consistency of the VILS test and vehicle test results. Section 6 presents the conclusion of the article along with future research implications.

2. Proposed PG-Based Vehicle-In-the-Loop Simulation System Structure

The composition of the proposed VILS system is divided into a virtual simulator and test vehicle, as shown in Figure 2. The virtual simulator comprises four components: virtual road generation, real–virtual synchronization, virtual traffic behavior generation, and perception sensor modeling. The test vehicle is a real vehicle equipped with an autonomous driving ECU which performs autonomous driving using the data generated in a virtual simulator.

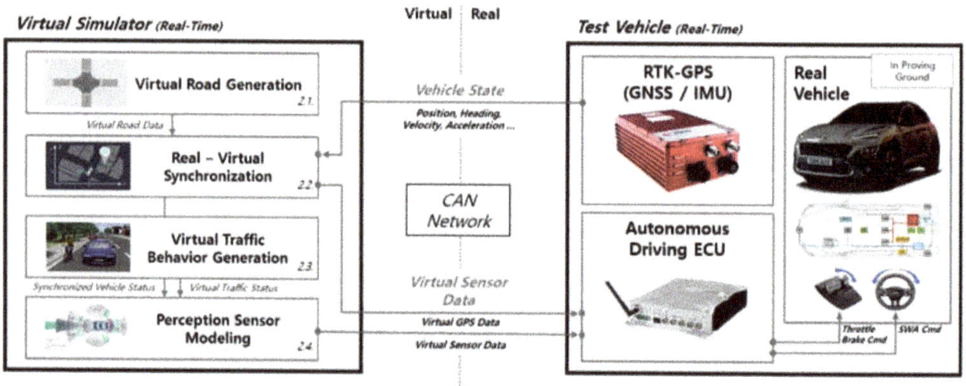

Figure 2. Proposed PG-based VILS System Structure.

Data transmission/reception between the virtual simulator, which generates the virtual environment, and the test vehicle, which operates in the actual proving ground environment, is performed using the CAN communication interface. The state information (position, heading, speed, etc.) which is measured by the RTK-GPS of the test vehicle is transmitted to the virtual simulator to enable interworking between the two environments. Subsequently, the virtual GPS and sensor data of the virtual simulator which are required for autonomous driving control are transmitted to the ECU of the test vehicle. The virtual simulator consists of four elements: Virtual Road Generation, Synchronization, Virtual Traffic Behavior generation, and perception sensor modeling. Virtual Road Data is used for synchronization between real and virtual. Perception sensor modeling is performed by utilizing the state information of the synchronized test vehicle and the behavior of virtual traffic. This study aims to perform a simulation similar to the vehicle test conditions through the proposed processes. This section explains the four elements that constitute the virtual simulator. The next, Section 3, explains the configuration of the test vehicle and presents a detailed explanation of the autonomous driving algorithm applied in this study.

2.1. Virtual Road Generation

To conduct a driving test in a virtual environment, a virtual road environment is required in which to drive the vehicle [22]. In this study, a virtual road was created using HD-MAP, which was provided by the National Geographic Information Institute (NGII) of South Korea. HD-MAP is an electronic map which was developed using the recognition and positioning sensor data installed in a Mobile Mapping System (MMS) vehicle; it provides precise locations and road object property information in centimeters. HD-MAP is essential for autonomous driving systems, which require precise control of the vehicle to perform operations such as lane changes and obstacle avoidance. It differs from navigation and ADAS maps, which can only distinguish the road units [23]. The virtual road environment was developed using the HD-MAP of K-City, an autonomous driving testbed at the Korea Automobile Testing and Research Institute (KATRI).

The road shape information provided by the HD-MAP is supported by the World Geodetic System 1984 (WGS84) coordinate system format. Here, the location on the surface of the earth is expressed in terms of latitude and longitude using units of [°], which makes it less useful and intuitive when implementing an autonomous driving system [24]. Therefore, the map must be converted to the Universal Transverse Mercator (UTM) coordinate system in meters, with precision expressed in terms of latitude and longitude. It is easy to express the distance and direction in the case of UTM because all the coordinates on the surface are expressed as X[m] and Y[m] coordinates based on a specific origin. The process of converting the WGS84 coordinate system into UTM coordinates is given as follows [25]:

$$X = k_0 N[A + (1 - T + C) * A^3/6 + (5 - 18T + T^2 + 72C - 58e'^2)A^5/120] \tag{1a}$$

$$\begin{aligned} Y = k_0[M - M_0 + N\tan\phi\,[A^2/2 + (5 - T + 9C + 4C^2)A^4/24 + \\ (61 - 58T + T^2 + 600C - 330e'^2)A^6/720]] \end{aligned} \tag{1b}$$

$$\begin{aligned} M = a\,[(1 - e^2/4 - 3e^4/256 - \ldots)\phi - (3e^2/8 + 3e^4/32 + 45e^6/1024 + \ldots)\sin 2\phi \\ +(15e^4/\ldots + 45e^6/1024 + \ldots)\sin 4\phi - (35e^6/3072 + \ldots)\sin 6\phi + \ldots] \end{aligned} \tag{1c}$$

$$\begin{aligned} k_0 = 0.9996,\ e^2 = (a^2 - b^2)/a^2,\ e'^2 = e^2/1 - e^2,\ N = a/\sqrt{1 - e^2\sin^2\phi}, \\ T = \tan^2\phi,\ C = e'^2\cos^2\phi,\ A = (\lambda - \lambda_0)\cos\phi \end{aligned} \tag{1d}$$

Here, k_0 represents the UTM Central meridian scale factor, and a and b represent the semi-major and semi-minor axes of Earth's ellipsoid, respectively. e denotes the first eccentricity and e' denotes the second eccentricity. ϕ and λ represent the latitude and longitude, respectively, and ϕ_0 and λ_0 represent the origin constants of latitude and longitude.M denotes the actual distance from the equator to latitude based on the central meridian, and M_0 denotes the value obtained by applying ϕ_0 to Equation (1c).

Figure 3 presents an example result obtained when the road shape information of the HD-MAP, which is provided as latitude and longitude in degrees units, is converted into the UTM coordinate system in meter units. The converted road shape information was applied to the IPG Carmaker Scenario Editor to build a virtual road environment.

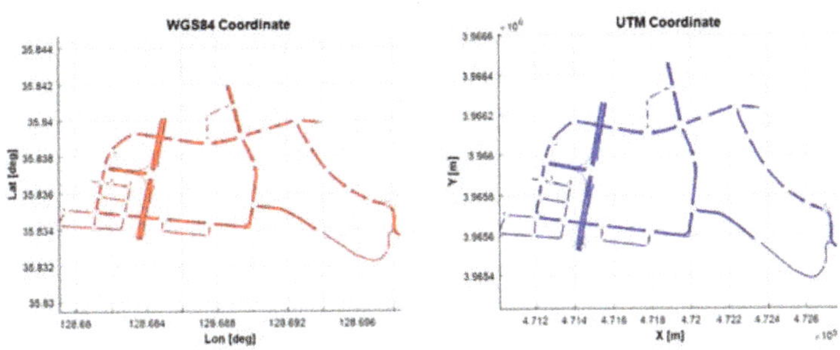

Figure 3. Example of WGS84 to UTM coordinate conversion.

2.2. Real–Virtual Synchronization

In the case of PG-based VILS, the actual location of the test vehicle and the location of the virtual road do not match. Therefore, the test vehicle must be positioned on the virtual road for the driving test, as shown in Figure 4.

The location information for the test vehicle at the start of the simulation and the location information (position, heading) errors for the starting point of the virtual road were applied to the location information for the test vehicle for each sample using the same process as in Equation (2).

$$X_{vir}(t) = X(t) - (X_{start} - X_{road}) \tag{2a}$$

$$Y_{vir}(t) = Y(t) - (Y_{start} - Y_{road}) \tag{2b}$$

$$\psi_{vir}(t) = \psi(t) - (\psi_{start} - \psi_{road}) \tag{2c}$$

In this equation, $(\ )_{road}$ represents the starting point of the virtual road and $(\ )_{start}$ represents the point where the test vehicle is located at the start of the simulation. $(\ )_{vir}$ represents the virtual location information, where X, Y, and ψ represent the X/Y global position and heading value of the vehicle, respectively.

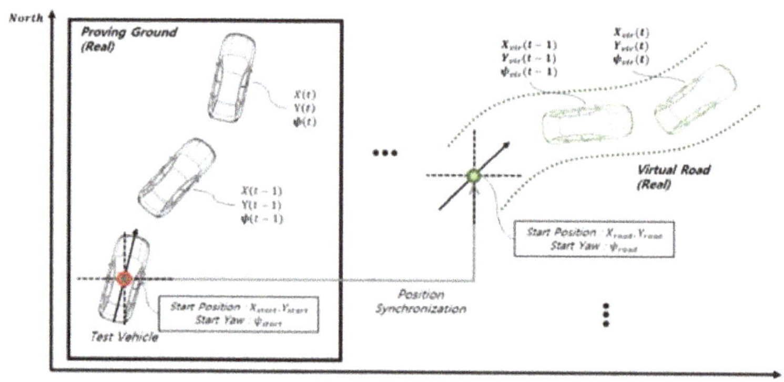

Figure 4. Concept of position synchronization between real and virtual environments.

Thus, the location information of the test vehicle is converted into virtual location information to ensure that the vehicle can be driven on a virtual road regardless of the actual location. Since the location information measured by the RTK-GPS of the test vehicle is provided in terms of latitude and longitude, the coordinate system transformation described in Section 2.1 is performed.

The measurement period (10 Hz–50 Hz) [26–28] of the RTK-GPS sensor differs from the period in which the virtual simulator operates (>100 Hz) [29–31]. This causes the test vehicle behavior to be updated continuously in the virtual simulator. To achieve an accurate simulation, location information for the test vehicle which matches the cycle of the virtual simulator must be updated continuously. In this study, location information was generated using the extrapolation method based on the operation period of the virtual simulator, as shown below [32]:

$$\begin{aligned} v_{x\_ext}(t) &= v_{x\_ext}(t-1) + \left\{ a_{x\_mea} + \frac{(x_{mea} - x_{ext}(t-1)) * \kappa}{\Delta t_{sim}{}^2} \right\} * \Delta t_{sim}, \\ x_{ext}(t) &= x_{ext}(t-1) + \left\{ v_{x\_ext}(t) + \frac{(x_{mea} - x_{ext}(t-1)) * \kappa}{\Delta t_{sim}} \right\} * \Delta t_{sim} \end{aligned} \tag{3a}$$

$$\begin{aligned} v_{y\_ext}(t) &= v_{y\_ext}(t-1) + \left\{ a_{y\_mea} + \frac{(y_{mea} - y_{ext}(t-1)) * \kappa}{\Delta t_{sim}{}^2} \right\} * \Delta t_{sim}, \\ y_{ext}(t) &= y_{ext}(t-1) + \left\{ v_{y\_ext}(t) + \frac{(y_{mea} - y_{ext}(t-1)) * \kappa}{\Delta t_{sim}} \right\} * \Delta t_{sim} \end{aligned} \tag{3b}$$

$$\psi_{ext}(t) = \psi_{ext}(t-1) + \left\{ \gamma_{mea} + \frac{(\psi_{mea} - \psi_{ext}(t-1)) * \kappa}{\Delta t_{sim}} \right\} * \Delta t_{sim} \tag{3c}$$

Here, $(\ )_{ext}$ and $(\ )_{mea}$ represent the extrapolation value and the measured value, respectively, κ represents the filtering constant, and Δt_{sim} represents the operation cycle of the simulation. The parameters x, y, v_x, v_y, a_x, a_y, ψ, γ represent the global position, longitudinal/lateral speed, acceleration, heading, and heading change rate of the test vehicle, respectively.

The period of the RTK-GPS data measurement value is lower than the period of the simulation, due to which the data are represented in a step shape, as shown in Figure 5. However, continuous position information for the test vehicle which is suitable for the simulation period is generated using the extrapolation method.

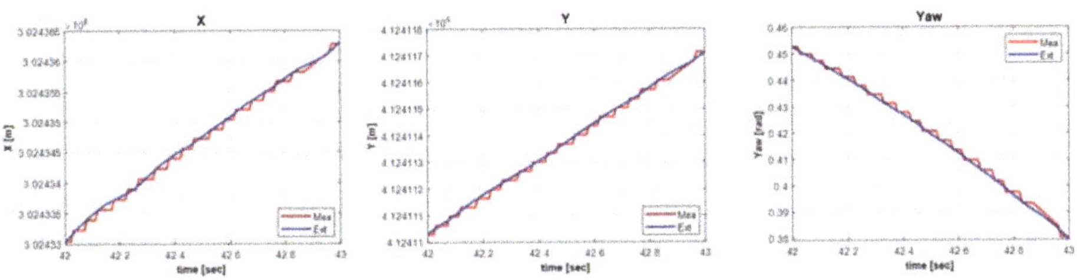

Figure 5. Result of extrapolation process.

2.3. *Virtual Traffic Behavior Generation*

Ensuring the accuracy of the behavior of the test vehicle used in the simulation test is important. However, in the case of an autonomous driving system simulation, it is also important to determine the behavior of the surrounding vehicles because the autonomous driving system must respond to various dangerous situations which can arise from the surrounding vehicles on the road. Additionally, it is important to accurately simulate the behavior of the target vehicle at a given time to replicate dangerous situations such as emergency steering and braking which can occur during the vehicle testing in the simulation and to verify the performance of the algorithm [33].

Figure 6 depicts the implementation of the virtual traffic behavior. Firstly, the location, speed, and heading of the RTK-GPS installed in the target vehicle are logged during the vehicle test. The measurement data were logged based on the GPS time provided by the RTK-GPS, and information on the target vehicle was obtained based on the vehicle test time. The location of the target vehicle was measured in terms of latitude and longitude. Therefore, the location of the target vehicle is converted into meter units by performing the UTM conversion, as explained in Section 2.1.

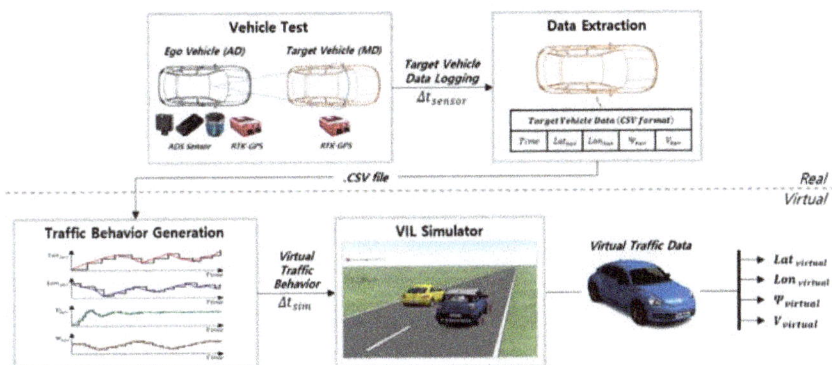

Figure 6. Procedure of virtual vehicle behavior generation.

Since the RTK-GPS data of the target vehicle are updated at a frequency lower than that used in the simulation, discontinuous vehicle behavior is observed in the virtual environment. To solve this problem, the logged target vehicle information was interpolated based on the simulation cycle and applied in the virtual simulator. The behavior of the target vehicle was interpolated using the Matlab Interpolation Tool, as shown in Figure 7. The Piecewise Cubic Hermite-Interpolating-Polynomial (PCHIP) method was used for the interpolation; this method interpolates the data most similar to the measured target vehicle data [34,35].

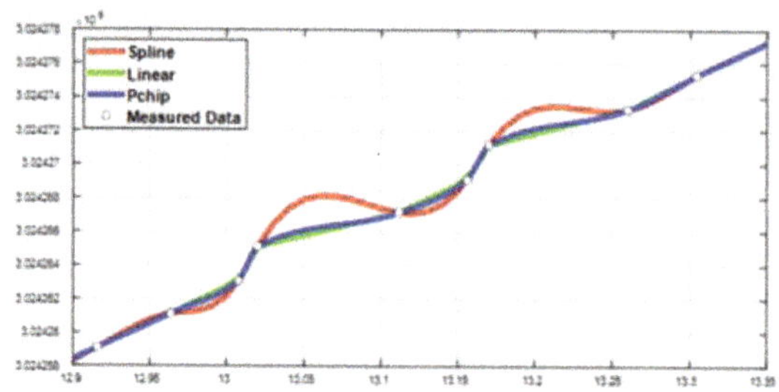

Figure 7. Interpolation result of measured target vehicle data.

2.4. Perception Sensor Modeling

The virtual traffic implemented as shown above cannot be measured through an actual perception sensor; therefore, a virtual perception sensor model is required to measure this traffic. Perception sensor models can be largely classified as using raw-data or object-list modeling. A raw-data sensor model provides a virtual image or point cloud similar to the real one, and can implement the driving environment in detail. However, it involves a large computational overhead. Conversely, the object-list sensor model can directly provide the data required for vehicle control to reduce the computational overhead and can provide realistic data by replicating physical characteristics such as sensor noise [36–39]. In this study, the relative distance and speed of the virtual traffic required for the autonomous driving algorithm were generated through object-list-based perception sensor modeling.

The relative distance between the test vehicle and the virtual traffic was determined based on the global location data of the vehicles. If the relative distance is calculated directly through the global location of each vehicle, the generated value is different from the relative distance of the actual perception sensor. This results in a difference in behavior between the vehicle test and the autonomous driving test performed using the VILS, making it difficult to ensure the repeatability and reproducibility.

Three aspects must be reflected to generate a value similar to the relative distance of an actual perception sensor. The distance between the test vehicle's RTK-GPS and the perception sensor mounting point, the RTK-GPS mounting position of the actual target vehicle used to implement the virtual traffic, and the point at which the actual perception sensor detects the target vehicle must be considered. Figure 8 depicts the method used to generate a relative distance which is similar to that of a real perception sensor in the VILS test by considering the aforementioned three factors. The relative distance between the test vehicle and virtual traffic is calculated as follows:

$$\begin{bmatrix} RD_x \\ RD_y \end{bmatrix} = \begin{bmatrix} X_{mea} - X_{sensor} \\ Y_{mea} - X_{sensor} \end{bmatrix} \begin{bmatrix} \cos(-\psi_{ego}) & -\sin(-\psi_{ego}) \\ \sin(-\psi_{ego}) & \cos(-\psi_{ego}) \end{bmatrix} \tag{4a}$$

$$\begin{bmatrix} X_{sensor} \\ Y_{sensor} \end{bmatrix} = \begin{bmatrix} d_x \\ d_y \end{bmatrix} \begin{bmatrix} \cos(\psi_{ego}) & -\sin(\psi_{ego}) \\ \sin(\psi_{ego}) & \cos(\psi_{ego}) \end{bmatrix} + \begin{bmatrix} X_{ego} \\ Y_{ego} \end{bmatrix} \tag{4b}$$

$$\begin{bmatrix} X_{mea} \\ Y_{mea} \end{bmatrix} = \begin{bmatrix} -l_x \\ -l_y \end{bmatrix} \begin{bmatrix} \cos(\psi_{traf}) & -\sin(\psi_{traf}) \\ \sin(\psi_{traf}) & \cos(\psi_{traf}) \end{bmatrix} + \begin{bmatrix} X_{traf} \\ Y_{traf} \end{bmatrix} \tag{4c}$$

Here, the subscripts $(\,)_{sensor}$ and $(\,)_{mea}$ represent the perception sensor mounted on the test vehicle and the point detected by the perception sensor, respectively. The parameter RD represents the relative distance based on the test vehicle; d denotes the distance between the RTK-GPS mounted on the test vehicle and the recognition sensor; and l denotes the distance between the RTK-GPS mounted on the target vehicle and the point detected by the perception sensor of the test vehicle.

The relative speed can be calculated based on the absolute speed difference between the virtual traffic and the test vehicle as shown below:

$$RV = V_{traf} - V_{ego} \tag{4d}$$

Here, the subscripts $(\ )_{ego}$ and $(\ )_{traf}$ represent each test vehicle and virtual traffic vehicle, RV represents the relative speed, and V represents the vehicle's absolute speed.

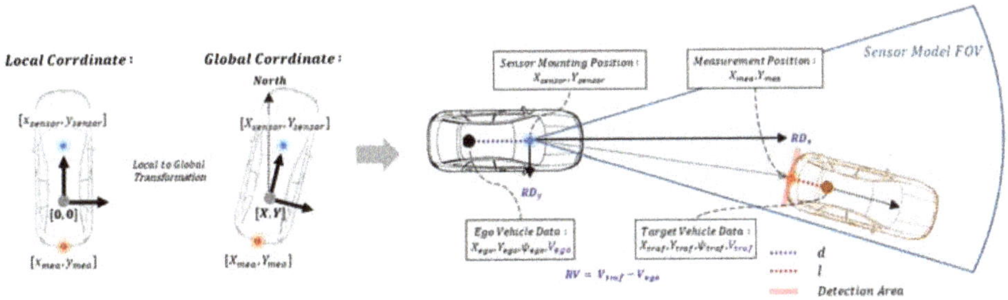

Figure 8. Concept of perception sensor modeling.

Even if the relative distance and speed of the test vehicle and the virtual traffic are obtained using the above method, the noise component generated by the actual perception sensor cannot be implemented in the virtual environment. Consequently, the RTK-GPS date and perception sensor data which were obtained during the vehicle tests were utilized to implement the sensor noise, which is similar to that of the actual conditions. Firstly, the RTK-GPS-based relative distance and relative speed of the test vehicle and target vehicle measured during the vehicle test were calculated using Equation (4). The error was calculated by comparing the calculated RTK-GPS-based relative distance and speed with the actual perception sensor data, and then calculating the mean and variance of the error using Equations (5a,b). Subsequently, a normal distribution was generated with the mean and variance of the error by using Equation (5c). The relative distance and velocity similar to the actual perception sensor data were generated by inserting the generated normal-distributed noise into the perception sensor model, as follows [40–42]:

$$\mu = \frac{1}{n}\sum_{k=1}^{n} X_k \tag{5a}$$

$$\sigma^2 = \frac{\sum (X-\mu)^2}{n} \tag{5b}$$

$$N(\mu,\sigma^2) = \frac{1}{\sqrt{2\pi\sigma^2}} \times e^{-((X-\mu)^2/2\sigma^2)} \tag{5c}$$

Here, μ denotes the mean of the data, σ^2 denotes the variance, and n denotes the total amount of data. $N(\mu,\sigma^2)$ represents a normal distribution with mean and variance as inputs.

Figure 9a presents the results generated by using the perception sensor modeling process. In the case of the relative distance calculated by using only the RTK-GPS data between each vehicle without sensor modeling (represented by the red point), an error was observed in comparison to the actual recognition sensor data that is sufficient to affect the vehicle control. In the case of the relative distance and speed (represented by the green point) generated through sensor modeling, it can be observed that the trend is similar to the actual perception sensor data (represented by the blue point). Figure 9b presents the distribution of the errors by comparing the data before and after performing sensor modeling using the actual perception sensor data. When sensor modeling is performed, the longitudinal and lateral relative distances and relative speeds present an average error distribution close to 0 when compared to the data obtained before performing sensor modeling.

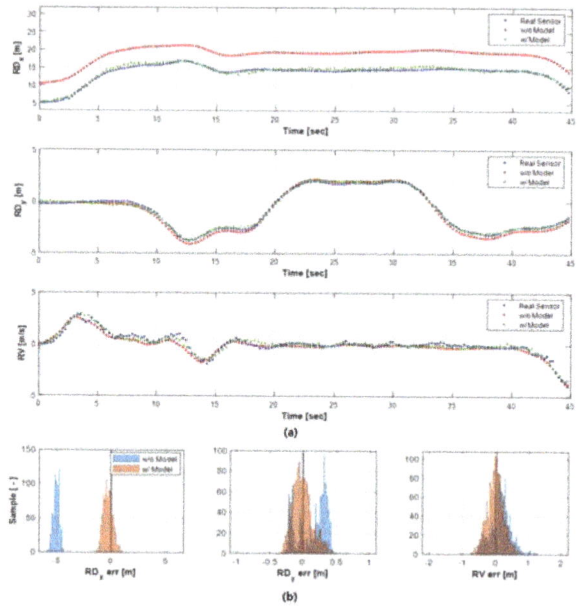

Figure 9. (**a**) Result of perception sensor modeling, (**b**) distribution of errors.

3. Test Autonomous Driving System

In the VILS, a test vehicle with an autonomous driving function is required to verify the performance, similar to in the vehicle test. The test vehicle comprised an autonomous driving ECU, RTK-GPS equipment, and a control target vehicle. The autonomous driving ECU is used to control the vehicle based on the information obtained in the virtual driving environment, and the real-time performance is verified. The RTK-GPS is used to measure the position, speed, acceleration, and rotation information of the test vehicle. The information obtained in the RTK-GPS is transferred to the synchronization module of the virtual simulator to implement the actual vehicle movement in the virtual environment, as explained in Section 2.2. The vehicle to be controlled is the actual vehicle to which the autonomous driving system is applied, and, in the case of the PG-VILS, the vehicle is driven on a proving ground.

The autonomous driving algorithm for the PG-VILS consists of longitudinal control that maintains the vehicle's distance from the preceding vehicle and lateral control that follows the designated path. Figure 10 presents the system architecture of the autonomous driving used in this study. In longitudinal control, relative distance and relative speed are used to maintain a distance from a preceding vehicle. In lateral direction control, path-following is performed using global coordinates measured by the RTK-GPS. The output values of the longitudinal and lateral controllers are transferred to the lower controller that controls the actuator of the vehicle. The lower controller uses PID control to generate throttle/brake and steering wheel angle commands for vehicle acceleration/deceleration and steering. Since the lower controller based on PID control has an intuitive and simple structure, a detailed description will be omitted in this paper.

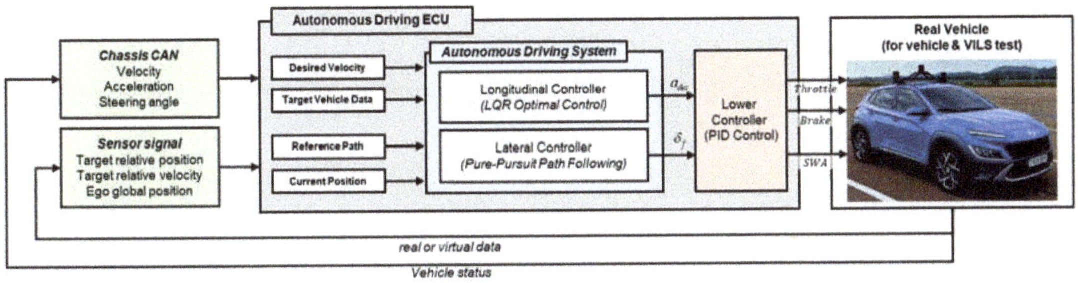

Figure 10. The architecture of the Test Autonomous Driving System.

3.1. Longitudinal Controller

An Adaptive Cruise Control (ACC) system was designed based on LQR optimal control for the longitudinal control. Figure 11 depicts the correlation between the test vehicle and the preceding vehicle in the ACC system, which can be expressed as a state equation, as shown below [43,44]:

$$\dot{x} = Ax + Bu = \begin{bmatrix} 0 & -1 \\ 0 & 0 \end{bmatrix} x + \begin{bmatrix} 0 \\ -1 \end{bmatrix} u \tag{6a}$$

$$x = \begin{bmatrix} x_1 \\ x_2 \end{bmatrix} = \begin{bmatrix} c_d - c \\ v_p - v_e \end{bmatrix} \tag{6b}$$

Here, u denotes the acceleration of the host vehicle, and c and c_d denote the relative distance and target relative distance to the preceding vehicle, respectively. v_e and v_p represent the speed of the test vehicle and the preceding vehicle, respectively, and c_0 represents the safe braking distance. The state variable, x_1, represents the error between the target relative distance, c_d, and the current relative distance, c, and x_2 represents the error between the speed of the preceding vehicle and the test vehicle. c_d represents the safe braking distance, time gap, and speed of the preceding vehicle.

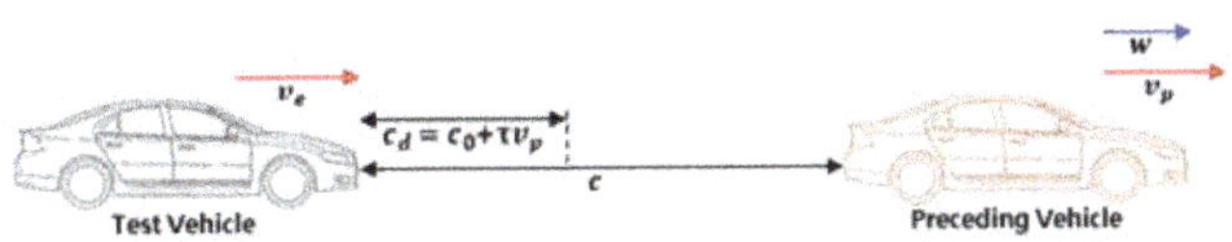

Figure 11. Correlation between test vehicle and preceding vehicle.

To obtain the target acceleration of the test vehicle, a cost function is defined and a control input, u, is generated as a value which minimizes it, as shown below:

$$J = \int_0^\infty (x^T Q x + u^T R u) dt \tag{7}$$

Here, Q and R represent the weights for the state variable and the input, respectively.

The gain, K, which represents the control input, is calculated using the Ricatti equation as follows:

$$u = a_{des} = -Kx \tag{8a}$$

$$A^T P + PA - PBR^{-1}B^T + Q = 0 \tag{8b}$$

$$K = R^{-1} B^T P \tag{8c}$$

Here, a_{des} represents the target acceleration and P represents the solution of the Ricatti equation. In the case of the calculated a_{des}, the maximum and minimum values are limited as shown in Equation (9a) to prevent a sudden acceleration/deceleration of the vehicle.

When the Time-To-Collision (TTC) (set to 0.8 s in this study) with the preceding vehicle expressed in Equation (9b) lies within the set time, a specific acceleration value is used for emergency braking [45].

$$a_{des} = \begin{cases} a_{\max}\ if & a_{des} > a_{\max} \\ a_{\min}\ else & a_{des} < a_{\min} \\ a_{des}\ else & \\ a_{AEB}\ if\ TTC < TTC_{\min} & \end{cases} \quad (9a)$$

$$TTC = \frac{c}{v_p - v_e} \quad (9b)$$

Here, $a_{\max}$ and $a_{\min}$ represent the limit values of the required acceleration and a_{AEB} represents the acceleration value used during the emergency braking.

3.2. Lateral Controller

The pure-pursuit algorithm, which is a path-following algorithm, was used for the lateral control [46]. The pure-pursuit algorithm generates the target steering angle to reach the look-ahead point, which is the target point on the path to be followed, as shown in Figure 12. The generated target steering angle is calculated as follows:

$$\delta_{des} = \frac{L}{r} = \frac{2\sin(\theta_{err})}{l_d} * L \quad (10a)$$

$$l_d = v_e * k \quad (10b)$$

Here, L represents the wheelbase of the vehicle, r represents the turning radius, θ_{err} represents the yaw angle between the vehicle heading and the look-ahead point, and l_d represents the distance to the look-ahead point. In the case of l_d, the vehicle speed, v_e, of the test vehicle and the linear coefficient, k, were used in proportion to the vehicle speed. Thus, an appropriate look-ahead point can be determined based on the vehicle speed, while achieving the path-following performance and steering stability.

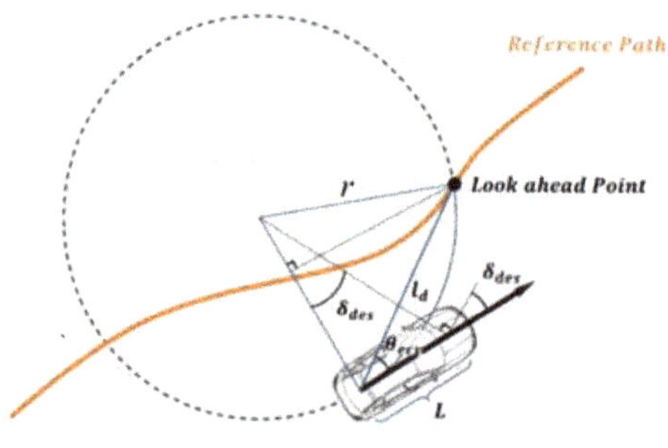

Figure 12. Geometry of pure-pursuit algorithm.

4. Consistency Validation Procedure and Methodologies

The VILS system must be capable of verifying the autonomous driving system and whether the test results obtained using the proposed VILS are similar to those of the vehicle test. Even if a VILS System with various components has been developed, it is impossible to replicate and reproduce the vehicle test scenario using the VILS system if the behavior of the vehicle is different from that observed during the vehicle test [47].

4.1. Consistency Validation Procedure

Figure 13 depicts the procedure used to verify the consistency between the vehicle test and the VILS test. Firstly, the target vehicle behavior required for the VILS scenario generation and the result data (Control Command, Vehicle Status, etc.) required for consistency verification are extracted from the various pieces of information obtained during the vehicle test. A VILS scenario is then generated using the extracted target vehicle behavior, and the VILS test result data are extracted. Time synchronization is performed based on the Time of Arrival (TOA), which is the time at which the VILS test data reaches a specific value, to compare the results obtained by performing the same VILS scenario several times [48]. The consistency was analyzed and the validity of the proposed VILS system was verified by comparing the time-synchronized VILS test results with the vehicle test results.

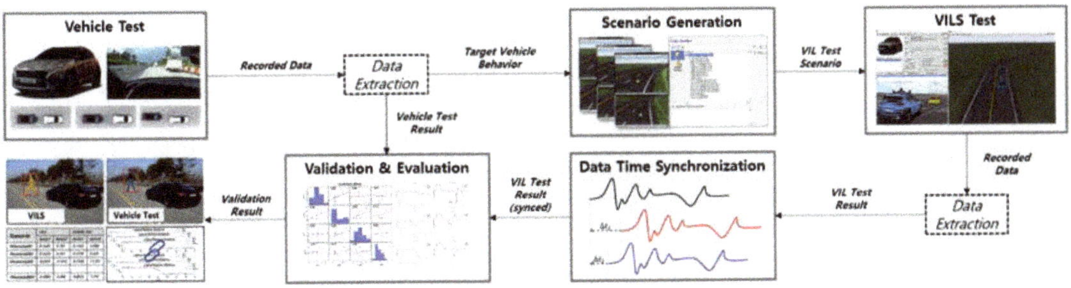

Figure 13. Vehicle Test—VILS Test Validation Procedure.

4.2. Consistency Validation Methodologies

In this study, a time-series comparison, which quantitatively verifies the consistency of the virtual and real data, and a scalar data comparison, used for data comparison at key points in the test, are utilized [49,50].

In the time-series comparison, NRMSE, which is an index indicating the error between datasets, and the Pearson correlation, which is an index indicating the linear correlation between the two datasets, were used [51]. The NRMSE is a value obtained by dividing the Root Mean Square Error (RMSE) value by the difference between the maximum and minimum data. It can determine the error normalized to a value of 0 to 100, and is expressed as follows [52]:

$$NRMSE = \frac{RMSE}{y_{real,\max} - y_{real,\min}} \times 100 = \frac{\sqrt{\frac{1}{N}\sum_i^N \left(y_{real,i} - y_{sim,i}\right)^2}}{y_{real,\max} - y_{real,\min}} \times 100 \tag{11}$$

Here, y_{real} and y_{sim} represent the real and simulated data, N represents the number of samples of the calculated data, and the subscripts, $(\,)_{\max}$ and $(\,)_{\min}$, represent the maximum and minimum values, respectively.

The Pearson correlation is a value obtained by dividing the absolute value of the covariance of two data points by the product of the standard deviation of each data point. It can be represented as a number between -1 and 1, where a value closer to 1 indicates a higher correlation between the data, and can be expressed as follows [53]:

$$r_{sim,real} = \frac{\left|\sum_i^N \left(y_{sim,i} - \overline{y}_{sim}\right) \times \left(y_{real,i} - \overline{y}_{real}\right)\right|}{\sqrt{\sum_i^N \left(y_{sim,i} - \overline{y}_{sim}\right)^2 \times \sum_i^N \left(y_{real,i} - \overline{y}_{real}\right)^2}} \tag{12}$$

Here, r denotes the Pearson correlation coefficient, y_{real} and y_{sim} represent the real and simulation data, respectively, and $\overline{y}$ denotes the average of the data, where N indicates the number of samples in the measured data.

In the case of scalar data comparison, the relative error between the virtual and real data is determined as a ratio by comparing the data peak values at a specific point in time. In this study, the peak value of the relative speed which is observed when the preceding vehicle arrives at a sudden stop is compared with the peak value of the longitudinal acceleration of the vehicle performing the AEB. The peak values of the yaw rate/lateral acceleration are obtained when driving on a turning road with a large curvature. The corresponding results are explained in Section 5, and the relative error between the actual and virtual peak data can be expressed as follows [54]:

$$PR = \frac{\left|y_{real,peak} - y_{sim,peak}\right|}{\left|y_{real,peak}\right|} \times 100 \tag{13}$$

Here,PR represents the ratio of the real simulation peak value, y_{real} and y_{sim} represent the real and simulated data, respectively, and the subscript $(\,)_{peak}$ represents the peak value of the data to be compared.

5. Field Test Configuration and Test Result

As explained in the previous section, the consistency of the VILS test results and vehicle test results is an important indicator to ensure that the proposed VILS system can achieve repeatability and reproducibility for a specific scenario. This is because VILS is implemented to replicate scenarios which are difficult to implement repeatedly in a vehicle test through a simulation and to replicate vehicle behaviors with similar tendencies. This section describes the vehicle configuration, test site, and scenario used for the actual vehicle tests and the VILS tests and analyzes the consistency of the test results from the two different environments.

5.1. Experimental Setup

An autonomous vehicle based on the Hyundai Kona was used as the test vehicle, as shown in Figure 14. Mobileye 630 was used as the perception sensor to detect the surrounding vehicles, and a real-time ECU loaded with the autonomous driving algorithm was configured using MicroAutobox3 (MAB3). Additionally, the actual position of the test vehicle was measured using RT3002-v2, and the Xpack4 RoadBox, a real-time computing device, was used as a virtual simulator equipped with the VILS system proposed in this study. In the case of the vehicle test, information on the surrounding vehicles measured in Mobileye and the location information of RT3002-v2 are transmitted to the real-time ECU to perform the longitudinal and lateral control in autonomous driving. In the case of the VILS, the virtual surrounding vehicles and virtual location information generated through the virtual simulator are transmitted to the Real-Time ECU.

Figure 14. Hyundai Kona vehicle equipment components.

The actual vehicle tests and VILS tests were conducted on the configured test vehicle in K-City (Autonomous Driving Testbed) and on the proving ground located inside the Korea Automobile Testing and Research Institute. The vehicle test was conducted on a suburban road inside K-City, as shown in Figure 15. The driving test was conducted on a selected target road (approximately 320 m in length), which includes curved sections with turning radii of 70 m, 80 m, and 70 m, respectively, among the suburban roads. A test that considered both the longitudinal and lateral dynamics along the target road was conducted. The VILS test was conducted by constructing a virtual road with the same shape as that of the target road selected for the vehicle test. Consequently, some areas within the proving ground that allowed for the target road size were selected for the VILS test.

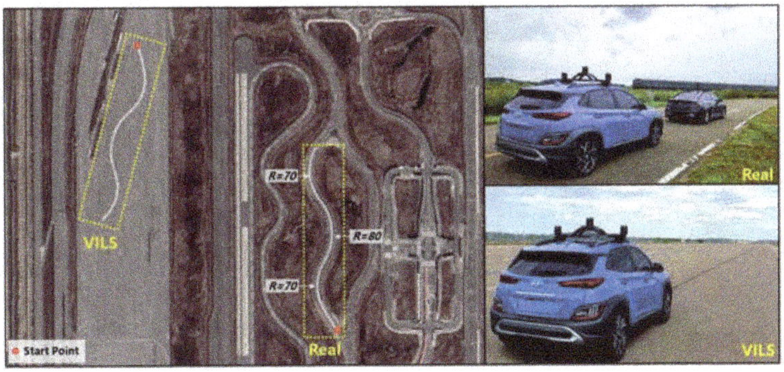

Figure 15. Test field KATRI K-City (Real) and proving ground (VILS).

Figure 16 presents the configuration and scenario of the vehicle test conducted using the test vehicle and the target road described above. A Hyundai Sonata was used as the target vehicle, and the ACC function, which maintains a longitudinal distance from the target vehicle, was implemented. The target vehicle was equipped with an autonomous driving function, along with manual driving; it was controlled to follow a set route, and a constant speed was maintained through cruise control. Additionally, it was configured to record the behavior measured using a mounted RTK-GPS. The recorded behavioral information of the target vehicle was used to implement the behavior of the virtual vehicle to be used in the VILS test. The vehicle test scenario was constructed for two cases, one in which the target vehicle was driven autonomously and one manually.

Figure 16. Real test and VILS test concept and scenarios.

In the first scenario, ACC control is performed on a target vehicle which is driven using autonomous driving to follow a predetermined route. In this scenario, the target vehicle maintains a steady-state speed of 30 km/h and stops when sudden braking is performed at the end of the last curved section (R = 70 m). In the second scenario, longitudinal and lateral control are implemented on a manually driven target vehicle. The vehicle is driven at a speed of 30–40 km/h, and, similarly to in Scenario 1, it is stopped when sudden braking is performed at the end of the last curved section. The speed of the target vehicle in the scenario reflected the width and curvature of the target road to set a safe speed for the test. In addition, in the case of a target vehicle that is manually driven by a real driver, it is difficult to drive the entire section of the target road at a constant speed; therefore, it is driven at a speed within a specific range. The ACC setting for the speed of the test vehicle was set to 50 km/h, which is higher than the driving speed of the target vehicle, in order to respond to the two scenarios.

The vehicle test was conducted using the scenario configured as described above, and the recorded behavior information of the target vehicle was applied to the virtual simulator. The VILS test was performed five times for each scenario, and the consistency of the behavior of the vehicle in the two environments was determined by analyzing the VILS test results and the vehicle test results.

5.2. Test Results

The test results of the two environments were compared by classifying the obtained data into perception sensor data, longitudinal behavior, and lateral behavior, and a Key Performance Index (KPI) was selected for each classification [55]. The KPI of the vehicle test is represented in black, and the KPI of the VILS test, which was performed five times, is represented in five different colors.

The consistency between the vehicle test and the VILS test was analyzed based on the results of each KPI. The NRMSE, Pearson correlation, and peak-ratio comparison methods, which were explained in Section 4, were used for the consistency analysis. The NRMSE is an index that indicates the error between datasets, and the Pearson correlation is an index that indicates a linear correlation between two datasets. The peak-ratio comparison is an indicator to determine whether the data peak values observed at a specific point in time, such as during emergency braking, are similar to each other in the vehicle test and VILS test.

5.2.1. Test Result #1—Sensor Data

In this section, the validity of the proposed sensor model of the VILS system is verified by analyzing the consistency between the virtual perception sensor data generated during the VILS and the actual sensor perception data. First, the longitudinal/lateral relative distance and relative speed of the target object, which are the sensor data necessary for the ACC performed by the test vehicle, were selected as the KPIs of the sensor data. To ensure high consistency with the vehicle test, the generated virtual perception sensor data of the VILS System must be similar to the vehicle test data. This is because, when the generated virtual perception sensor data are not similar to the real data, the control command value generated by the autonomous driving algorithm may exhibit a completely different behavior.

Figure 17a depicts Scenario 1 and Figure 17b depicts Scenario 2. In Scenario 1, the front target vehicle is autonomously driven at a fixed speed. Therefore, it exhibits a constant trend in the relative distance and relative speed after the steady state. In Scenario 2, the measured sensor data values are not constant when compared to those of Scenario 1 since the front target vehicle is manually driven.

The five VILS tests performed using these scenarios demonstrated that the results of the VILS tests were similar to the sensor information obtained during the vehicle test, as shown in Figure 17. A comparison of the errors of each VILS test and the vehicle test demonstrated that the NRMSE of the longitudinal/lateral relative distance and the relative speed presented an average value of approximately two percent, as shown in Table 2. Even in the case of the Pearson correlation, it can be observed that the VILS system, which was built with a high positive correlation close to one, generates sensor data similar to those of the practical conditions. Table 3 presents the comparison results of the relative speed of the front target vehicle, which was measured during the sudden braking at the end of the target road. This relative speed has an error rate of less than 2.5% on average when compared to the vehicle test in both scenarios. This comparison confirmed the validity of the virtual perception sensor data generated by the proposed VILS system.

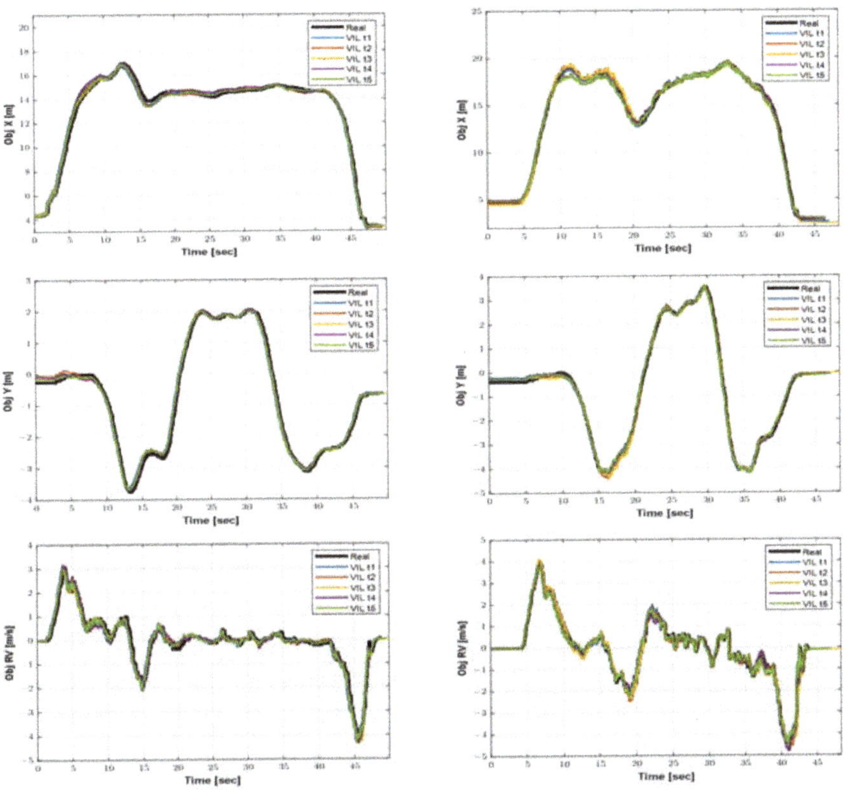

(a) Scen #1 **(b) Scen #2**

Figure 17. Result of sensor data KPI.

Table 2. Data comparison result—Sensor data (Object relative distance, Object relative velocity) KPI.

Real—VILS (Scen #1)	**NRMSE [%]**			**Pearson Correlation**		
	ObjX	**ObjY**	**ObjRV**	**ObjX**	**ObjY**	**ObjRV**
#1	2.13	2.41	2.71	0.99	0.99	0.98
#2	1.76	1.87	2.07	0.99	0.99	0.99
#3	1.98	2.28	2.68	0.99	0.99	0.99
#4	1.84	1.24	1.68	0.99	0.99	0.99
#5	1.60	1.80	2.30	0.99	0.99	0.99
Avg	1.80	1.80	2.18	0.99	0.99	0.99
Real—VILS (Scen #2)	NRMSE [%]			Pearson Correlation		
	ObjX	ObjY	ObjRV	ObjX	ObjY	ObjRV
#1	1.61	1.48	1.97	0.99	0.99	0.99
#2	1.56	0.99	1.33	0.99	0.99	0.99
#3	2.18	1.21	1.87	0.99	0.99	0.99
#4	2.04	1.34	2.30	0.99	0.99	0.99
#5	2.17	1.61	2.91	0.99	0.99	0.99
Avg	1.99	1.29	2.10	0.99	0.99	0.99

Table 3. Data Comparison Result—Object relative velocity Peak-Ratio.

Real—VILS (Scen #1)	*PR* [%] Obj RV	Real—VILS (Scen #2)	*PR* [%] Obj RV
#1	0.70	#1	1.89
#2	0.46	#2	1.05
#3	2.32	#3	0.42
#4	2.55	#4	0.63
#5	0.23	#5	4.42
Avg	1.25	Avg	1.68

5.2.2. Test Result #2—Longitudinal Behavior (Adaptive Cruise Control)

In this section, the longitudinal behavior consistency of the vehicle test and VILS test is analyzed. The desired acceleration, velocity, longitudinal acceleration, and AEB flag were selected as the KPIs corresponding to the longitudinal behavior, and a consistency analysis was conducted. Figure 18 depicts the longitudinal behavior of the KPIs for the two selected scenarios. In the case of Scenario 1, it can be observed that the values of the speed, longitudinal acceleration, and required acceleration of the test vehicle are constant, except for the acceleration section, since the front target vehicle drove the target road at a constant speed. In Scenario 2, the test vehicle controls the front target vehicle, which is manually driven, and the required acceleration value, which is the control command value, is not uniformly generated, unlike in Scenario 1. Therefore, the vehicle speed is maintained at a value between approximately 30 km/h and 40 km/h (approximately 8 to 11 m/s), and the longitudinal acceleration value also changes based on the required acceleration value.

In both scenarios, the front target vehicle performs sudden braking at the endpoint of the driving target road; the test vehicle also implements control to activate the AEB. If the set TTC condition is not satisfied, the autonomous driving controller outputs an AEB Flag signal even though the minimum required acceleration calculated by the autonomous driving controller of -4m/s^2 is applied to the vehicle. Simultaneously, the acceleration command (-8m/s^2) required for sudden braking is input to the vehicle actuator.

The five VILS tests performed using these scenarios demonstrated that they were implemented in a similar manner to the longitudinal behavior information obtained during the vehicle test.

The NRMSE for the longitudinal KPIs was within four percent on average, and the Pearson correlation exhibited a positive correlation close to one, as shown in Table 4. The AEB flag was also observed at a similar time to the time observed during the vehicle test, confirming the repeatability of the scenario in which the same emergency braking situation occurs at the same location and at the same time. Additionally, the longitudinal acceleration peak values in the vehicle test and VILS test, which occur in the AEB situation near the test endpoint, were compared as shown in Table 5. The comparison results demonstrate that, in Scenario 1, the average error rate was 6.18%, and, in Scenario 2, the average error rate was 3.06%. As the test vehicle was driven on a real proving ground instead of a virtual one, the VILS test result exhibits some error every time. However, since the error in each test is not at a level that exhibits a different trend from the vehicle test, the validity of the longitudinal control test performed using the proposed VILS system is demonstrated.

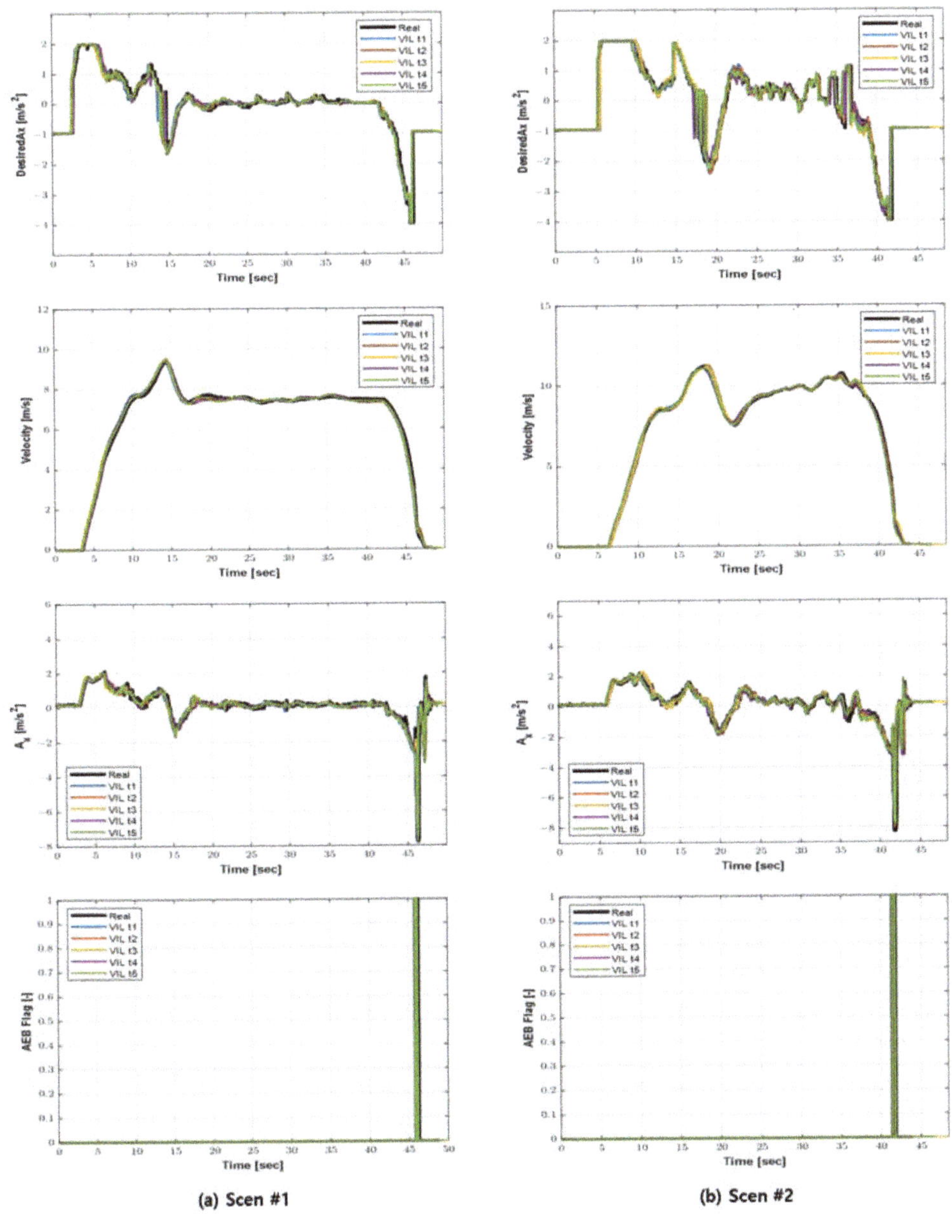

Figure 18. Result of Longitudinal behavior KPI.

Table 4. Data Comparison Result—Longitudinal behavior (Desired longitudinal acceleration, Velocity and Longitudinal acceleration) KPI.

Real—VILS (Scen #1)	**NRMSE [%] Desired_Ax**	**Vel**	**Ax**	**Pearson Correlation Desired_Ax**	**Vel**	**Ax**
#1	4.21	2.08	4.34	0.95	0.99	0.88
#2	5.71	1.16	3.23	0.97	0.99	0.93
#3	3.98	1.96	3.15	0.96	0.99	0.94
#4	3.04	0.31	2.26	0.98	0.99	0.97
#5	2.33	1.29	2.82	0.97	0.99	0.95
Avg	3.76	1.18	2.87	0.966	0.99	0.934
Real—VILS (Scen #2)	NRMSE [%] Desired_Ax	Vel	Ax	Pearson Correlation Desired_Ax	Vel	Ax
#1	2.49	1.13	3.34	0.97	0.99	0.95
#2	1.79	0.72	2.08	0.98	0.99	0.98
#3	2.40	1.18	3.86	0.97	0.99	0.94
#4	2.95	1.74	3.58	0.95	0.99	0.95
#5	2.70	0.60	4.21	0.96	0.99	0.93
Avg	2.46	1.06	3.43	0.966	0.99	0.95

Table 5. Data Comparison Result—Longitudinal acceleration Peak-Ratio.

Real—VILS (Scen #1)	***PR* [%] Ax**	**Real—VILS (Scen #2)**	***PR* [%] Ax**
#1	7.18	#1	3.25
#2	5.00	#2	3.37
#3	6.28	#3	2.53
#4	6.28	#4	5.66
#5	6.15	#5	0.48
Avg	6.18	Avg	3.06

5.2.3. Test Result #3—Lateral Behavior (Path Following)

The consistency of the lateral behavior of the test vehicle and the VILS test was analyzed by selecting the desired Steering Wheel Angle (SWA), yaw rate, and lateral acceleration as the related KPIs. In both scenarios described above, the test vehicle is driven in a three-section curved road (radius of curvature = 70 m, 80 m, and 70 m) on the target road through path-following control. Figure 19 presents the lateral KPI data of the vehicle test and VILS test conducted based on this scenario. The desired SWA is generated to travel on a set route, and the yaw rate and lateral acceleration generated as a result are presented. It can be observed that, for the desired SWA, the initial values of the vehicle and VILS tests are different. This error is attributed to the fact that the SWA arranged at the start of the test was different for each VILS test. When the test vehicle is driven using the autonomous driving algorithm, the desired SWA is generated in a manner similar to the vehicle test. It can be observed that the yaw rate and lateral acceleration also appear to be similar to the results of the vehicle test.

The NRMSE for the lateral KPIs was within 5.5% on average, and the Pearson correlation also exhibited a positive correlation close to 1, as shown in Table 6. However, it presents a higher NRMSE value than the sensor data KPI and longitudinal KPI. This is caused by the SWA of the initially aligned test vehicle, and, after the test vehicle starts driving, it does not cause a significant difference in the vehicle behavior from the vehicle test, as explained above. The yaw rate and lateral acceleration values generated when turning in a curved line were used for the comparison of the peak values. In Scenario 1, the peak value was observed on the first curved road (radius of curvature = 70 m) because this curved section was encountered when the test vehicle started driving and accelerated. In Scenario 2, the acceleration of the test vehicle occurred on the third curved road (radius of curvature = 70 m), resulting in a peak value. The peak ratio of each scenario was compared, and it was observed that the yaw rate ratios were 1.04% and 1.17%, and the lateral acceleration ratios were 3.31% and 2.25%, respectively, as shown in Table 7. In both scenarios, the peak value of the VILS test was observed to be similar to that of the vehicle test. The above results demonstrate the validity of the lateral control test performed using the proposed VILS system.

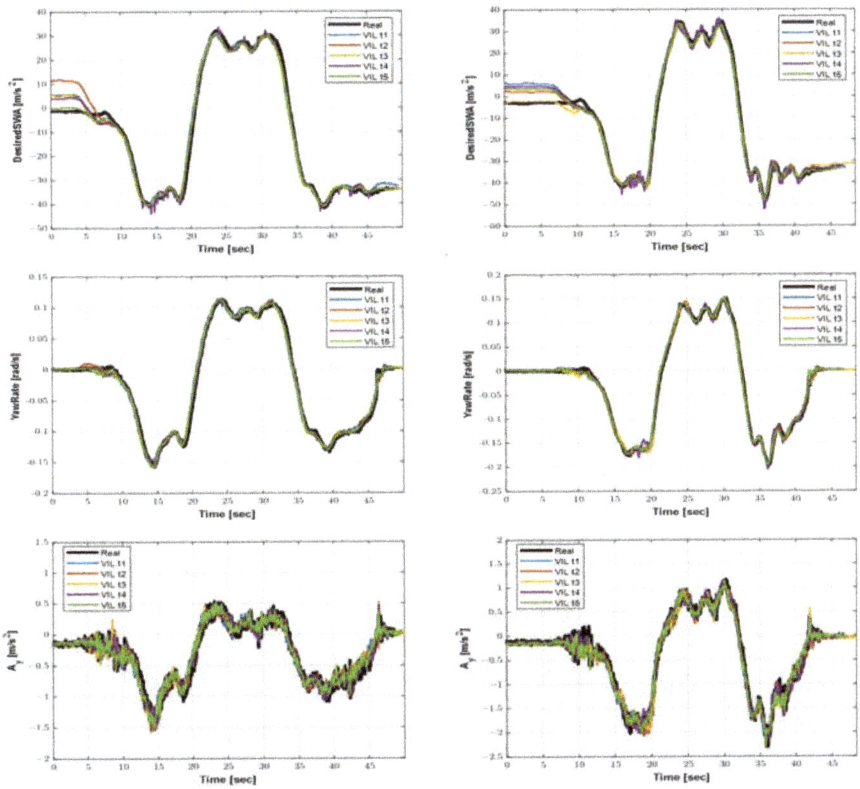

(a) Scen #1 **(b) Scen #2**

Figure 19. Results of Lateral behavior KPI.

Table 6. Data Comparison Result—Lateral behavior (Desired steering wheel angle, Yaw rate, Lateral acceleration) KPI.

Real—VILS (Scen #1)	NRMSE [%] Desired_SWA	NRMSE [%] YawRate	NRMSE [%] Ay	Pearson Correlation Desired_SWA	Pearson Correlation YawRate	Pearson Correlation Ay
#1	4.21	2.92	5.56	0.99	0.99	0.97
#2	5.71	1.78	5.03	0.99	0.99	0.97
#3	3.98	2.73	5.91	0.99	0.99	0.96
#4	3.04	1.6	5.24	0.99	0.99	0.97
#5	2.33	2.05	5.27	0.99	0.99	0.97
Avg	3.76	2.04	5.36	0.99	0.99	0.97
Real—VILS (Scen #2)	NRMSE [%] Desired_SWA	NRMSE [%] YawRate	NRMSE [%] Ay	Pearson Correlation Desired_SWA	Pearson Correlation YawRate	Pearson Correlation Ay
#1	4.88	1.6	3.57	0.99	0.99	0.99
#2	3	1.04	2.93	0.99	0.99	0.99
#3	2.13	1.39	3.45	0.99	0.99	0.99
#4	4.58	2.19	3.98	0.99	0.99	0.99
#5	3.86	2.23	3.95	0.99	0.99	0.99
Avg	3.39	1.71	3.58	0.99	0.99	0.99

Table 7. Data comparison result—Yaw rate, Lateral acceleration Peak-Ratio.

Real—VILS (Scen #1)	*PR* [%] YawRate	Ay	Real—VILS (Scen #2)	*PR* [%] YawRate	Ay
#1	0.33	2.65	#1	0.44	0.43
#2	1.77	3.31	#2	0.18	1.73
#3	1.33	5.30	#3	0.26	1.30
#4	1.11	2.65	#4	4.50	5.63
#5	0.66	2.65	#5	0.44	2.16
Avg	1.04	3.31	Avg	1.17	2.25

6. Conclusions

In this study, a VILS system which considers the consistency between the vehicle test and VILS test results was proposed. Four components of the VILS system were implemented to obtain simulation results similar to the behavior of a vehicle during vehicle testing. First, a virtual road was built, and the behavior of the test vehicle in the real environment and the virtual environment were synchronized. In addition, the behavior of the surrounding traffic measured during the vehicle test was implemented in the simulation. Finally, a virtual perception sensor that outputs similar values to the real perception sensor was modeled. The VILS test was conducted using the VILS system built through the above process.

The VILS test results and vehicle test results were analyzed using three types of coherence measures (i.e., NRMSE, Pearson correlation, data peak value comparison), which verified that the vehicle behavior in the two different environments was similar. In particular, as the peak values of the two environments show a small difference of around five percent, it can be seen that the vehicle behavior at a specific point is similarly repeated and reproduced. Furthermore, it was verified that the proposed VILS system can simulate not only general driving situations, but also dangerous driving situations, by applying a scenario in which the target vehicle behavior changes between autonomous driving and manual driving.

In the near future, the proposed VILS system will be expanded to be applied to various speeds, road types, and surrounding environments. Additionally, the proposed VILS system can be improved through virtual perception sensor advancement, which considers the characteristics and modeling methodologies of various perception sensors. With these improvements, the VILS system is expected to be able to verify various autonomous driving functions in addition to the ACC and path-following functions.

Author Contributions: Conceptualization, W.S.; investigation, S.W.; methodology, W.S.; resources, S.C.; software, W.S., Y.H., T.O.; validation, W.S., Y.H., T.O.; writing—original draft preparation, W.S.; writing—review and editing, J.Y.; visualization, W.S. All authors have read and agreed to the published version of the manuscript.

Funding: This work is supported by the Korea Agency for Infrastructure Technology Advancement (KAIA) grant funded by the Ministry of Land, Infrastructure, and Transport (Grant 22AMDP-C162182-02).

Data Availability Statement: Not applicable.

Conflicts of Interest: The authors declare no conflict of interest.

References

1. Abbasi, R.; Bashir, A.K.; Alyamani, H.J.; Amin, F.; Doh, J.; Chen, J. Lidar Point Cloud Compression, Processing and Learning for Autonomous Driving. *IEEE Trans. Intelligent Trans. Syst.* **2022**, 1–18. [CrossRef]
2. SAE On-Road Automated Vehicle Standards Committee. Taxonomy and definitions for terms related to on-road motor vehicle automated driving systems. *SAE Standard J.* **2014**, *3016*, 1–16.
3. Chen, Y.; Chen, S.; Xiao, T.; Zhang, S.; Hou, Q.; Zheng, N. Mixed Test Environment-based Vehicle-in-the-loop Validation—A New Testing Approach for Autonomous Vehicles. In Proceedings of the IEEE Intelligent Vehicles Symposium (IV), Las Vegas, NV, USA, 19 October 2020–13 November 2020; pp. 1283–1289. [CrossRef]

4. Wang, B.; Han, Y.; Wang, S.; Tian, D.; Cai, M.; Liu, M.; Wang, L. A Review of Intelligent Connected Vehicle Cooperative Driving Development. *Mathematics* **2022**, *10*, 3635. [CrossRef]
5. Szalay, Z. Next Generation X-in-the-Loop Validation Methodology for Automated Vehicle Systems. *IEEE Access* **2021**, *9*, 35616–35632. [CrossRef]
6. Riedmaier, S. Validation of x-in-the-loop Approaches for Virtual Homologation of Automated Driving Functions. In Proceedings of the 11th Graz Symposium Virtual Vehicle, Graz, Austria, 15–16 May 2018.
7. Koopman, P.; Wagner, M. Challenges in Autonomous Vehicle Testing and Validation. *SAE Int. J. Transp. Saf.* **2016**, *4*, 15–24. [CrossRef]
8. Kihong, P.; Son, W. Research trends of vehicle-in-the-loop-simulation for automated driving. *AUTO J. J. Korean Soc. Automotive Eng.* **2022**, *44*, 21–25.
9. VMAD-26th SG2 Session–Transport–Vehicle Regulations–UNECE Wiki. Available online: wiki.unece.org/display/trans/VMAD-26th+SG2+session. (accessed on 6 November 2022).
10. Gietelink, O.; Ploeg, J.; De Schutter, B.; Verhaegen, M. Development of advanced driver assistance systems with vehicle hardware-in-the-loop simulations. *Veh. Syst. Dyn.* **2006**, *44*, 569–590. [CrossRef]
11. Galko, C.; Rossi, R.; Savatier, X. Vehicle-Hardware-in-the-Loop System for ADAS Prototyping and Validation. In Proceedings of the 2014 International Conference on Embedded Computer Systems: Architectures, Modeling, and Simulation (SAMOS XIV), Agios Konstantinos, Greece, 14–17 July 2014.
12. Siegl, S.; Ratz, S.; Düser, T.; Hettel, R. Vehicle-in-the-Loop at Testbeds for ADAS/AD Validation. *ATZelectronics Worldw.* **2021**, *16*, 62–67. [CrossRef]
13. Diewald, A.; Kurz, C.; Kannan, P.; Gießler, M.; Pauli, M.; Göttel, B.; Kayser, T.; Gauterin, F.; Zwick, T. Radar Target Simulation for Vehicle-in-the-Loop Testing. *Vehicles* **2021**, *3*, 257–271. [CrossRef]
14. Gao, Y.; Zhao, X.; Xu, Z.; Cheng, J.; Wang, W. An Indoor Vehicle-in-the-Loop Simulation Platform Testing Method for Autonomous Emergency Braking. *J. Adv. Transp.* **2021**, *2021*, 1–12. [CrossRef]
15. Zhang, Y.; Lu, S.; Yang, Y.; Guo, Q. Internet-Distributed Vehicle-in-the-Loop Simulation for HEVs. *IEEE Trans. Veh. Technol.* **2018**, *67*, 3729–3739. [CrossRef]
16. Park, C.; Chung, S.; Lee, H. Vehicle-in-the-Loop in Global Coordinates for Advanced Driver Assistance System. *Appl. Sci.* **2020**, *10*, 2645. [CrossRef]
17. Hyundai MOBIS: Augmented Reality. dSPACE. Available online: www.dspace.com/en/pub/home/applicationfields/stories/hyundai-mobis-augmented-reali.cfm (accessed on 6 November 2022).
18. Tettamanti, T.; Szalai, M.; Vass, S.; Tihanyi, V. Vehicle-In-the-Loop Test Environment for Autonomous Driving with Microscopic Traffic Simulation. In Proceedings of the 2018 IEEE International Conference on Vehicular Electronics and Safety (ICVES), Madrid, Spain, 12–14 September 2018; pp. 1–6. [CrossRef]
19. Lee, D.; Lee, S.; Chen, Z.; Park, B.B.; Shim, D.H. Design and field evaluation of cooperative adaptive cruise control with unconnected vehicle in the loop. *Transp. Res. Part C Emerg. Technol.* **2021**, *132*, 103364. [CrossRef]
20. Kim, D.; Kim, G.; Kim, H.; Huh, K. A hierarchical motion planning framework for autonomous driving in structured highway environ-ments. *IEEE Access* **2022**, *10*, 20102–20117. [CrossRef]
21. Solmaz, S.; Rudigier, M.; Mischinger, M. A Vehicle-in-the-Loop Methodology for Evaluating Automated Driving Functions in Virtual Traffic. In Proceedings of the IEEE Intelligent Vehicles Symposium (IV), Las Vegas, NV, USA, 19 October 2020–13 November 2020; pp. 1465–1471. [CrossRef]
22. A Comprehensive Approach for the Validation of Virtual Testing Toolchains. IAMTS. Available online: https://www.iamts.org/storage/app/media/Publications/iamts0001202104.pdf (accessed on 6 November 2022).
23. Kang, J.M.; Yoon, T.S.; Kim, E.; Park, J.B. Lane-Level Map-Matching Method for Vehicle Localization Using GPS and Camera on a High-Definition Map. *Sensors* **2020**, *20*, 2166. [CrossRef]
24. Gacoki, T.G.; Aduol, F.W.O. Transformation between GPS coordinates and local plane UTM coordinates using the excel spreadsheet. *Surv. Rev.* **2002**, *36*, 449–462. [CrossRef]
25. Snyder, J.P. *Map Projections: A Working Manual*; U.S. Government Printing Office: Washington, DC, USA, 1987. [CrossRef]
26. Feng, Y.; Wang, J. GPS RTK Performance Characteristics and Analysis. *J. Glob. Position. Syst.* **2008**, *7*, 1–8. [CrossRef]
27. Takasu, T.; Yasuda, A. Development of the Low-Cost RTK-GPS Receiver with an Open Source Program Package RTKLIB. In Proceedings of the International Symposium on GPS/GNSS, International Convention Center, Jeju, Korea, 4–6 November 2009.
28. Schall, G.; Wagner, D.; Reitmayr, G.; Taichmann, E.; Wieser, M.; Schmalstieg, D.; Hofmann-Wellenhof, B. Global Pose Estimation using Multi-Sensor Fusion for Outdoor Augmented Reality. In Proceedings of the 8th IEEE International Symposium on Mixed and Augmented Reality, Orlando, FL, USA, 19–22 October 2009; pp. 153–162. [CrossRef]
29. Boge, T.; Ma, O. Using Advanced Industrial Robotics for Spacecraft Rendezvous and Docking simulation. In Proceedings of the IEEE International Conference on Robotics and Automation, Shanghai, China, 9–13 May 2011; pp. 1–4. [CrossRef]
30. Shah, S.; Dey, D.; Lovett, C.; Kapoor, A. AirSim: High-Fidelity Visual and Physical Simulation for Autonomous Vehicles. In *Field and Service Robotics, Proceedings of the 11th Conference on Field and Service Robotics, Zürich, Switzerland, 13–15 September 2017*; Springer: Cham, Switzerland; pp. 621–635. [CrossRef]
31. Ma, O.; Wang, J.; Misra, S.; Liu, M. On the Validation of SPDM Task Verification Facility. *J. Robot. Syst.* **2004**, *21*, 219–235. [CrossRef]

32. Miquet, C. New test method for reproducible real-time tests of ADAS ECUs: "Vehicle-in-the-Loop" connects real-world vehicles with the virtual world. *Proceedings* **2014**, 575–589. [CrossRef]
33. Roth, E.; Dirndorfer, T.; Neumann-Cosel, K.V.; Fischer, M.O.; Ganslmeier, T.; Kern, A.; Knoll, A. Analysis and validation of perception sensor models in an integrated vehicle and environment simulation. In Proceedings of the 22nd Enhanced Safety of Vehicles Conference, Washington, DC, USA, 13–16 June 2011.
34. Carlson, R.E.; Fritsch, F.N. Monotone Piecewise Bicubic Interpolation. *SIAM J. Numer. Anal.* **1985**, *22*, 386–400. [CrossRef]
35. Fritsch, F.N.; Kahaner, D.; Moler, C.; Nash, S. Numerical Methods and Software. *Math. Comput.* **1990**, *55*, 865. [CrossRef]
36. Ponn, T.; Müller, F.; Diermeyer, F. Systematic analysis of the sensor coverage of automated vehicles using phenomeno-logical sensor models. In Proceedings of the IEEE Intelligent Vehicles Symposium (IV), Las Vegas, NV, USA, 19 October 2020–13 November 2020.
37. Schöner, H.-P. Simulation in development and testing of autonomous vehicles. In Proceedings of the 18th Internationales Stuttgarter Symposiium, Wiesbaden, Germany; 2018; pp. 1083–1095. [CrossRef]
38. Stolz, M.; Nestlinger, G. Fast generic sensor models for testing highly automated vehicles in simulation. *Elektrotechnik Informationstechnik* **2018**, *135*, 365–369. [CrossRef]
39. Schlager, B.; Muckenhuber, S.; Schmidt, S.; Holzer, H.; Rott, R.; Maier, F.M.; Saad, K.; Kirchengast, M.; Stettinger, G.; Watzenig, D.; et al. State-of-the-Art Sensor Models for Virtual Testing of Advanced Driver Assistance Systems/Autonomous Driving Functions. *SAE Int. J. Connect. Autom. Veh.* **2020**, *3*, 233–261. [CrossRef]
40. Genser, S.; Muckenhuber, S.; Solmaz, S.; Reckenzaun, J. Development and experimental validation of an intelligent camera model for automated driving. *Sensors* **2021**, *21*, 7583. [CrossRef]
41. Hanke, T.; Hirsenkorn, N.; Dehlink, B.; Rauch, A.; Rasshofer, R.; Biebl, E. Generic architecture for simulation of ADAS sensors. In Proceedings of the 16th International Radar Symposium (IRS), Dresden, Germany, 24–26 June 2015; pp. 125–130. [CrossRef]
42. Hirsenkorn, N.; Hanke, T.; Rauch, A.; Dehlink, B.; Rasshofer, R.; Biebl, E. A non-parametric approach for modeling sensor behavior. In Proceedings of the 16th International Radar Symposium (IRS), Dresden, Germany, 24–26 June 2015; pp. 131–136. [CrossRef]
43. Moon, S.; Moon, I.; Yi, K. Design, tuning, and evaluation of a full-range adaptive cruise control system with collision avoidance. *Control Eng. Pr.* **2009**, *17*, 442–455. [CrossRef]
44. Jiang, Y. Modeling and simulation of adaptive cruise control system. *arXiv preprint*. arXiv:2008.02103, 2020.
45. Funk Drechsler, M.; Sharma, V.; Reway, F.; Schütz, C.; Huber, W. Dynamic vehicle-in-the-loop: A novel method for testing automated driving functions. *SAE Int. J. Connect Autom. Veh.* **2022**, *5*, 367–380. [CrossRef]
46. Coulter, R.C. *Implementation of the Pure Pursuit Path Tracking Algorithm*; Carnegie-Mellon UNIV Pittsburgh PA Robotics INST: Pittsburgh, PA, USA, 1992.
47. Groh, K.; Wagner, S.; Kuehbeck, T.; Knoll, A. Simulation and Its Contribution to Evaluate Highly Automated Driving Functions. *SAE Int. J. Adv. Curr. Prac. Mobility* **2019**, *1*, 539–549. [CrossRef]
48. Mongiardini, M.; Ray, M.; Anghileri, M. Acceptance criteria for validation metrics in roadside safety based on repeated full-scale crash tests. *Int. J. Reliab. Saf.* **2010**, *4*, 69. [CrossRef]
49. VMAD-31st SG2 Session–Transport–Vehicle Regulations–UNECE Wiki. Available online: wiki.unece.org/display/trans/VMAD-31st+SG2+session (accessed on 6 November 2022).
50. Dona, R.; Ciuffo, B. Virtual Testing of Automated Driving Systems. A Survey on Validation Methods. *IEEE Access* **2022**, *10*, 24349–24367. [CrossRef]
51. Dona, R.; Vass, S.; Mattas, K.; Galassi, M.C.; Ciuffo, B. Virtual Testing in Automated Driving Systems Certification. A Longitudinal Dynamics Validation Example. *IEEE Access* **2022**, *10*, 47661–47672. [CrossRef]
52. Feng, D.; Feng, M.Q.; Ozer, E.; Fukuda, Y. A vision-based sensor for noncontact structural displacement measurement. *Sensors* **2015**, *15*, 16557–16575. [CrossRef] [PubMed]
53. Kamal, M.; Srivastava, G.; Tariq, M. Blockchain-Based Lightweight and Secured V2V Communication in the Internet of Vehicles. *IEEE Trans. Intell. Transp. Syst.* **2020**, *22*, 3997–4004. [CrossRef]
54. Guha, S.; Shim, K. A Note on Linear Time Algorithms for Maximum Error Histograms. *IEEE Trans. Knowl. Data Eng.* **2007**, *19*, 993–997. [CrossRef]
55. Schyr, C.; Inoue, H.; Nakaoka, Y. Vehicle-in-the-Loop Testing—A Comparative Study for Efficient Validation of ADAS/AD Functions. In Proceedings of the International Conference on Connected Vehicle and Expo (ICCVE), Lakeland, FL, USA, 7–9 March 2022; pp. 1–8. [CrossRef]

electronics

Article

Design and Analysis of the Trajectory of an Overtaking Maneuver Performed by Autonomous Vehicles Operating with Advanced Driver-Assistance Systems (ADAS) and Driving on a Highway

Josue Ortega [1,2,*], Henrietta Lengyel [2] and Jairo Ortega [3]

1 Logistics and Transportation Department, Universidad Técnica Particular de Loja, San Cayetano Street, Loja 110107, Ecuador
2 Department of Vehicle Technology, Faculty of Transportation Engineering and Vehicle Engineering, Budapest University of Technology and Economics, Műegyetem rkp. 3, 1111 Budapest, Hungary
3 Department of Transport Technology and Economics, Faculty of Transportation Engineering and Vehicle Engineering, Budapest University of Technology and Economics, Műegyetem rkp. 3, 1111 Budapest, Hungary
* Correspondence: jdortega19@utpl.edu.ec

Abstract: Overtaking is a maneuver that consists of passing another vehicle traveling on the same trajectory, but at a slower speed. Overtaking is considered one of the most dangerous, delicate and complex maneuvers performed by a vehicle, as it requires a quick assessment of the distance and speed of the vehicle to be overtaken, and also the estimation of the available space for the maneuver. In particular, most drivers have difficulty overtaking a vehicle in the presence of vehicles coming on other trajectories. To solve these overtaking problems, this article proposes a method of performing safe, autonomous-vehicle maneuvers through the PreScan simulation program. In this environment, the overtaking-maneuver scenario (OMS) is composed of highway infrastructure, vehicles and sensors. The proposed OMS is based on the solution of minimizing the risks of collision in the presence of any oncoming vehicle during the overtaking maneuver. It is proven that the overtaking maneuver of an autonomous vehicle is safe to perform through the use of advanced driver-assistance systems (ADAS) such as adaptive cruise control (ACC) and technology-independent sensors (TIS) that detect the driving environment of the maneuver.

Keywords: autonomous vehicle; communication; ADAS; model predictive control; ACC

Citation: Ortega, J.; Lengyel, H.; Ortega, J. Design and Analysis of the Trajectory of an Overtaking Maneuver Performed by Autonomous Vehicles Operating with Advanced Driver-Assistance Systems (ADAS) and Driving on a Highway. *Electronics* **2023**, *12*, 51. https://doi.org/10.3390/electronics12010051

Academic Editors: Calin Iclodean, Bogdan Ovidiu Varga and Felix Pfister

Received: 9 November 2022
Revised: 19 December 2022
Accepted: 20 December 2022
Published: 23 December 2022

1. Introduction

Over time, the process of integrating automation systems into commercial vehicles is becoming more and more evident. Among the triggers for this integration are, in particular, traffic accidents, and specifically, collisions, which are mostly related to human mistakes when performing high-risk maneuvers, such as overtaking and obstacle avoidance. Basically, overtaking is passing another vehicle traveling on the same trajectory at a lower speed [1]. To avoid a collision, the driver performing the maneuver calculates the time remaining before colliding with the vehicle approaching from the opposite direction. Performing these calculations in real time is difficult, because it means that the human driver must take into account situations such as traffic or the speed of other vehicles [2].

Therefore, different manufacturers and research groups around the world are working on the development of systems to improve driving in urban areas and highways [3]. Autonomous vehicles (AVs) are part of this development, due to their role as an intelligent transportation system (ITS), in which control and communication techniques such as advanced driver-assistance systems (ADAS) are applied [4]. The research into cooperative maneuvers such as overtaking between AVs is one of the most challenging areas in the field of ITS and ADAS systems, since overtaking is a high-risk maneuver. Decisions made in the execution of overtaking maneuvers depend not only on the state of the vehicle, but also on many other variables, such as environmental factors, surrounding conditions, the distance of the vehicle to curves and intersections, the position and speed of other vehicles, etc.

The ADAS systems also provide control systems such as adaptive cruise control (ACC); this system allows the vehicle to drive in an autonomous mode once a cruising speed and a safety distance have been set. Thanks to this device, the vehicle accelerates and brakes in accordance with the traffic conditions present, and the driver has the opportunity to relax in the driver's seat while keeping the situation under control [5,6].

Several studies have suggested applications associated with ADAS systems to perform an overtaking maneuver. Similarly, many studies have explored the use of control systems such as model predictive control (MPC), algorithms, and sensors such as cameras, the Global Positioning System (GPS) and RADAR, to perform an autonomous overtaking maneuver.

However, no study has focused solely on the performance of overtaking maneuvers with ADAS systems such as ACC and technology-independent sensors (TIS), whose operations are a combination of RADAR and LIDAR.

In this study, in order to provide parity between the different elements that constitute the overtaking situation, an overtaking-maneuver scenario (OMS) with optimal trajectory generation for autonomous vehicle overtaking without collisions, was developed. This OMS is proposed for a three-lane highway, in which several elements, such as the highway infrastructure, sensors, and vehicles are combined. PreScan was used to build and recreate the OMS simulation features for the overtaking maneuver.

This study explores how an autonomous vehicle overtakes another vehicle traveling io the same trajectory at a lower speed. This includes comprehending what elements are involved and how sensors, highway infrastructure, vehicles or their combination contribute to the successful performance of the overtaking maneuver. Furthermore, it implies understanding in the near future how the ADAS system will adapt to real-life interaction with AVs.

This article is structured as follows. Section 2 discusses the overtaking maneuver and the different control methods to improve active safety. Section 3 describes the OMS. Section 4 presents the OMS results. Section 5, together with the limitations of the study, discusses and analyses the results obtained. Finally, the last section presents the conclusions of the study.

2. Literature Review

Driving systems have become more complex with time, leading to the development of new technologies in order to improve driving performance and stability [7]. This is also the case with AVs, which provide numerous benefits to people who, for some reason, cannot or should not drive. Such a solution could lead to improved access to education, employment opportunities, and increased productivity, by reducing the burden on the families and friends of people who cannot drive [8,9].

The AVs also have numerous advantages, such as reducing traffic congestion, reducing fuel consumption, improving public transportation services, and facilitating maneuvers [10]. The application of new technologies in maneuvers implies that interactions between AVs and environmental sensing may improve safety, efficiency and sustainability [11,12].

One of the most challenging maneuvers is overtaking, because the maneuver does not depend only on the state of the vehicle but also on many other variables, such as environmental conditions, the distance of the vehicle to curves or intersections, the type of highway, the traffic signals, and the position and speed of other vehicles. Overtaking, considered a common practice between vehicles, involves the consideration of a series of factors by the users. These factors include the dimensions of the highway, the speed of both the vehicle performing the maneuver and the vehicle to be overtaken, and also the applicable traffic laws and regulations. [13].

Regarding autonomous overtaking, the main issue is to determine the onboard system carried by the AVs. The onboard system determines how decision making is planned during the operation of the AVs on the highway. The systems are also responsible for providing safety in various scenarios. To achieve successful overtaking, the vehicle with the onboard system must calculate parameters such as speed, distance and time [14].

One of the most popular onboard systems used in AVs is the adaptive cruise control (ACC). The ACC system bases its operation on sensors and RADARs, which are used to control the speed and position of the AVs automatically. A notable feature of ACC systems is the possibility of detecting and analyzing avoidable collisions with other vehicles in a specific environment [15]. The emergency brake is activated if the system detects that the collision is inevitable even with evasive maneuvers [16].

Yi et al. [17] proposed a control scheme that uses a full-range ACC with braking, for a collision-avoidance system. The algorithm can perform user-like driving in high and low-speed gears. Moon et al. [18] also used an algorithm comprised of full-range ACC for collision avoidance. With this algorithm, the vehicle in question operates in three modes: (1) comfort, (2) high, and (3) severe braking.

Another method for improving overtaking maneuvers is the model predictive control (MPC). MPC is a control algorithm that calculates in real time a sequence of future actions of a system, in order to provide a basis for optimizing future control actions [19,20]. Wang et al. [21] used the MPC as a method for autonomous navigation of unmanned surface vessels (USVs). MPC creates commands for precise tracking control of the USVs, to avoid collisions with objects detected around. The study carried out by Wang et al. [22] proposed a method to minimize the risk of collisions between a group of vehicles. This method uses an MPC controller and a coordinated brake control (CBC) that further helps the vehicle to reduce the impact when a collision with another vehicle is unavoidable. Lin et al. [23] presented a collision-avoidance system that predicts the obstacle trajectory using a trajectory-replanning controller integrated with driver characteristics, longitudinal control and vehicle speed. Wnag et al. [24] studied a new method that consists of the development of a trajectory-tracking controller and a path planner to avoid collisions of autonomous vehicles, based on active front-steering systems (AFS). The MPC can also be used as a 3D tracking-control tool. This is done through a symplectic pseudospectral optimal-control method that performs tracking of the reference trajectory [25].

Bae and Lee [26] suggested an adaptive overtaking-assistance system (VAOAS) based on vehiclar ad hoc networks (VANET). The VAOAS system calculates the safe passing distance and the passing sight-distance, based on other vehicles' location and speed, and provides either a warning or an overtaking-approval message. Mei et al. [27] presented a robust motion-detection and planning method for avoiding vehicles parked on the roadside. The method uses lidar and long-range radar sensors in a complementary manner for obstacle detection. Zhu et al. et al. [28] developed a system that uses a fusion algorithm and dynamic scenes to detect overtaking vehicles. Their modelling technique for the identification of overtaking vehicles combines dynamic-scene modeling, hypothesis testing, multi-scale means-displacement-based information fusion, and bandwidth density. The combination of these techniques results in a reliable model for detecting overtaking vehicles [29].

A novel technique for performing overtaking maneuvers is through a heuristic algorithm. With this algorithm, the AVs are able to determine an optimal distance and speed based on the relative position of other vehicles involved in the overtaking maneuver [14].

Yang et al. [30] designed and validated a decision-making system to generate overtaking decisions by using a deep neural network (DNN). With this system, the DNN can learn the decisions of humans in complex situations, and implement them in the AVs. An automatic overtaking-maneuver can be performed through vision sensors (cameras, GPS or a global navigation satellite system (GNSS) and the inertial measurement unit), and longitudinal and lateral control- systems based on fuzzy controllers [31,32]. The study by Anindyaguna et al. [33] also proposed using a fuzzy controller, but with the difference that it used a camera combined with two components: GPS and radio control (RC).

Petrov et al. [34] proposed a nonlinear adaptive controller using third-order reference trajectories for automated two-vehicle overtaking maneuvers. The trajectories and the derived kinematic relationships between the vehicles were used to establish a rationale for designing a feedback control based on kinematics. Andersen et al. [35] developed a method to perform overtaking by generating a trajectory planner. With this method, the planner is guided by a real-time non-linear system that captures trajectory information, so that AVs can overtake safely.

A modern approach to vehicle identification during an overtaking maneuver is through the use of a Kinect system. Through red, green and blue, plus depth (RGB-D) data collected within traffic scenes, the action of the detected vehicle is identified and tracked at the moment of the overtaking maneuver [36].

Olaverri-Monreal et al. [37] proposed the see-through system (STS). With STS, AVs can have a perspective view of another vehicle. To obtain an even clearer view, video transmission technology and VANET are used to stream video between the involved vehicles in the overtaking maneuver.

The scientific literature on AVs and the different systems for improving driving performance and stability shows a gap in the development of scenarios involving overtaking maneuvers with AVs. This article aims to develop an OMS through a simulation program of a safe overtaking-maneuver that combines highway infrastructure, vehicles and systems and technology-independent sensors (TIS).

3. Modeling Approach

This section presents in detail the simulation conditions to be fulfilled in an overtaking-maneuver scenario (OMS) with autonomous vehicles. Subsequently, a description of the construction of the OMS is included. Finally, the auxiliary settings implemented with MATLAB/SIMULINK software are explained, in order to establish the simulation parameters of the OMS.

Figure 1 shows the flowchart of our study. This flowchart illustrates the procedure that the OMS follows to carry out an overtaking maneuver.

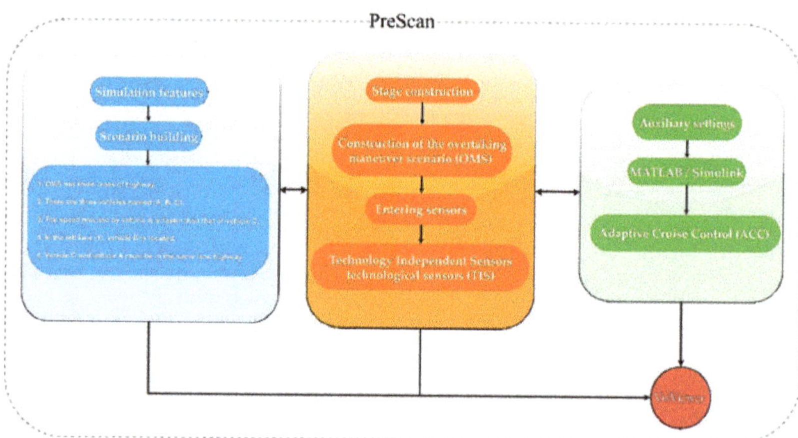

Figure 1. The flowchart of the method developed for the overtaking maneuver in PreScan.

3.1. Simulation Features

The first step in the scenario-building process is to specify the conditions of the simulation. Next, the conditions necessary for the OMS simulation with autonomous vehicles are specified:

1. The OMS has three lanes of highway;
2. There are three vehicles, named A, B, and C;
3. The speed reached by vehicle A is faster than that of vehicle C;
4. In the left lane (1), vehicle B is located;
5. Vehicle C and vehicle A must be in the same lane of the highway.

Once the simulation conditions are known, a virtual experiment can be constructed, using the PreScan simulation function, in which different simulation elements such as the highway infrastructure, vehicles, and sensors are combined. The following subsections explain in detail how an OMS is constructed and which control systems are used to carry out the simulation.

3.1.1. Stage Construction

The first step in the construction of the OMS is to determine the number of lanes of the highway. In this OMS, a three-lane highway was chosen, with lanes designated as the left lane (1), the center lane (2) and the right lane (3) (see Figure 2).

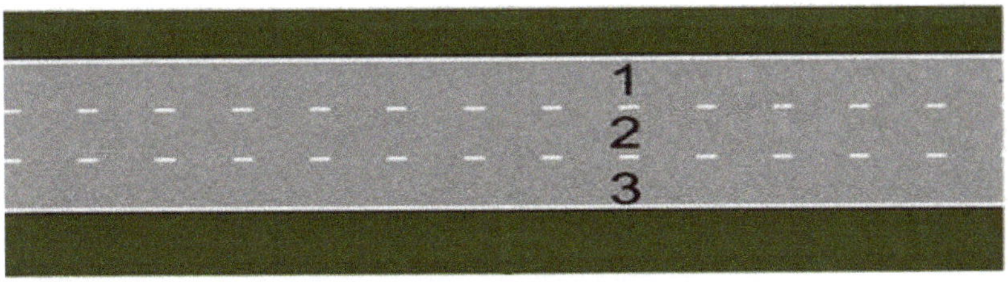

Figure 2. Scheme and layout of OMS highway infrastructure in PreScan.

This type of layout was chosen because in this type of lane, vehicles travelling in both directions can use the center lane for passing, and thus approaching traffic may intend to make the same maneuver at the same time.

In lane 2 there are two vehicles: vehicle A, which is responsible for carrying out the overtaking maneuver, and vehicle C, which is ahead of vehicle A. Since both vehicles (A and C) are located in lane 2, both vehicles move io the same trajectory, which goes from X to Z (see Figure 3).

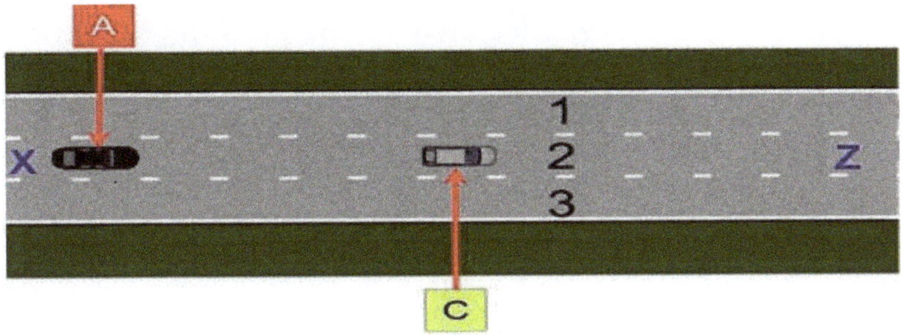

Figure 3. OMS with vehicles located in the center lane (2).

In lane 1, there is one vehicle, vehicle B, which has a trajectory that goes from Z to X. As shown in Figure 4, the trajectory of vehicle B going from Z to X is the opposite of that of vehicles A and C, which goes from X to Z: i.e., the trajectory in lane 1 is different from the trajectory in lane 2.

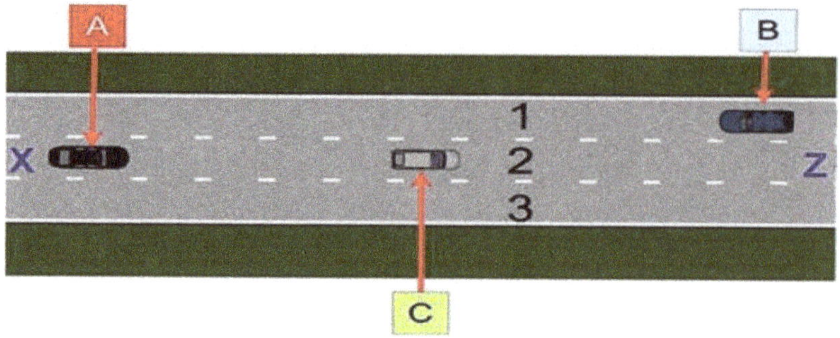

Figure 4. OMS with vehicles located in the left lane (1) and in the center lane (2).

Once the highway infrastructure and the vehicles (A, C, B) that interact with the OMS have been determined, it is possible to determine the sensors that operate with the autonomous vehicle in the OMS.

Vehicle A, being responsible for performing the overtaking maneuver, is equipped with two technology-independent sensors (TIS), and each of these sensors is labeled with a color in order to identify it in the OMS. The first TIS sensor is labeled yellow (TIS1), while the other TIS sensor is labeled red (TIS2) (see Figure 5).

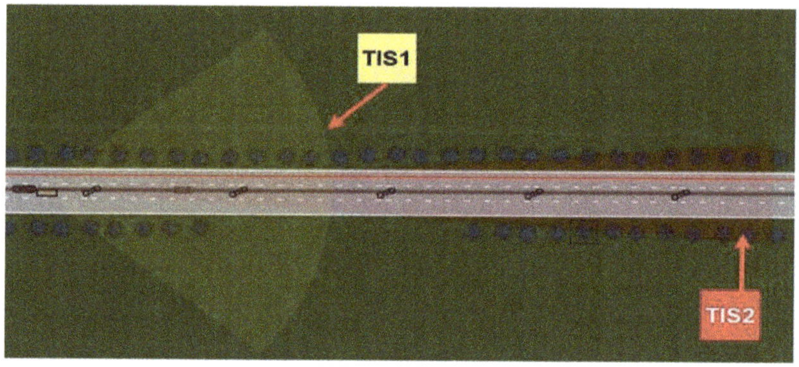

Figure 5. Short-range sensor (TS1) and long-range sensor (TS2) in vehicle A.

The operation of the TIS sensor is based on the combination of the technology of two sensors: RADAR and LIDAR. This allows the sensor to have a general active scanning and not be limited to a specific technology (such as radar, lidar or laser scanner).

TIS sensors can help to use and understand other sensors for active scanning, as they have a strong scene-scanning performance. This allows them to receive information about the environment and the obstacles.

As vehicle A moves along the highway path, it tracks and detects the vehicles that are in the OMS. The range distance assigned for short-range detection is given by TS1, which has a range of 60 m. For long-range detection, the TS2 is designated. This sensor has a range of 150 m.

3.1.2. Auxiliary Settings

After building the OMS with all the necessary elements (highway infrastructure, vehicles and sensors), it is necessary to execute auxiliary adjustments; i.e., using a graphical programming software such as MATLAB/Simulink, an analysis of the OMS is carried out through a graphical user interface (GUI). This is achieved through the interactive use of the block-diagram systems that make up the OMS (see Figure 6).In addition, through the GUI it is possible to add ADAS, such as ACC.

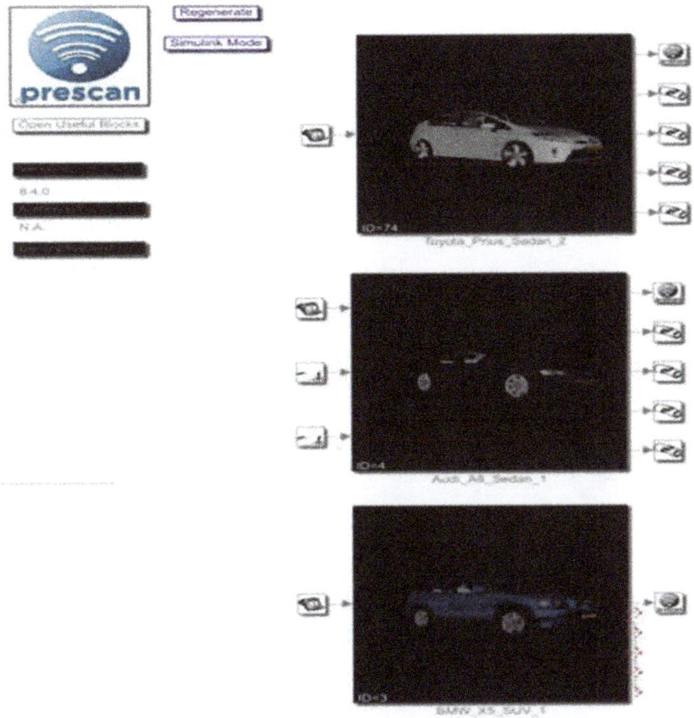

Figure 6. Compilation sheet in MATBLAB/Simulink.

With ACC technology, vehicle A may be able to modify its speed, depending on whether there are vehicles ahead traveling at a lower speed. In the case of failure, the user can also enter new values for the identified problem areas.

If the OMS contains all the road infrastructure, vehicles and sensors needed for the simulation, a compilation sheet is created, which allows for the launching of the experiment. Figure 7 Shows a 3D scene during simulation, through a PreScan window called VisViewer.

Figure 7. Viewpoints on PreScan.

Through VisViewer, the OMS scene can be handled from three types of viewpoints, such as: (1) the default viewpoints, (2) the PreScan libraries viewpoints, and (3) the VisViewer viewpoint (see Figure 7). Each of these viewpoints provides a view of the overtaking maneuver, which results in a better visualization of the OMS analysis.

PreScan provides a preprocessing system for defining the OMS, and an execution environment to carry it out. An accurate and complete simulation was performed to achieve an OMS that fulfills all the guidelines established by the simulation conditions. Furthermore, an analysis was performed through the GUI to optimize the OMS simulation.

4. Results

The results set out in this section consist of three phases. Each of these phases includes a description of the elements that compose the OMS (see Figure 8).

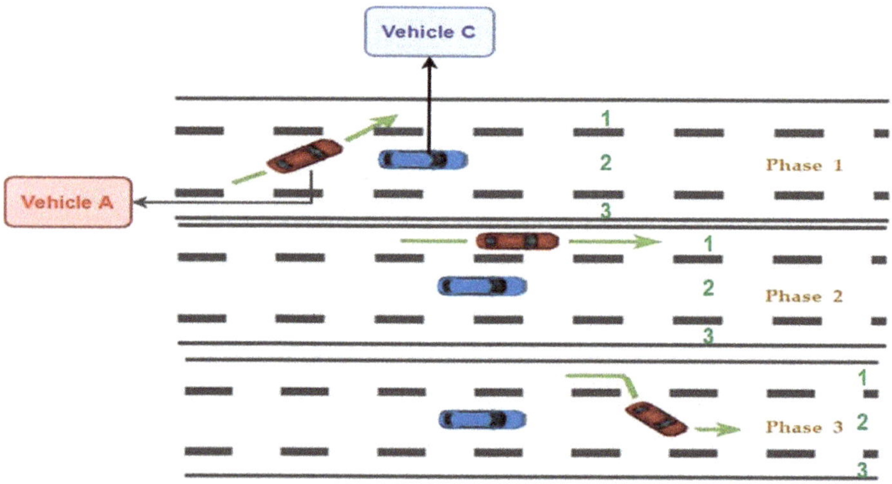

Figure 8. Scheme of overtaking-maneuver phases.

Phase 1 shows the behavior of vehicle A upon starting to carry out the overtaking maneuver. In addition, during Phase 1, vehicle A is analyzed when detecting other vehicles (B and C) through the TIS sensors and the ACC controller.

Phase 2 shows the change of lane made by vehicle A at the end of Phase 1. In this change of lane, which is from lane 2 to lane 1, vehicle A moves until it manages to pass vehicle C. As a result, Phase 2 ends and Phase 3 begins.

Finally, Phase 3 describes how vehicle A returns to its original lane, which is lane 2. This results in the end of the overtaking maneuver.

4.1. Phase 1

In Phase 1, vehicle A, which is located in lane 2, starts moving along its trajectory from X to Z at an initial speed of 54 km/h (see Figure 9). As vehicle A moves along its trajectory, (X–Z), its speed progressively increases until it eventually encounters another vehicle ahead, which would cause it to slow down.

In phase 1, both the TIS sensors and the ACC controller in vehicle A continuously monitor the OMS.

Figure 10 shows two TIS sensors in vehicle A. Each of the TIS sensors is labeled with a color: the first TIS sensor is labeled red, and has a range of 60 m; d the other TIS sensor is labeled blue, and has a range of 150 m.

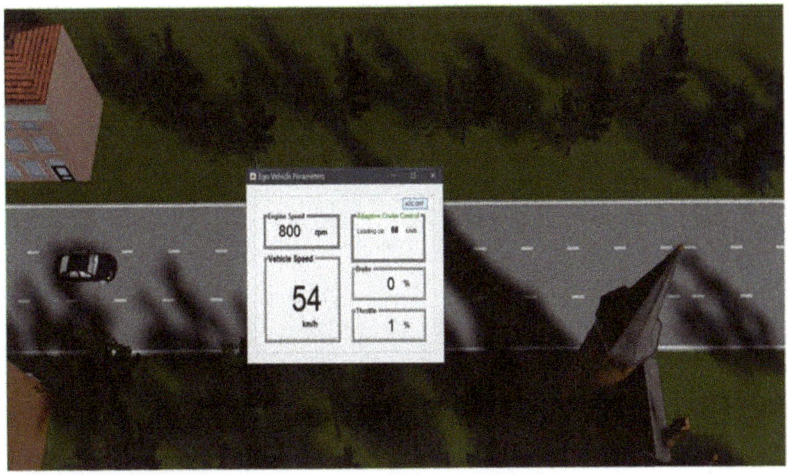

Figure 9. Ego-vehicle parameters showing lower speed of the vehicle A.

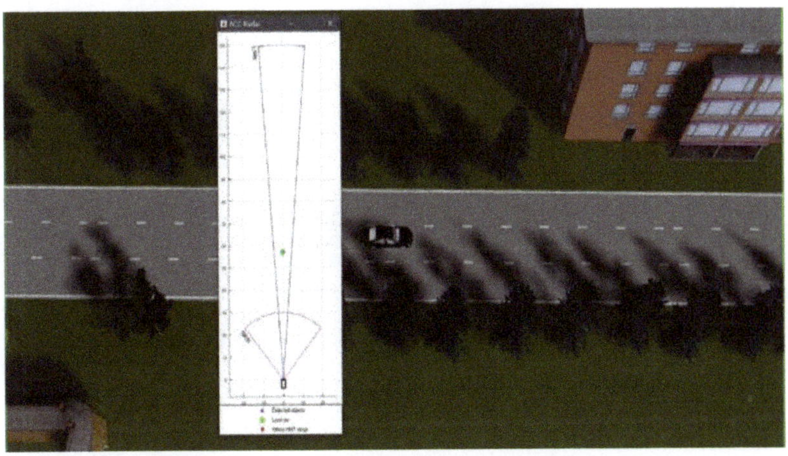

Figure 10. Vehicle A with TIS sensors in PreScan.

Vehicle C is represented by a green dot when it is detected by the TIS sensors in vehicle A.

Once the vehicles have been identified through the TIS sensors, a signal is sent to the ACC controller, which enters demanded-headway-time (HWT) mode (see Figure 11), i.e., vehicle A brakes and starts to reduce its speed, and thus avoids a collision with vehicle C.

4.2. Phase 2

Regarding the ACC controller and TIS sensors, as mentioned before, they also start to work when vehicle A starts to move along its trajectory (X–Z). Figure 12 shows how vehicle A starts to reduce its speed from 59 km/h to 58 km/h when it approaches vehicle C. This speed reduction is made possible by the TIS sensors on vehicle A, which detect the vehicles ahead (B and C) and thus provide an overview of the OMS environment.

Figure 11. Vehicle A with TIS sensors and ACC controller.

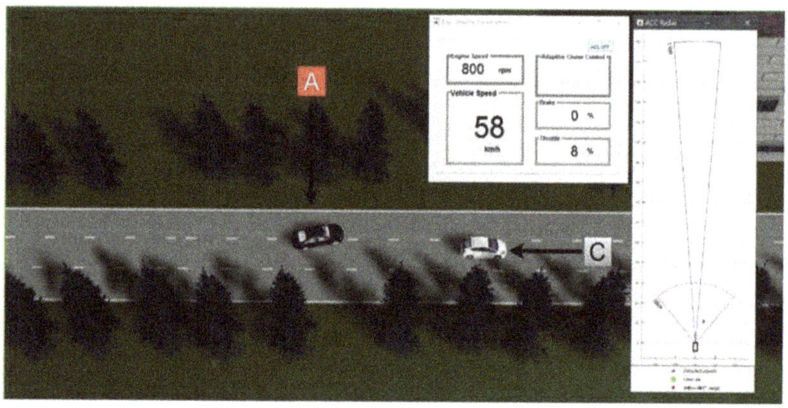

Figure 12. Vehicle A moving from lane 2 to lane 1.

In Phase 2, vehicle A is behind vehicle C, while vehicle B continues to move along its path (Z–X) (see Figure 4). In Phase 2, vehicle A still continues detecting both vehicles C and B through the TIS sensors and ACC control.

As shown in Figure 12, vehicle A starts to change lane (from lane 2 to lane 1), and thus begins the overtaking maneuver. This change of lane happens only when vehicle A, through the TIS sensors and the ACC controller, detects that there is no danger of collision with vehicle C, and vehicle B is already ahead of vehicle C and A.

Vehicle A, once in lane 1, is still following the same trajectory (X–Z). In Phase 2, vehicle A, which is moving in lane 1, starts to accelerate, which causes the speed increase until it overtakes vehicle C. As shown in Figure 13, vehicle A increases to a speed of 98 km/h in order to overtake vehicle C. Throughout the lane change, both the TIS sensors and the ACC controller keep working continuously to detect and act, in case there are vehicles ahead of vehicle A.

In Phase 2, when vehicle A detects that there are no vehicles ahead, it again sends a signal to the ACC controller, which causes the HWT mode to be deactivated.

4.3. Phase 3

Finally, in Phase 3, vehicle A, once it has overtaken vehicle C, can return to its original lane 2 (see Figure 14) to complete the overtaking maneuver. At the end of the overtaking maneuver, vehicle A can start accelerating, causing the speed to increase to 119 km/h, until it progressively moves away from vehicle C (see Figure 15).

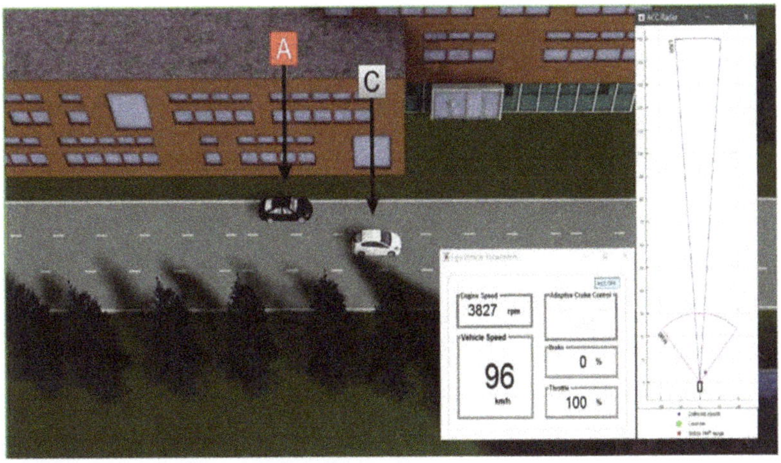

Figure 13. Second phase of the overtaking maneuver.

Figure 14. Vehicle A changing phase, from left lane to center lane.

Figure 15. Third phase of the overtaking maneuver.

5. Discussion

Single-lane highway overtaking is a common OMS that is present on most highways, and hence most studies that have been conducted on overtaking with autonomous vehicles opt for this type of OMS. However, when dealing with an OMS with three lanes of highway and with vehicles that have a different trajectory than the vehicle performing the overtaking maneuver, the operation of typical overtaking on a single-lane highway is not guaranteed. To solve this problem, in this article an analysis of the overtaking maneuver with autonomous vehicles on a three-lane highway is presented, with a vehicle on a trajectory opposite to that of the vehicle performing the overtaking maneuver.

The results obtained demonstrate that the PreScan software is a versatile tool, which is able to build scenarios with highway infrastructure, AVs and ADAS systems. This allows obtaining data close to reality without the need to use real vehicles or physical infrastructure. The OMS simulation showed how vehicle A starts to slow down gradually when it approaches vehicle C. This speed change is due to the assistance of the ACC system. The results also showed that with the help of TIS sensors, vehicle A can continuously monitor its area to avoid a collision with both vehicle C and vehicle B.

From the results obtained in the OMS, for an autonomous vehicle to perform an overtaking maneuver of another vehicle that is in the same lane of the highway moving at a lower speed, and at the same time having to deal with vehicles that are in an opposite lane to the vehicle performing the maneuver, it is necessary that all the analyzed phases (1, 2, 3) fulfill the conditions of the simulation, in order for sensors such as TIS and the ACC controller to detect the driving environment. Under these conditions, an autonomous vehicle can perform the overtaking maneuver.

Figure 9 shows vehicle A with an initial speed of 54 km/h increasing its speed to a maximum speed of 59 km/h (see Figure 11) to catch up with vehicle C. Once vehicle A catches up with vehicle C, it starts the overtaking maneuver. The speed of vehicle A is progressively reduced, while vehicle C maintains a speed of 58 km/h, as shown in Figure 12. Vehicle A performs the lane-change maneuver (lane 2 to lane 1). To perform this lane-change maneuver, vehicle A increases its speed to 96 km/h (see Figure 13). When vehicle A achieves the overtaking maneuver with respect to vehicle C, vehicle A returns to lane 2, and once the overtaking maneuver has been performed, vehicle A increases its speed to 119 km/h, to move away from vehicle C. This can be compared to an overtaking maneuver performed in the literature by Né-meth et al. [38], who designed a scenario with three lanes and with the overtaking maneuver in lane 2. The overtaking scenario proposed by the author starts with vehicle A, whose initial speed is 110 km/h, which is reduced once it reaches vehicle C, and maintains similar speeds of 90 km/h, subsequently increasing to a speed higher than that of vehicle C, to perform the overtaking maneuver. This means that in our research, as in Nemeth's, in order to perform the overtaking maneuver, it is necessary that the vehicle that performs the overtaking maneuver (vehicle A) performs speed changes (acceleration and deceleration), and these speed differences depend on the speed of the vehicle to be overtaken, which is vehicle C.

The limitations of this study are that the speed and lateral movement of vehicles B and C are constant in the scenario simulation, thereby not generating speed changes. In addition, the scenario constructed in this work only considers the use of three vehicles on a three-lane highway, and does not cover the inclusion of other elements such as traffic signs, pedestrians or more vehicles. These topics could offer a fascinating continuation of this work.

6. Conclusions

An autonomous vehicle is a vehicle with a complex system, in which many technologies and different disciplines such as artificial vision, mechanics, and even geopositioning, are involved. AVs convert and expand the present circumstances through small and subtle details entirely into the future, and modify various aspects already known in vehicles, to achieve better safety or comfort. This article aims to present a method of performing safe maneuvers with AVs through the PreScan simulation program, in which the OMS is composed of highway infrastructure, vehicles and sensors.

The study presented in this article is governed by modeling approaches such as stage construction and auxiliary settings. Each of these models employs a program for the development of the scenario. For scenario construction, a simulation program is used, in order to create highway infrastructure. For auxiliary settings, a graphical programming program (MATLAB/Simulink) was applied, to modify and interpret a graphical user interface (GUI). Moreover, each methodology is governed by simulation conditions that must be fulfilled in order to perform the overtaking maneuver with AVs.

The proposed OMS can incorporate a safe overtaking driving-system in the front of vehicles that have an opposite trajectory to the vehicle that carries out the maneuver. The proposed OMS can be employed as a fully automated system, using ADAS systems such as the ACC controller and TIS sensors, to detect the driving environment for performing the overtaking maneuver.

The OMS complies with EuroNCAP tests. These tests mention that the autonomous vehicle-assistance evaluations consist of an ACC system that has a lateral assistance to control the position of the vehicle when moving in a fully marked lane and at a desired test speed.

An autonomous vehicle that has an ACC controller and TIS sensors can perform an overtaking maneuver while controlling its speed and recognizing vehicles on the same trajectory or on the opposite trajectory.

Considering further research, it is proposed that the scenario used here should be complemented with different elements on the highway, such as traffic signs, traffic signals, pedestrians, or more vehicles, to study the behavior of the maneuver when interacting with more elements. Another interesting extension could be the implementation of several sensors in more vehicles and for the highway infrastructure to have a continuous data network that can assist the autonomous vehicle to perform and carry out different actions based on the information from the environment. In addition, adding other ADAS systems that help to analyze the behavior of the vehicle which has systems that help to perform different maneuvers.

Author Contributions: Conceptualization, J.O. (Jairo Ortega); Methodology, J.O. (Josue Ortega); Formal analysis, J.O. (Josue Ortega) and H.L.; Investigation, H.L.; Resources, H.L.; Data curation, J.O. (Josue Ortega); Writing—original draft, J.O. (Josue Ortega) and J.O. (Jairo Ortega); Writing—review & editing, H.L. and J.O. (Jairo Ortega). All authors have read and agreed to the published version of the manuscript.

Funding: This research received no external funding.

Acknowledgments: Thanks to the Department of Transport Technology and Economics of the Faculty of Transportation Engineering and Vehicle Engineering of Budapest University of Technology and Economics, and Universidad Técnica Particular de Loja from the Republic of Ecuador for their collaboration and support.

Conflicts of Interest: The authors declare no conflict of interest.

References

1. Singh, S. *Critical Reasons for Crashes Investigated in the National Motor Vehicle Crash Causation Survey*; National Center for Statistics and Analysis: Washington, DC, USA, 2018.
2. Hegeman, G.; Van Der Horst, R.; Brookhuis, K.A.; Hoogendoorn, S.P. Functioning and Acceptance of Overtaking Assistant Design Tested in Driving Simulator Experiment. *Transp. Res. Rec.* **2007**, *2018*, 45–52. [CrossRef]
3. Milanés, V.; Naranjo, J.E.; González, C.; Alonso, J.; García, R.; de Pedro, T. Sistema de Posicionamiento Para Vehiculos Autonomos. *Rev. Iberoam. Automática e Informática Ind. RIAI* **2008**, *5*, 36–41. [CrossRef]
4. Pérez, J.; Milanés, V.; Alonso, J.; Onieva, E.; de Pedro, T. Adelantamiento Con Vehiculos Autónomos En Carreteras de Doble Sentido. *Rev. Iberoam. Automática e Informática Ind. RIAI* **2010**, *7*, 25–33. [CrossRef]
5. Van Arem, B.; Van Driel, C.J.G.; Visser, R. The Impact of Cooperative Adaptive Cruise Control on Traffic-Flow Characteristics. *IEEE Trans. Intell. Transp. Syst.* **2006**, *7*, 429–436. [CrossRef]
6. Wang, C.S.; Liu, D.Y.; Hsu, K.S. Simulation and Application of Cooperative Driving Sense Systems Using Prescan Software. *Microsyst. Technol.* **2018**, *7*, 1201–1210. [CrossRef]

7. Mazur, C.; Offer, G.J.; Contestabile, M.; Brandon, N.B. Comparing the Effects of Vehicle Automation, Policy-Making and Changed User Preferences on the Uptake of Electric Cars and Emissions from Transport. *Sustainability* **2018**, *10*, 676. [CrossRef]
8. Khan, J. Using ADAS Sensors in Implementation of Novel Automotive Features for Increased Safety and Guidance. In Proceedings of the 2016 3rd International Conference on Signal Processing and Integrated Networks (SPIN), Noida, India, 11–12 February 2016; pp. 753–758.
9. Seet, M.; Harvy, J.; Bose, R.; Dragomir, A.; Bezerianos, A.; Thakor, N. Differential Impact of Autonomous Vehicle Malfunctions on Human Trust. *IEEE Trans. Intell. Transp. Syst.* **2020**, *23*, 548–557. [CrossRef]
10. Lin, Y.C.; Lin, C.L.; Huang, S.T.; Kuo, C.H. Implementation of an Autonomous Overtaking System Based on Time to Lane Crossing Estimation and Model Predictive Control. *Electronics* **2021**, *10*, 2293. [CrossRef]
11. Lengyel, H.; Szalay, Z. Traffic Sign Anomalies and Their Effects To the Highly Automated and Autonomous Vehicles. In *Advanced Manufacturing and Repair Technologies in Vehicle Industry Monograph*; Budapest University of Technology and Economics: Budapest, Hungary, 2018; pp. 193–204, ISBN 978-83-945647-1-1.
12. Jing, P.; Huang, H.; Ran, B.; Zhan, F.; Shi, Y. Exploring the Factors Affecting Mode Choice Intention of Autonomous Vehicle Based on an Extended Theory of Planned Behavior—A Case Study in China. *Sustainability* **2019**, *11*, 1155. [CrossRef]
13. Vanholme, B.; Gruyer, D.; Lusetti, B.; Glaser, S.; Mammar, S. Highly Automated Driving on Highways Based on Legal Safety. *IEEE Trans. Intell. Transp. Syst.* **2013**, *14*, 333–347. [CrossRef]
14. Easa, S.M.; Diachuk, M. Optimal Speed Plan for the Overtaking of Autonomous Vehicles on Two-Lane Highways. *Infrastructures* **2020**, *5*, 44. [CrossRef]
15. Okuda, R.; Kajiwara, Y.; Terashima, K. A Survey of Technical Trend of ADAS and Autonomous Driving. In Proceedings of the Technical Program—2014 International Symposium on VLSI Technology, Systems and Application (VLSI-TSA), Hsinchu, Taiwan, 28–30 April 2014; pp. 0–3. [CrossRef]
16. Kaempchen, N.; Schiele, B.; Dietmayer, K. Situation Assessment of an Autonomous Emergency Brake for Arbitrary Vehicle-to-Vehicle Collision Scenarios. *IEEE Trans. Intell. Transp. Syst.* **2009**, *10*, 678–687. [CrossRef]
17. Yi, K.; Moon, S.W.; Lee, I.S.; Um, J.Y.; Moon, I. Design of a Full-Range ACC with Collision Avoidacne/Mitigation Braking. *IFAC Proc. Vol.* **2007**, *5*, 127–134. [CrossRef]
18. Moon, S.; Moon, I.; Yi, K. Design, Tuning, and Evaluation of a Full-Range Adaptive Cruise Control System with Collision Avoidance. *Control Eng. Pract.* **2009**, *17*, 442–455. [CrossRef]
19. Camacho, E.F.; Bordons, C. Control Predictivo: Pasado, Presente y Futuro. *Rev. Iberoam. Automática E Inf. Ind.* **2010**, *1*, 5–28. [CrossRef]
20. Liang, Y.; Li, Y.; Khajepour, A.; Huang, Y.; Qin, Y.; Zheng, L. A Novel Combined Decision and Control Scheme for Autonomous Vehicle in Structured Road Based on Adaptive Model Predictive Control. *IEEE Trans. Intell. Transp. Syst.* **2022**, *23*, 16083–16097. [CrossRef]
21. Wang, X.; Liu, J.; Peng, H.; Qie, X.; Zhao, X.; Lu, C. A Simultaneous Planning and Control Method Integrating APF and MPC to Solve Autonomous Navigation for USVs in Unknown Environments. *J. Intell. Robot. Syst. Theory Appl.* **2022**, *105*, 1–16. [CrossRef]
22. Wang, J.Q.; Li, S.E.; Zheng, Y.; Lu, X.Y. Longitudinal Collision Mitigation via Coordinated Braking of Multiple Vehicles Using Model Predictive Control. *Integr. Comput. Aided. Eng.* **2015**, *22*, 171–185. [CrossRef]
23. Lin, F.; Wang, K.; Zhao, Y.; Wang, S. Integrated Avoid Collision Control of Autonomous Vehicle Based on Trajectory Re-Planning and V2v Information Interaction. *Sensors* **2020**, *20*, 1079. [CrossRef]
24. Wnag, C.Y.; Zhao, W.Z.; Xu, Z.J.; Zhou, G. Path Planning and Stability Control of Collision Avoidance System Based on Active Front Steering. *Sci. China Technol. Sci.* **2017**, *60*, 1231–1243. [CrossRef]
25. Wang, X.; Liu, J.; Zhang, Y.; Shi, B.; Jiang, D.; Peng, H. A Unified Symplectic Pseudospectral Method for Motion Planning and Tracking Control of 3D Underactuated Overhead Cranes. *Int. J. Robust Nonlinear Control* **2019**, *29*, 2236–2253. [CrossRef]
26. Bae, I.H.; Lee, J.K. Design and Performance Evaluation of a VANET-Based Adaptive Overtaking Assistance System. *Lect. Notes Electr. Eng.* **2016**, *354*, 59–69. [CrossRef]
27. Mei, X.; Nagasaka, N.; Okumura, B.; Prokhorov, D. Detection and Motion Planning for Roadside Parked Vehicles at Long Distance. *IEEE Intell. Veh. Symp. Proc.* **2015**, *2015*, 412–418. [CrossRef]
28. Zhu, Y.; Comaniciu, D.; Pellkofer, M.; Koehler, T. Reliable Detection of Overtaking Vehicles Using Robust Information Fusion. *IEEE Trans. Intell. Transp. Syst.* **2006**, *7*, 401–414. [CrossRef]
29. Wang, J.; Bebis, G.; Miller, R. Overtaking Vehicle Detection Using Dynamic and Quasi-Static Background Modeling. In Proceedings of the 2005 IEEE Computer Society Conference on Computer Vision and Pattern Recognition (CVPR'05)—Workshops, San Diego, CA, USA, 21–23 September 2006; p. 64. [CrossRef]
30. Yang, J.; Lee, S.; Lim, W.; Sunwoo, M. Human-like Decision-Making System for Overtaking Stationary Vehicles Based on Traffic Scene Interpretation. *Sensors* **2021**, *21*, 6768. [CrossRef]
31. Milanés, V.; Llorca, D.F.; Villagrá, J.; Pérez, J.; Fernández, C.; Parra, I.; González, C.; Sotelo, M.A. Intelligent Automatic Overtaking System Using Vision for Vehicle Detection. *Expert Syst. Appl.* **2012**, *39*, 3362–3373. [CrossRef]
32. Kan, S.; Lyu, W.; Zhao, S. Evaluation of the Environmental Effect of Automated Vehicles Based on IVIULWG Operator Development. *Sustainability* **2022**, *14*, 9669. [CrossRef]

33. Anindyaguna, K.; Basjaruddin, N.C.; Saefudin, D. Overtaking Assistant System (OAS) with Fuzzy Logic Method Using Camera Sensor. In Proceedings of the 2016 2nd International Conference of Industrial, Mechanical, Electrical, and Chemical Engineering (ICIMECE), Yogyakarta, Indonesia, 6–7 October 2017; pp. 89–94. [CrossRef]
34. Petrov, P.; Nashashibi, F. Modeling and Nonlinear Adaptive Control for Autonomous Vehicle Overtaking. *IEEE Trans. Intell. Transp. Syst.* **2014**, *15*, 1643–1656. [CrossRef]
35. Andersen, H.; Schwarting, W.; Naser, F.; Eng, Y.H.; Ang, M.H.; Rus, D.; Alonso-Mora, J. Trajectory Optimization for Autonomous Overtaking with Visibility Maximization. In Proceedings of the 2017 IEEE 20th International Conference on Intelligent Transportation Systems (ITSC), Yokohama, Japan, 16–19 October 2017; pp. 1–8. [CrossRef]
36. Xia, Y.; Wang, C.; Shi, X.; Zhang, L. Vehicles Overtaking Detection Using RGB-D Data. *Signal Process.* **2015**, *112*, 98–109. [CrossRef]
37. Olaverri-Monreal, C.; Gomes, P.; Fernandes, R.; Vieira, F.; Ferreira, M. The See-Through System: A VANET-Enabled Assistant for Overtaking Maneuvers. In Proceedings of the 2010 IEEE Intelligent Vehicles Symposium, La Jolla, CA, USA, 21–24 June 2010; pp. 123–128.
38. Németh, B.; Gáspár, P.; Hegeds, T. Optimal Control of Overtaking Maneuver for Intelligent Vehicles. *J. Adv. Transp.* **2018**, *2018*, 1–11. [CrossRef]

electronics

MDPI

Article

Neurofuzzy Data Aggregation in a Multisensory System for Self-Driving Car Steering

Antonio Luna-Álvarez [1], Dante Mújica-Vargas [1,*], Arturo Rendón-Castro [1], Manuel Matuz-Cruz [2] and Jean Marie Vianney Kinani [3]

1 Department of Computer Science, Tecnológico Nacional de México/CENIDET, Interior Internado Palmira S/N, Palmira, Cuernavaca 62490, Mexico
2 Departamento de Sistemas Computacionales, Tecnológico Nacional de México/ITTapachula, Tapachula Chiapas 30700, Mexico
3 Unidad Profesional Interdisciplinaria de Ingeniería Campus Hidalgo, Instituto Politécnico Nacional, Pachuca 07738, Mexico
* Correspondence: dante.mv@cenidet.tecnm.mx

Abstract: In the self-driving vehicles domain, steering control is a process that transforms information obtained from sensors into commands that steer the vehicle on the road and avoid obstacles. Although a greater number of sensors improves perception and increases control precision, it also increases the computational cost and the number of processes. To reduce the cost and allow data fusion and vehicle control as a single process, this research proposes a data fusion approach by formulating a neurofuzzy aggregation deep learning layer; this approach integrates aggregation using fuzzy measures μ as fuzzy synaptic weights, hidden state using the Choquet fuzzy integral, and a fuzzy backpropagation algorithm, creating a data processing from different sources. In addition, implementing a previous approach, a self-driving neural model is proposed based on the aggregation of a steering control model and another for obstacle detection. This was tested in an ROS simulation environment and in a scale prototype. Experimentation showed that the proposed approach generates an average autonomy of 95% and improves driving smoothness by 9% compared to other state-of-the-art methods.

Keywords: self-driving; deep learning; neurofuzzy data processing; fuzzy backpropagation; self-driving simulation; scale prototype

Citation: Luna-Álvarez, A.; Mújica-Vargas, D.; Rendón-Castro, A.; Matuz-Cruz, M.; Kinani, J.M.V. Neurofuzzy Data Aggregation in a Multisensory System for Self-Driving Car Steering. *Electronics* **2023**, *12*, 314. https://doi.org/10.3390/electronics12020314

Academic Editors: Calin Iclodean, Bogdan Ovidiu Varga and Felix Pfister

Received: 12 December 2022
Revised: 2 January 2023
Accepted: 3 January 2023
Published: 7 January 2023

1. Introduction

In systems, the use of multiple data sources allows an actuator to have a broad view of the environment it is seeking to control, thus allowing it to perform this task with greater precision and to tolerate anomalies in data from some sources [1]. This method is used in physical and robotic systems to widen the field of perception and avoid blind spots, while the redundancy of the data allows the controller system to detect anomalies and filter them [2]. Mobile robots use this method to perceive their environment from different sensors that provide spatial, inertial, and visual information [3]; this is most notable in the Autonomous Vehicles application. The research area for vehicles control is divided into sub areas such as: environment perception, object and vehicles recognition, behaviors, planning and route selection, lane maintenance, signal detection, steering and speed control, among others [4]. This research focuses on the sensor data fusion for steering control.

Various strategies have been proposed in the literature to control the self-driving vehicle direction, just to name a few: starting with the fuzzy inference systems [5,6], which controls the direction by means of inference rules. In simulations, the predictive models [7,8] are recurrently used to trace a route to follow within the path, as well as its combination with fuzzy logic [9]. The aforementioned methods share the characteristic of requiring knowledge transfer from the algorithm designer, other similar strategies are path detection [10], knowledge transfer classifiers [11], and semantic segmentation [12]. On the other hand, self-tuning methods are the most widely used, from strategies based on reinforcement learning [13–16], which are generally executed in simulations since they require learning based on trial and error or mistakes. Another type of reinforcement learning is based on vision [17,18], although there are other methods based on pure vision [19,20]. The aforementioned strategies mainly focus on keeping the vehicle on track; however, the current trend for steering control is mainly focused on deep learning. This deep neural domain can be divided into control strategies through vision and detection with convolutional networks [21–23]. Other alternatives within the same paradigm are recurrent networks for a visual–temporal relationship [24,25]. The previously mentioned works of the deep learning paradigm present good results; however, they use a single data source such as camera vision. In the literature, it is possible to notice that the methods that present the best results are those based on multisensory systems, for example those that combine visual and spatial information [26–28], the a priori data fusion methods by object detection [29], and fusion 3D [30]. Returning to the deep learning paradigm, another approach to process multiple data sources is to create parallel architectures that process each source separately and integrate the outputs [31–33], as these are the ones with the highest computational cost. To reduce computational cost, deep learning [34,35] fusion methods are used, although they depend on a second process to perform post-fusion control.

Although the aforementioned works present good results, they have some weaknesses. In parallel models, the computational cost increases according to the increase in sensors to be processed, making this option viable only in simulation, since it is too much load for a system on board the vehicle or embedded at scale. Similarly, a priori fusion methods represent a double computational cost since they must fit the data sources as well as filter and extract features that will be processed by the control algorithm. A controller-integrated fusion method lowers the computational cost, particularly deep learning-based fusion models performed using a neural layer. Ideally they are formulated from the adjustment of synaptic weights W, the hidden states h, activation functions $\sigma(\cdot)$, and the backpropagation algorithm $\Delta W(\nabla L)$. However, in Ref. [34] it depends on a PID controller coupled to the neural network to perform the direction and speed adjustment, stopping the network process and processing the data fusion separately from the neural model. In Ref. [35] the multiview aggregation approach uses perspective transformation to project n feature maps according to the corresponding $1 - n$ camera settings. The N projected feature maps are concatenated and a convolution is used to obtain the result. Like these examples, many others [36–38] perform control using a methodology composed of parallel or sequential processing blocks that interact with the neural model, creating a different dependency on the methods, thus increasing response time, fusion quality, and computational cost.

As an alternative, this article proposes a layer with a neurofuzzy aggregation approach, expressed as a neuronal model layer. The main contributions are the following: (a) the neurofuzzy aggregation layer allows extracting and relating features from different structural shapes sources; (b) it has generalization capacity, so it is viable for data fusion in any application through model training; (c) it allows for multiple dimensions inputs unlike the known neural models; (d) it can be added to an existing model as one more layer; and (e) the implementation is compatible with the TensorFlow framework, so it can be parallelizable on the GPU and is not processed as an external program. This proposal was evaluated in a self-driving vehicle steering control task; the experimentation was carried out in a simulated environment in the Robotic Operating System (ROS), as well as in a scale prototype in a controlled environment.

The rest of the document is organized as follows. Section 2 details the proposed layer formulation, as well as the neural architecture for self-driving. Section 3 details the experiments performed, as well as their validation and comparative analysis with other current methods in the literature. Finally, in Section 4 a synthesis of results is made concluding with a brief discussion and future work.

2. Materials and Methods

The main contribution is the data aggregation neural layer formulation, so to understand the proposal it is necessary to know the basic operation of the neural paradigm. This performs only data aggregation; for a more complex task such as self-driving, a more complex model composed of several layers is required. For this reason, this section details the basic concepts, the proposal formulation, and its adaptation for self-driving.

2.1. General Neural Layer Formulation

In the ANNs paradigm, a model is understood as a set of algorithms grouped into layers that perform a specific task. A neural layer consists of three main elements: for the synaptic weights, given a X data set structured in tensors of size n, there will be a weight W for each X such that $\forall x \in X, \exists w \in W : |X| = |W| \wedge x \neq \emptyset$. The hidden state h obtained by a two Euclidean magnitudes $\mathcal{N}(\cdot,\cdot)$ product function, is defined as a linear algebraic product operation of order n according to the dimensionality of X and W. The parameters of such a function are $\mathcal{N}(X, W)$ which generate a continuous h output. Furthermore, the activation function $\sigma(h)$ is understood as a linear, non-linear, binary, or probabilistic function that scales the h state. The output y can be discrete or continuous, depending on $\sigma(\cdot)$ domain and the purpose of the layer application. Depending on the layer application, it is defined as a feature extraction layer, memory, recurring, etc.; however, most of them depend on these 3 essential components. For the layer components to correctly process the data, it is necessary to train the parameters using the backpropagation algorithm. To adjust the weights W it is necessary to know the empirical error generated by the propagation in the training stage, based on the expected outputs y' known as ground truth. The error can be calculated by a distance measure known as loss function $L(W)$, which can be:

$$L(W) = \sum_{i=1}^{I} ||\sigma(h_i) - y_i'||_2^2 \tag{1}$$

although depending on the application it can be considered an absolute, polynomial, or radial Euclidean measure. The general method for minimize the error is to iteratively update the parameters by adding an increment ΔW to the current value: $W := W + \Delta W$. Conversely, if a function $\mathcal{N}(x, W)$ is used to approximate the output values and it is differentiable with respect to W, it is possible to use the Gradient Descending method as a learning algorithm:

$$\Delta W = -\alpha \frac{\partial E(W)}{\partial W} \tag{2}$$

where $0 < \alpha < 1$ is a parameter known as the learning rate, which regulates the updating ΔW in the error gradient ∇. In this way the general algorithm for the neural layers is described; graphically it is represented by the scheme of Figure 1.

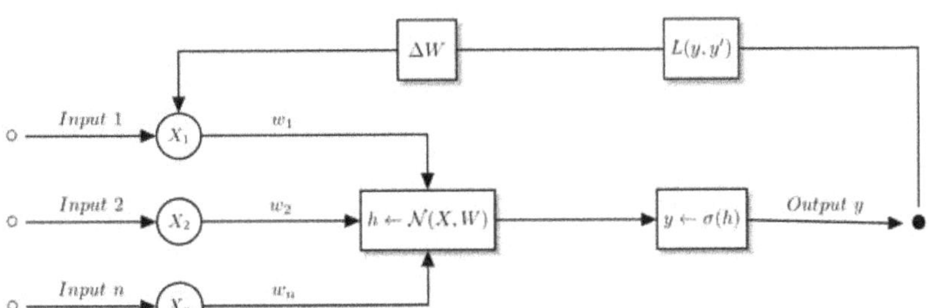

Figure 1. General neural layer formalization.

From this base, it is possible to reformulate the algorithm to give a different objective to the classification or characteristics extraction. To perform the aggregation, the neural layer was transformed to the fuzzy domain and its basic components were reformulated to be compatible with fuzzy aggregation measures μ, in addition to working with variable dimensionality inputs. This is described below.

2.2. *Neurofuzzy Layer Formulation*

To summarize the algorithmic adaptations required to formulate the neurofuzzy layer, the proposal components are listed below. Inputs X: whether it is a tensor of order 2 or higher, an input of different order is allowed for each element of the first dimension, that is, the signal $x_{1,}$ is structurally different from the one $x_{2,}$, allowing us to operate 2D and 1D signals in the same tensor. The inputs X are structured in an asymmetric tensor x_s of maximum dimensionality $x_s \in \mathbb{R}^{d\times(m\times n)}$, according to the highest order tensor.

Fuzzy weights W_μ: from a fuzzy measure μ defined as a transformation function within the fuzzy domain $\mu(x) : [0,1]^\eta \mapsto [0,1]$, and interpreted as a weighting function such that $\mu(\emptyset) = 0$, $\mu(X) = 1$, the fuzzy weights W_μ are defined for each $x \in X$ such that $\sum_{i=0}^{|X|} w_{\mu_i} = 1$. At the same time, there are non-fuzzy weights W_s for each ordered entry x_s and that are adjusted by them, independent of the fuzzy aggregation. Hidden state h: depends on the implemented operator, for the neuron generality an algebraic function of the Euclidean magnitudes product can be used. Therefore, the Choquet fuzzy integral defined as:

$$\int f \circ \mu = \sum_{i=1}^{\eta}[f(x_{s(i)}) - f(x_{s(i-1)})] \cdot \mu(x_{s(i)}) \tag{3}$$

where $f(\cdot)$ indicates the ordered data fuzzy transformation and $\mu(\cdot)$ the aggregation using the fuzzy measures μ. Given that $x \in X$ is a tensor of order greater than 0, a reduction is necessary using an operator such as the inner product $\mathcal{N}(\cdot,\cdot)$. This in relation to the introduction of non-fuzzy weights W_s for each x_s. Thus, the state h is detailed as the extension of the Choquet integral:

$$h = \sum_{i=1}^{\eta}[\mathcal{N}(x_{s(i)}, W_{s(i)}) - \mathcal{N}(x_{s(i-1)}, W_{s(i-1)})] \cdot \mu(X, W_\mu) \tag{4}$$

in such a way that the fuzzy integral $\mu(\cdot,\cdot)$ depends on the linear product $\mathcal{N}(\cdot,\cdot)$, which in the case must be an outer product $\times$ due to the asymmetry of the input tensors. Activation $\sigma(h)$: As this neuron is not dedicated to classification, a probabilistic function such as Softmax is not used, so a non-linear or rectifying function must be used like any hidden layer. Training algorithm: the propagation of the error within a complete model is carried out in the normal way except for this layer, for which it is necessary to update both the W_s and the fuzzy W_μ weights. Regarding Momentum, its fuzzy version resides in the scaling of the gradient ∇, this is modified by the scalar value of the fuzzy integral obtained from the aggregation product:

$$m_{\mu_{i+1}} = \alpha m_{\mu_i} + \eta \nabla \mu(x_s, w_s) \tag{5}$$

where $\nabla \in [0,1]$ and the diffuse momentum $m_\mu \in [-1,1]$ are restricted. In this way the adjustment ΔW_μ is given by:

$$\Delta W_\mu = W_\mu - \alpha(y \cdot m_\mu) - \alpha(1-y))\nabla\mu(X, W_\mu). \tag{6}$$

On the other hand, the loss function must remain in the fuzzy domain, which is why the MSE can generate alterations to the adjustment by being able to obtain values greater than 1. For this reason, the Logarithm of the Hyperbolic Cosine is used:

$$L(W) = \sum_{i=0}^{n} \log\left(\frac{e^{(y'-y)} + e^{-(y'-y)}}{2}\right) \tag{7}$$

In summary, in the Algorithm 1 the propagation and backpropagation algorithm of this proposal is presented.

Algorithm 1 Neurofuzzy ΔW_μ training

Require: $e \in \mathbb{Z}, \alpha \in \mathbb{R}, X_1, X_2, X_n \in \mathbb{R}^+$
Ensure: y
1: $W_s \leftarrow random([-1,1])$
2: $W_\mu \leftarrow 0$
3: $x_s \leftarrow X_1 \times X_2 \times X_n : m \times N = \{n \in \mathbb{N} | \, |N| = m\}$
4: **while** converge **do**
5: **for** $i \leftarrow 0$ **to** $i = n$ **do**
6: $h_{s_i} \leftarrow \mathcal{N}(w_s, x_{s_i})$
7: $h_{\mu_i} \leftarrow \mu(x_{s_i}, w_\mu)$
8: $h_i = \sum_{i=1}^{\eta} [h_{s_i} - h_{s_i-1}] \cdot h_\mu$
9: $y_e \leftarrow \sigma(h_i)$
10: **end for**
11: $L(W)_e = \sum_{i=0}^{n} \log\left(\frac{e^{(y'-\sigma(h_\mu))} + e^{-(y'-\sigma(h_\mu))}}{2}\right)$
12: $m_{\mu_{i+1}} = \alpha m_{\mu_i} + \eta \nabla \mu(x_s, w_s)$
13: $\Delta W_\mu = W_\mu - \alpha(\sigma(h) \cdot m_\mu) - \alpha(1 - \sigma(h)) \nabla \mu(X, W_\mu)$
14: $\Delta W_s = W_s - \alpha(y_e \cdot m_j) - \alpha(1 - y_e) \nabla L(W)_e$
15: $e \leftarrow e + 1$
16: **end while**
return y

In the Algorithm 1 in step 3, the asymmetric tensor structuring is used. This transformation generates a single structure that respects each data source shape, grouping in an additional m dimension. The algorithm requires two product calculations, one linear h_s and one based on fuzzy measure h_μ, later they are unified by means of the fuzzy integral in step 8 in the output structure h. It should be noted that an output y is not generated from this algorithm since this belongs to the classification part used in the neural model to be implemented; however, it is necessary for the adjustment of the weights W_s. Therefore the neurofuzzy model is defined in Figure 2.

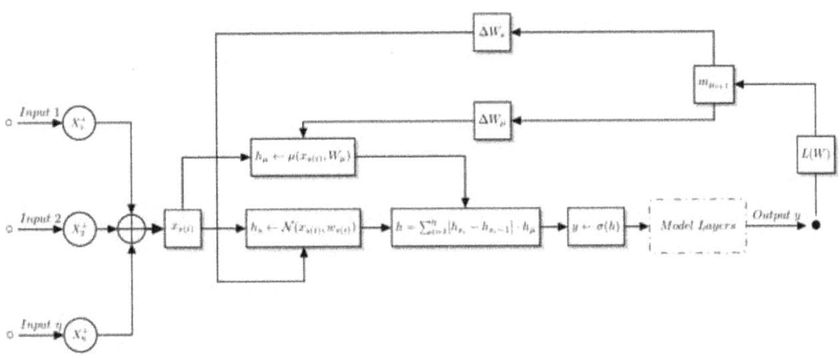

Figure 2. Neurofuzzy aggregation layer scheme.

In the above description, the hidden state function h is capable of performing the aggregation of two signals, using only an activation function σ rectifier. However, for an application such as self-driving, the model requires greater complexity as well as a greater number of layers with different activations. For this reason, Figure 2 represents the layers of the model in a dotted block, implying that after the aggregation layer there are dense layers, convolutions, or any other item that performs different tasks with information from unified sources. The resources used to adapt the self-driving model with neurofuzzy aggregation are detailed below.

2.3. Self-Driving Model

The proposed fuzzy aggregation layer purpose is to combine multiple signals to reduce the data within a neural model. Thus, it is necessary to have a driving model to guarantee performance and subsequently reduce it. For this, the time-distributed Chauffeur model shown in Figure 3 is used.

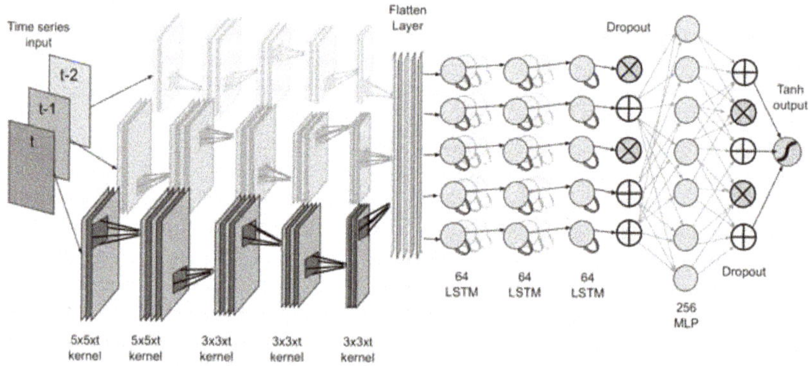

Figure 3. Time-distributed Chauffeur model [39].

This model receives information in a sequence of images form in the YUV color space, so it is structured as a 4D tensor of size $t \times 3 \times 200 \times 320$, where t represents the frames per instance, 3 are the color channels, and 200×320 is the size of each image. The model output is reduced to a scalar obtained from the output layer activation function σ, given by a Hyperbolic Tangent function that generates outputs in the interval $[-1, 1]$. These values represent degrees of turn in the direction of the vehicle. However, the td-Chauffeur model cannot detect objects. To carry out the object detection from visual information, the Mobilenet model was used as a base, as shown in Figure 4.

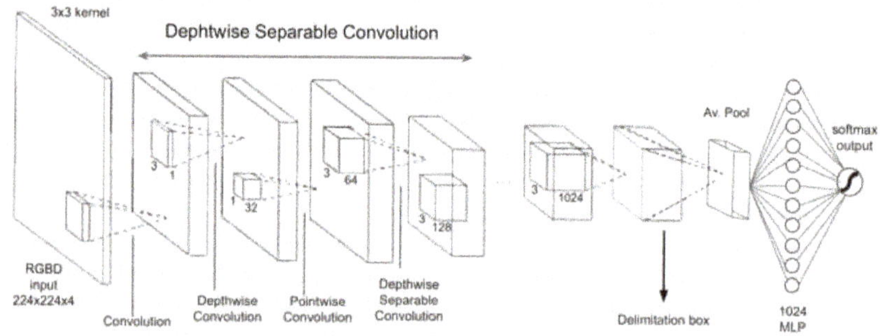

Figure 4. Mobilenet model for object detection [40].

This mobile model uses depth separable convolutions. Significantly reduces the number of parameters compared to networks with regular convolutions and same depth, resulting in a lightweight deep neural network. The model basis is to use depth convolution and point convolution layers, so the model can be adapted to the information to be processed. Originally it receives a $224 \times 224 \times 3$ 3D tensor that represents an RGB image; however, for this application it has been adapted to process a $224 \times 224 \times 4$ 3D tensor that is interpreted as an RGBD image. As an output it obtains the bounding boxes and distance from the center of the objects.

In a simple methodology, it is possible to perform the task by running both models in different threads and processing both outputs in a third algorithm. The limitation of that is the computational cost, in a computer with processing capacity it is possible to run using most of the resources; however, in an embedded system this is limited by: (1) the GPU and CPU processing capacity is with less capacity than a desktop computer, (2) the increased heating and power consumption due to the overuse of resources make the implementation unfeasible, and (3) the models synchronization is a point to consider even on a workstation. For this reason, the integration of both models through an asymmetric tensor distribution is proposed, generating a single multifunction model with layers of neurofuzzy aggregation to reduce the complexity of the model. This proposal can be seen visually in Figure 5.

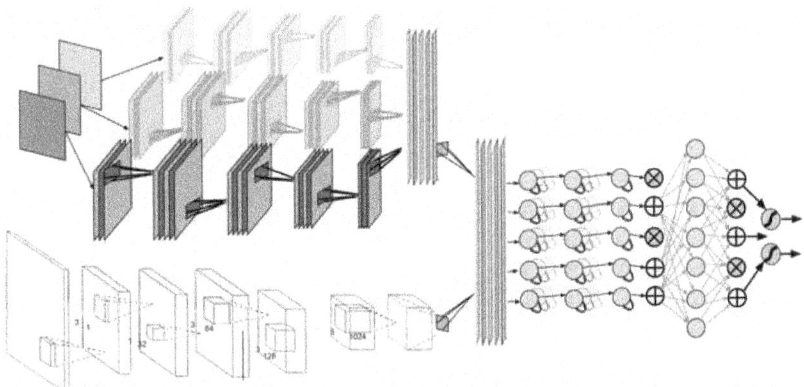

Figure 5. Self-driving operational model.

The graphical representation makes it appear that the model is a parallel structure of two neural networks. At the tensor level it is interpreted as two layers of the outermost dimension n; however, in the dimension $n-1$ the tensor shape changes in such a way that the input is defined as $I = \{\{224 \times 224 \times 3 \times t\}, \{224 \times 224 \times 4\}\}$. The asymmetry of the input of each neural layer is notorious; however, the reduction in processing threads is carried out by distributing the GPU process only. Where model shrinking is observed is in the A_{W_μ} layer where the bounding boxes are obtained, the temporal shrinking of the *Flatten* layer and added along with the translation and rotation $I_p = \{\{x^t, y^t, z^t\}, \{x^r, y^r, z^r\}\}$. Finally, the output is extended to two TanH functions to control direction and acceleration.

3. Experimental Results

To validate the proposal performance, two experiments were carried out: a simulation in ROS with variables from a real urban environment and a scale prototype tested in a controlled environment. In both cases, scenarios with obstacles and free paths are considered, as well as the use of more than one sensor. To quantify performance, supervised and unsupervised metrics were used, which specialized in measuring the quality, precision, and autonomy of self-driving. Both experiments were evaluated by the following metrics.

3.1. Metrics

First, the supervised evaluation was carried out to find out the driving model similarity with the actions carried out by a human driver. For this a ground truth y' was created from capturing steering commands during manual driving on the test path. From this reference it was possible to measure the difference in distance terms between what is obtained and what is expected, for this reason the Mean Square Error is used as evaluation metric, in the same way the Cosine Distance is used:

$$\cos(\theta) = \frac{\sum_{i=0}^{n} y_i \cdot y'_i}{\sqrt{\sum_{i=0}^{n} y_i^2} \cdot \sqrt{\sum_{i=0}^{n} y_i'^2}} \tag{8}$$

the output range spans $[-1, 1]$, where 0 indicates zero error; therefore, 1 and -1 indicate full left and right error. The MSE provides an error rate between both types of driving, regardless of the variations produced by time and the speed of movement of the vehicle. For this reason the Driving Behavior metric is used:

$$DB = \left\| \sqrt{\frac{\sum_{i=1}^{n}(y_i - \bar{y})}{n}} - \sqrt{\frac{\sum_{i=1}^{n}(y'_i - \bar{y}')}{n}} \right\| \tag{9}$$

this metric evaluates the entire route by calculating the deviation between both lines; thus, time does not take part in the evaluation and thus providing a speed-invariant distance index. In an unsupervised way, driving is measured by Path Smoothness, i.e., this refers to the angles amplitude that are described while the vehicle is moving:

$$\kappa = \frac{1}{n} \sum_{i=2}^{n} \left[\arctan\left(\frac{y_{i+1} - y_i}{x_{i+1} - x_i} \right) - \arctan\left(\frac{y_{i-1} - y_i}{x_{i-1} - x_i} \right) \right] \tag{10}$$

where x_i and y_i represent the position of the vehicle in a specific trajectory segment. The metric obtains the angle between two consecutive segments of the path. A lower smoothness value indicates a smoother path. On the other hand, to evaluate the ability to operate independently of a driver, use is made of the autonomy metric proposed by [41]:

$$autonomy = \left(1 - \frac{interventions \cdot 6}{elapsed\ time} \right) \cdot 100 \tag{11}$$

as well as the Absolute Autonomy Time metric [42], which measure the interventions of a human driver as an error in the system, which is the autonomy time the metric oriented to absolute intervention:

$$AAT = \left(1 - \frac{tiempo\ de\ intervención}{tiempo\ transcurrido} \right). \tag{12}$$

To complement the experimentation, some methods known from the literature for self-driving were evaluated under the same conditions. Specifically they were the Pilotnet model [41], Donkeycar self driving library [43], and Matlab Automated Driving Toolbox. For these, self-driving models were taken and replicated in the ROS system using it as an intermediary.

3.2. Experiment 1: ROS Self-Driving Simulation

This task was carried out based on the free access ROS design, CAT Vehicle Testbed, which consists of a Ford Escape vehicle with the actual physical measurements. The package includes sensors built into the vehicle; however, they were modified to suit this application. Originally featuring a Velodyne LiDAR sensor with a 2000 point cloud, this was modified to produce a 12,000 point cloud with a horizontal aperture of 90° and a vertical aperture of 33.67°. In the same way, the package has two cameras positioned on the vehicle at angles of $-45°$ and 45° with origin in front of the vehicle, these capture RGB images of size 800 × 800 pixels. For this application, a single camera pointing to the origin with a size of 640 × 320 pixels was required. To evaluate the vehicle autonomy it was necessary to integrate city environments with different characteristics such as intersections, houses and buildings, closed roads, and obstacles in the way. In order to have a variety of possible scenarios, the Vehicle and City Simulation was integrated, which fully represents a small city. Some fragments of the road with obstacles were taken to carry out the self-driving evaluation. These are shown in Figure 6.

Figure 6. ROS simulated environments.

Figure 7 shows the sketches with the shapes and measurements of the simulation segments in which the tests were performed. To validate the effectiveness of each method, two short scenarios and two more complex ones were used, all with objects partially obstructing the path and one with incorrect paths. For all the scenarios there was a trajectory to reach the goal.

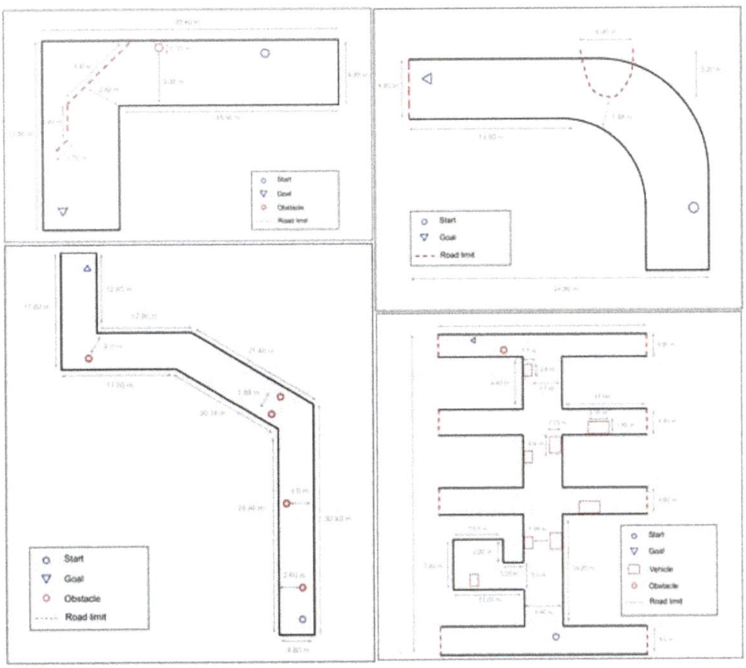

Figure 7. Real-scale simulated environments maps for self-driving tests.

Figure 8 shows the conduction trajectories generated by each evaluated method. It can be seen that in the simplest scenarios all the methods can reach the goal; however, it should be noted that the proposal performs the smoothest movements, approaching what is performed by a human, as shown in the ground truth. In the most complex environment it can be observed that a method takes the wrong path, ending the evaluation for it. For the cases in which the algorithm cannot complete the route, it was necessary to enable manual driving to resume the path, an action necessary to calculate the Autonomy and AAT metrics. This intervention is also carried out intentionally when the displacement moves away from the ground truth, thus obtaining an evaluation of autonomy. Quantitatively, the results are summarized in Table 1.

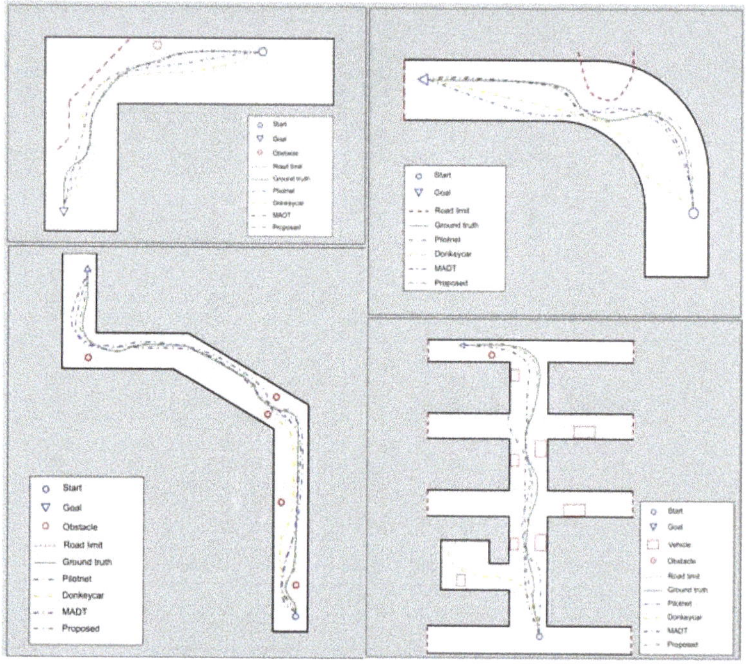

Figure 8. Trajectories obtained during self-driving in simulated environments.

According to what is shown in Table 1 and with respect to Figure 8 trajectories, it is observed that the proposal shows the best autonomy metrics are obtained by the proposed method. A range security of 96% is guaranteed, while the worst performance offers only 71%, as long as timely interventions are made, otherwise a collision would result. In simulation, an autonomy of 96.6% is obtained, which indicates that it depends on human intervention for only 3.4% of route, compared to the second best method, which is 8.6% more dependent. In road smoothness, the proposal obtains a metric of 0.256, which is 0.72 lower than the mean of the other methods. This metric indicates that the proposed approach offers a ride that is up to 2.8 times smoother and more comfortable, without any hard rocking.

Table 1. Quantitative results of simulation experimentation.

Method	MSE	Cosine Distance	Behavior	Path Smoothness	Autonomy	AAT
Pilotnet	0.106	0.221, −0.076	0.030	0.682	86.4%	92.9%
Donkeycar	0.385	0.380, −0.510	0.236	1.826	71.3%	79.8%
MADT	0.077	0.112, −0.014	0.024	0.442	88.0%	94.7%
Proposed	0.021	0.011, −0.044	0.003	0.256	96.6%	97.4%

To complement the experiments, each conductions method was compared with respect to a human conduction. The metrics showed that the proposed method obtains a mean error of 0.021, where the lowest showed difference of 0.056 compared to the second best. This error indicates that the method provides up to 3.6 times better than human-like driving than another method in this comparison. This measurement is seconded by the driving behavior metric, in which the minimum error of 0.003 is obtained, which indicates that the speed is similar to that of a human driving, without affecting the steering commands smoothness. Finally, the cosine distance showed an imbalance in the lateral displacement error, which was −0.044 and 0.011, so it presented more errors in left turns. This is due to the notable imbalance in the turns of the tracks that can be seen on the maps shown. Still, it showed the best balance of ≈1–3, compared to MADT which was ≈1–9. Although the Donkeycar method had a relationship ≈1–1.3, the error was up to 10× higher.

As is known, in simulation environments the results are not affected by environmental effects such as lighting, ground textures, obstructions, etc. For this reason, to complement the experimentation, a small scale prototype was designed that integrates in an embedded model light version to control the vehicle direction.

3.3. Experiment 2: Scale Prototype Steering Control

In the first instance, the base vehicle was acquired, which includes the mechanics and motors in operation. Its dimensions and appearance shown in Figure 9 represent a 1:16 scale of a compact vehicle in length and width.

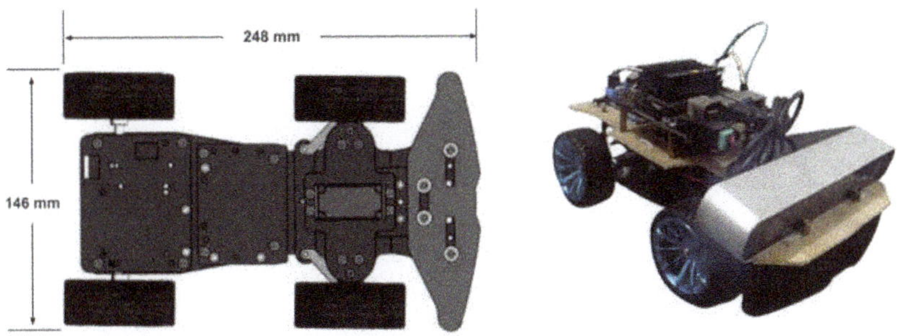

Figure 9. Scale prototype designed for self-driving physical tests.

Inspired by the Jetracer prototype, a Jetson Nano Dev Kit 4G embedded card was used as the control module due to its versatility in terms of high-level and low-level programming, as well as its GPU-embedded computing capability on CUDA under Ubuntu. This card includes a 4-core @ 1.43 GHz ARMv7 processor, 4 GB of 1600 MHz RAM, 16 GB eMMC 5.1, GPIO with L2C and PWM support for geared and servo motors, and Nvidia Maxwell 128 CUDA core GPU. In conjunction with the Jetson Nano card, some electronic components were used in the periphery: PCA9685 module, connected to the GPIO ports dedicated to the L2C type connection of the Jetson Nano. This device is used to control the motors by transforming the digital signals to PWM signals that give precision to the servo motor angular speed and position. The device can work at 3.3 V with direct power from the Jetson Nano; however, an external 5 V power is required to move the servo motor. The L298N module serves to magnify the voltage at the 3.3 V input from the PCA9685 to 12 V output. It is used to power the geared motor that pushes the vehicle in a channel.The Lm2596 dc-dc voltage regulator is used to convert 12 V input voltage to 5 V to power the PCA9685 module and the Jetson Nano. Due to the prototype size and power limitations, it used the ZED camera since it allows for obtaining visual information as well as spatial information from the generation of point clouds, in a similar way to the LiDAR sensor. In this way, two types of signals are obtained from the same sensor.

The embedded system works in a similar way to the simulation; however, the processing is conducted on the card, so it was required to interpret the model in the TensorFlow Lite version. Since the information is only obtained from the ZED camera, an acquisition system was designed in Python using the camera drivers. This program acquires the point clouds $I_d = \{M \cdot N \cdot D\}$ and creates the image sequences in tensors of T times at $I_t = \{M \times N \times C \times T\}$. Rotation and translation information is also acquired from the camera controllers. This generates as input three tensors: 4D, 1D, and 2D. Figure 10 shows the implemented system scheme.

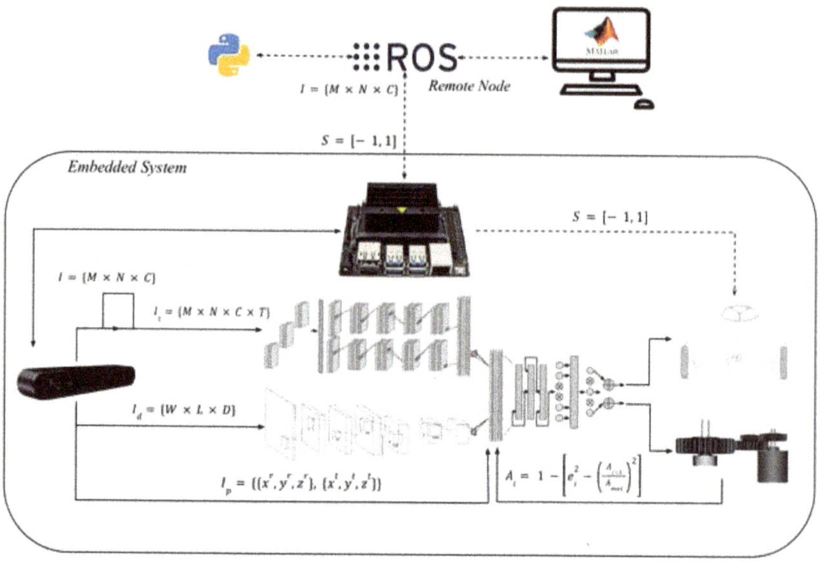

Figure 10. Embedded self-driving system operational diagram and remote interaction for comparison methods evaluation.

Due to the prototype characteristics, it was evaluated in simpler road scenarios than the simulation urban environments. The designed paths are inspired by the BlueRaven tracks for Jetbot and some obstacles were added to validate the ability to evade and correct the trajectory. Figure 11 details the designed roads in real-scale measurements.

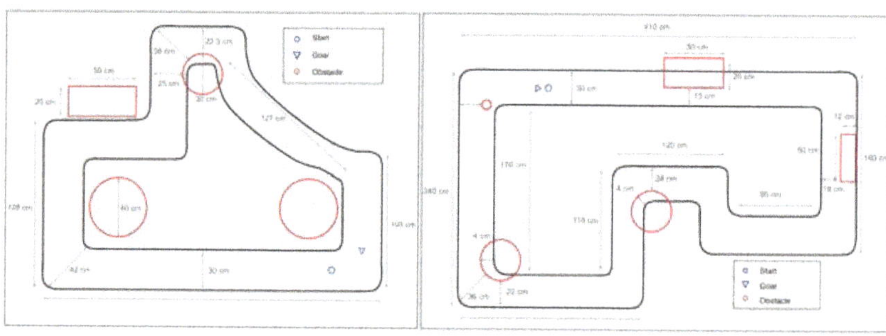

Figure 11. Real-scale track maps for physical self-driving tests.

In Figure 11, it can be seen that the paths have a width of 0.30 m, while in the simulation maps of Figure 7 these are exactly 4.8 m, so the tracks designed for this experiment are at 1:16 scale. In accordance with previously mentioned measurements of the prototype shown in Figure 9, both the test tracks and the vehicle are on the same scale, so the experimentation for the self-driving test it is considered viable in non-urban environment conditions at a scale of 1:16. In order for the displacement to be consistent with the track scale, the maximum speed was limited to 0.8 m/s, limiting the acceleration to:

$$A_i = 1 - \left[e_i^2 - \left(\frac{A_{i-1}}{A_{max}} \right)^2 \right] \tag{13}$$

where A_{i-1} is the acceleration at the previous instant emitted by the neural model, on average at 50 Hz frequency. e_i^2 represents the rotation command of the current instant and

A_{max} the maximum speed of 0.8 m/s, which in full scale would represent a maximum of 50 km/h. From the above, it is possible to observe the relationship between both experiments and their viability when evaluating the proposal and the comparison methods in the same way. Therefore, the results are discussed below.

Figure 12 shows the paths traveled on the test tracks. To increase experimentation, on the first track the obstacles location was modified to validate the evasion capacity. In the first scenario, all the methods were able to finish the course; however, when adding the obstacles, it was observed that Donkeycar did not overcome the second obstacle. On the more complex track both the Donkeycar and the Pilotnet model were unable to complete the full course, the latter evading all three obstacles but went off the road. In this case, MADT also goes beyond the path limits; however, it manages to resume autonomously. It can also be seen that both MADT and the proposed method generate very similar trajectories in the first scenarios. To make the comparison, a metrics summary is shown in Table 2.

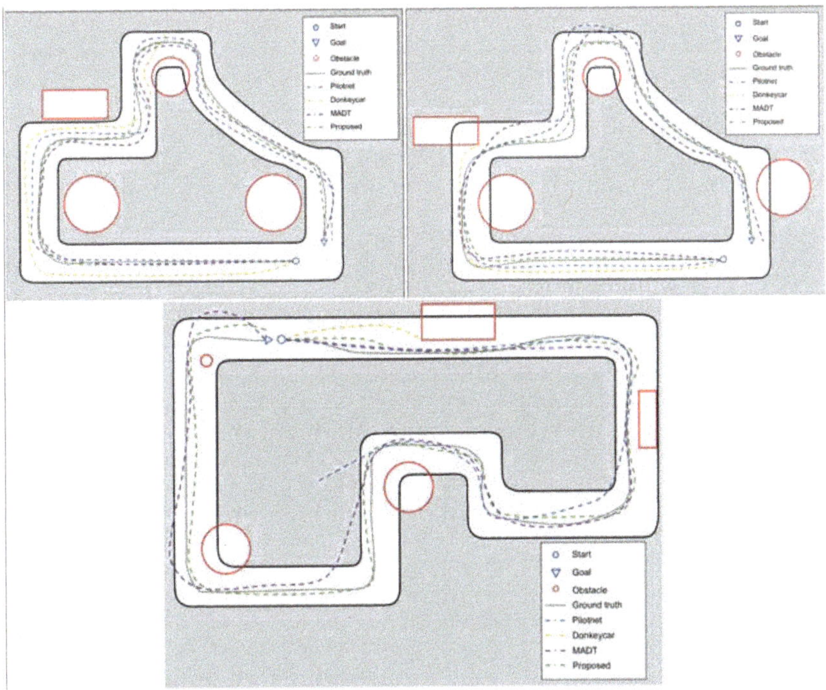

Figure 12. Trajectories obtained during self-driving in physical environments.

Table 2 shows the AAT of 75% for the Pilotnet model, this indicates that 25% of the path depends on human intervention, unlike the proposed method that depends only 5.6% if perfect driving is needed. Similarly, MADT receives an AAT of 89.8% as it went off track twice and needs to be corrected. In this experimentation, an autonomy of 89% was obtained, a decrease of 7.6% with respect to the autonomy in simulation is due to the fact that in the physical environment there are factors that alter perception such as the lighting changes, sensor noise, and affectations to control such as inertia and motors cumulative error. However, the proposed method is the least affected by this domain change, since the other methods decrease an average of 9.73% and up to 12.9% in the worst case. It can also be observed that the smoothness is reduced since a metric of 0.598 is obtained due to the path and physical effects complexity; however, the method is autonomous by 89%, which is better by ≈17.7% than the average of the other methods.

Table 2. Quantitative results of the experimentation on the scale prototype.

Method	MSE	Cosine Distance	Behavior	Path Smoothness	Autonomy	AAT
Pilotnet	0.185	0.171, −0.441	0.051	1.216	73.5%	75.5%
Donkeycar	0.205	0.322, −0.571	0.087	2.948	63.8%	66.9%
MADT	0.106	0.132, −0.100	0.030	0.834	78.3%	89.8%
Proposed	0.081	0.148, −0.075	0.027	0.598	89.0%	94.4%

Compared to human driving, the proposed method obtains a mean error of 0.081 and a smoothness of 0.589. For the aforementioned reasons, the metrics show higher performance and therefore lower performance. Even so, it is shown as the best by 0.025 with respect to the maximum and 0.084 with respect to the average, which is up to 2× better than the rest of the comparison methods.

An important observation is that the proposed method can complete the path without interventions, compared to the Pilotnet model and the Donkeycar method, which require forced interventions to stay on track. From these experimental observations it can be inferred that the multiple data sources used improves the quality of self-driving, as well as that the neurofuzzy aggregation layer helps to decrease computational consumption making it possible to operate in a lightweight prototype. Furthermore, in this experimentation, the proposed method can be trusted by 89% in the worst case and 95% in a normal case. With these specific observations, the conclusions of this investigation can be reached.

4. Conclusions

Two main contributions are made in this work. The first is the formulation of a neurofuzzy aggregation layer for deep learning neural networks, which allows for composing new features from the several data sources fusion, reducing the computational cost while maintaining the precision of a multisensory system. The second contribution is the application of the proposed layer in the creation of a multisensory model for steering control. Additionally, the proposed model was tested in a simulated urban environments in ROS and in a scale prototype. The experimentation was carried out equally with our proposal and other methods of the state of the art. From this experimentation the following was observed: in simulation, an autonomy of 96.6% was obtained, which indicates that it depends on human intervention only 3.4% of the time. Regarding the smoothness of the road, a metric of 0.256 was obtained, demonstrating that the proposed method offers driving that is up to 2.8× smoother and more comfortable. Regarding human driving, a difference of 0.021 is shown, which is 3.6 times more similar than the other methods. In the experimentation with the scale prototype, an autonomy of 89% was obtained. However, the proposed method is the one with the best performance, since the other methods only obtain an autonomy of 78.3% in the best case and 63.8% in the worst case. Compared with human driving, the proposed method obtains a mean error of 0.081 and a smoothness of 0.589, so it is shown to be the best by 0.025 with respect to the maximum and 0.084 with respect to the average, which is up to 2× better.

In conclusion, the following advantages can be highlighted from this work: a model was created that performs data fusion and self-driving in a single process, based on the use of the proposed neurofuzzy aggregation approach, thus reducing the computational cost and allowing for its operation in real time. The proposed model is 95% reliable on average, with 2.8 better movement smoothness and 92% similarity to human behavior. The autonomous model completed the paths in 100% of the experiments. All this is because the two main contributions of this work offer the benefits of a multi-sensory system of several processes, but they are unified in a single one and reduce the cost to the point of being viable for use in real time in an embedded system. In future work, the model will be adapted to operate with more and different sensors for tasks such as pedestrians and traffic signs detection, as well as to integrate the capacity of planning and following routes. It is also intended to increase experimentation in different environments to demonstrate the feasibility of being implemented in a full-size prototype.

Author Contributions: Conceptualization, A.L.-Á., D.M.-V., M.M.-C., A.R.-C. and J.M.V.K.; methodology, A.L.-Á., D.M.-V., A.R.-C. and J.M.V.K.; software, A.L.-Á., M.M.-C. and A.R.-C.; validation, A.L.-Á., D.M.-V. and J.M.V.K.; formal analysis, A.L.-Á., D.M.-V. and J.M.V.K.; investigation, A.L.-Á., D.M.-V., M.M.-C. and J.M.V.K.; resources, A.L.-Á., D.M.-V., M.M.-C. and A.R.-C.; writing—original draft A.L.-Á. and D.M.-V.; writing—review and editing, A.L.-Á., A.R.-C., M.M.-C. and J.M.V.K.; visualization, A.L.-Á., A.R.-C. and J.M.V.K.; supervision, A.L.-Á., D.M.-V., M.M.-C. and J.M.V.K. All authors have read and agreed to the published version of the manuscript.

Funding: This research received no external funding.

Institutional Review Board Statement: Not applicable.

Informed Consent Statement: Not applicable.

Data Availability Statement: Not applicable.

Acknowledgments: The authors thank CONACYT, as well as Tecnológico Nacional de México/Centro Nacional de Investigación y Desarrollo Tecnológico for their support.

Conflicts of Interest: The authors declare no conflict of interest.

References

1. Xu, X.; Zhang, L.; Yang, J.; Cao, C.; Wang, W.; Ran, Y.; Tan, Z.; Luo, M. A Review of Multi-Sensor Fusion SLAM Systems Based on 3D LIDAR. *Remote Sens.* **2022**, *14*, 2835. [CrossRef]
2. Schinkel, W.; van der Sande, T.; Nijmeijer, H. State estimation for cooperative lateral vehicle following using vehicle-to-vehicle communication. *Electronics* **2021**, *10*, 651. [CrossRef]
3. Yan, Y.; Zhang, B.; Zhou, J.; Zhang, Y.; Liu, X. Real-Time Localization and Mapping Utilizing Multi-Sensor Fusion and Visual–IMU–Wheel Odometry for Agricultural Robots in Unstructured, Dynamic and GPS-Denied Greenhouse Environments. *Agronomy* **2022**, *12*, 1740. [CrossRef]
4. Badue, C.; Guidolini, R.; Carneiro, R.V.; Azevedo, P.; Cardoso, V.B.; Forechi, A.; Jesus, L.; Berriel, R.; Paixao, T.M.; Mutz, F.; et al. Self-driving cars: A survey. *Expert Syst. Appl.* **2021**, *165*, 113816. [CrossRef]
5. Nguyen, A.T.; Rath, J.; Guerra, T.M.; Palhares, R.; Zhang, H. Robust set-invariance based fuzzy output tracking control for vehicle autonomous driving under uncertain lateral forces and steering constraints. *IEEE Trans. Intell. Transp. Syst.* **2020**, *22*, 5849–5860. [CrossRef]
6. Chen, L.; Hu, X.; Tang, B.; Cheng, Y. Conditional DQN-based motion planning with fuzzy logic for autonomous driving. *IEEE Trans. Intell. Transp. Syst.* **2020**, *23*, 2966–2977. [CrossRef]
7. Lazcano, A.M.R.; Niu, T.; Akutain, X.C.; Cole, D.; Shyrokau, B. MPC-based haptic shared steering system: A driver modeling approach for symbiotic driving. *IEEE/ASME Trans. Mechatronics* **2021**, *26*, 1201–1211. [CrossRef]
8. Liang, Y.; Yin, Z.; Nie, L. Shared steering control for lane keeping and obstacle avoidance based on multi-objective MPC. *Sensors* **2021**, *21*, 4671. [CrossRef]
9. Awad, N.; Lasheen, A.; Elnaggar, M.; Kamel, A. Model predictive control with fuzzy logic switching for path tracking of autonomous vehicles. *ISA Trans.* **2022**, *129*, 193–205. [CrossRef]
10. Alhussan, A.A.; Khafaga, D.S.; El-Kenawy, E.S.M.; Ibrahim, A.; Eid, M.M.; Abdelhamid, A.A. Pothole and Plain Road Classification Using Adaptive Mutation Dipper Throated Optimization and Transfer Learning for Self Driving Cars. *IEEE Access* **2022**, *10*, 84188–84211. [CrossRef]
11. Sumanth, U.; Punn, N.S.; Sonbhadra, S.K.; Agarwal, S. Enhanced Behavioral Cloning-Based Self-driving Car Using Transfer Learning. In *Data Management, Analytics and Innovation*; Springer: Berlin/Heidelberg, Germany, 2022; pp. 185–198.
12. Sharma, S.; Ball, J.E.; Tang, B.; Carruth, D.W.; Doude, M.; Islam, M.A. Semantic segmentation with transfer learning for off-road autonomous driving. *Sensors* **2019**, *19*, 2577. [CrossRef] [PubMed]
13. García Cuenca, L.; Puertas, E.; Fernandez Andrés, J.; Aliane, N. Autonomous driving in roundabout maneuvers using reinforcement learning with Q-learning. *Electronics* **2019**, *8*, 1536. [CrossRef]
14. Chopra, R.; Roy, S.S. End-to-end reinforcement learning for self-driving car. In *Advanced Computing and Intelligent Engineering*; Springer: Berlin/Heidelberg, Germany, 2020; pp. 53–61.
15. Nagasai, L.; Sriprasath, V.; SajithVariyar, V.; Sowmya, V.; Aniketh, K.; Sarath, T.; Soman, K. Electric Vehicle Steering Design and Automated Control Using CNN and Reinforcement Learning. In *Soft Computing and Signal Processing*; Springer: Berlin/Heidelberg, Germany, 2021; pp. 513–523.
16. Liang, X.; Liu, Y.; Chen, T.; Liu, M.; Yang, Q. Federated transfer reinforcement learning for autonomous driving. In *Federated and Transfer Learning*; Springer: Berlin/Heidelberg, Germany, 2023; pp. 357–371.
17. de Morais, G.A.; Marcos, L.B.; Bueno, J.N.A.; de Resende, N.F.; Terra, M.H.; Grassi, V., Jr. Vision-based robust control framework based on deep reinforcement learning applied to autonomous ground vehicles. *Control Eng. Pract.* **2020**, *104*, 104630. [CrossRef]
18. Cai, P.; Wang, H.; Huang, H.; Liu, Y.; Liu, M. Vision-based autonomous car racing using deep imitative reinforcement learning. *IEEE Robot. Autom. Lett.* **2021**, *6*, 7262–7269. [CrossRef]

19. Lin, H.Y.; Dai, J.M.; Wu, L.T.; Chen, L.Q. A vision-based driver assistance system with forward collision and overtaking detection. *Sensors* **2020**, *20*, 5139. [CrossRef] [PubMed]
20. Dewangan, D.K.; Sahu, S.P.; Sairam, B.; Agrawal, A. VLDNet: Vision-based lane region detection network for intelligent vehicle system using semantic segmentation. *Computing* **2021**, *103*, 2867–2892. [CrossRef]
21. Kocić, J.; Jovičić, N.; Drndarević, V. An end-to-end deep neural network for autonomous driving designed for embedded automotive platforms. *Sensors* **2019**, *19*, 2064. [CrossRef]
22. Boloor, A.; Garimella, K.; He, X.; Gill, C.; Vorobeychik, Y.; Zhang, X. Attacking vision-based perception in end-to-end autonomous driving models. *J. Syst. Archit.* **2020**, *110*, 101766. [CrossRef]
23. Lee, D.H.; Liu, J.L. End-to-end deep learning of lane detection and path prediction for real-time autonomous driving. *Signal Image Video Process.* **2022**, 1–7 . [CrossRef]
24. Kim, C.J.; Lee, M.J.; Hwang, K.H.; Ha, Y.G. End-to-end deep learning-based autonomous driving control for high-speed environment. *J. Supercomput.* **2022**, *78*, 1961–1982. [CrossRef]
25. Kortli, Y.; Gabsi, S.; Voon, L.F.L.Y.; Jridi, M.; Merzougui, M.; Atri, M. Deep embedded hybrid CNN–LSTM network for lane detection on NVIDIA Jetson Xavier NX. *Knowl.-Based Syst.* **2022**, *240*, 107941. [CrossRef]
26. Elgharbawy, M.; Schwarzhaupt, A.; Frey, M.; Gauterin, F. A real-time multisensor fusion verification framework for advanced driver assistance systems. *Transp. Res. Part F Traffic Psychol. Behav.* **2019**, *61*, 259–267. [CrossRef]
27. Chiang, K.W.; Tsai, G.J.; Li, Y.H.; Li, Y.; El-Sheimy, N. Navigation engine design for automated driving using INS/GNSS/3D LiDAR-SLAM and integrity assessment. *Remote Sens.* **2020**, *12*, 1564. [CrossRef]
28. Han, J.H.; Park, C.H.; Kwon, J.H.; Lee, J.; Kim, T.S.; Jang, Y.Y. Performance evaluation of autonomous driving control algorithm for a crawler-type agricultural vehicle based on low-cost multi-sensor fusion positioning. *Appl. Sci.* **2020**, *10*, 4667. [CrossRef]
29. Bocu, R.; Bocu, D.; Iavich, M. Objects Detection Using Sensors Data Fusion in Autonomous Driving Scenarios. *Electronics* **2021**, *10*, 2903. [CrossRef]
30. Xu, J.; Liang, Z. Multiview Fusion 3D Target Information Perception Model in Nighttime Unmanned Intelligent Vehicles. *J. Funct. Spaces* **2022**, *2022*, 9295395. [CrossRef]
31. Bonnard, J.; Abdelouahab, K.; Pelcat, M.; Berry, F. On building a CNN-based multi-view smart camera for real-time object detection. *Microprocess. Microsyst.* **2020**, *77*, 103177. [CrossRef]
32. Deng, J.; Zhou, W.; Zhang, Y.; Li, H. From multi-view to hollow-3D: Hallucinated hollow-3D R-CNN for 3D object detection. *IEEE Trans. Circuits Syst. Video Technol.* **2021**, *31*, 4722–4734. [CrossRef]
33. Xiong, H.; Liu, H.; Ma, J.; Pan, Y.; Zhang, R. An NN-based double parallel longitudinal and lateral driving strategy for self-driving transport vehicles in structured road scenarios. *Sustainability* **2021**, *13*, 4531. [CrossRef]
34. Huang, Z.; Lv, C.; Xing, Y.; Wu, J. Multi-modal sensor fusion-based deep neural network for end-to-end autonomous driving with scene understanding. *IEEE Sens. J.* **2020**, *21*, 11781–11790. [CrossRef]
35. Liu, Y.; Han, C.; Zhang, L.; Gao, X. Pedestrian detection with multi-view convolution fusion algorithm. *Entropy* **2022**, *24*, 165. [CrossRef] [PubMed]
36. Almalioglu, Y.; Turan, M.; Trigoni, N.; Markham, A. Deep learning-based robust positioning for all-weather autonomous driving. *Nat. Mach. Intell.* **2022**, *4*, 749–760. [CrossRef]
37. Xiao, Y.; Codevilla, F.; Gurram, A.; Urfalioglu, O.; López, A.M. Multimodal end-to-end autonomous driving. *IEEE Trans. Intell. Transp. Syst.* **2020**, *23*, 537–547. [CrossRef]
38. Roszyk, K.; Nowicki, M.R.; Skrzypczyński, P. Adopting the YOLOv4 Architecture for Low-Latency Multispectral Pedestrian Detection in Autonomous Driving. *Sensors* **2022**, *22*, 1082. [CrossRef]
39. Luna-Alvarez, A.; Mújica-Vargas, D.; Matuz-Cruz, M.; Kinani, J.M.V.; Ramos-Díaz, E. Self-driving through a Time-distributed Convolutional Recurrent Neural Network. In Proceedings of the 2020 17th International Conference on Electrical Engineering, Computing Science and Automatic Control (CCE), Mexico City, Mexico, 11–13 November 2020; pp. 1–6.
40. Howard, A.G.; Zhu, M.; Chen, B.; Kalenichenko, D.; Wang, W.; Weyand, T.; Andreetto, M.; Adam, H. Mobilenets: Efficient convolutional neural networks for mobile vision applications. *arXiv* **2017**, arXiv:1704.04861.
41. Bojarski, M.; Yeres, P.; Choromanska, A.; Choromanski, K.; Firner, B.; Jackel, L.; Muller, U. Explaining how a deep neural network trained with end-to-end learning steers a car. *arXiv* **2017**, arXiv:1704.07911.
42. Mújica-Vargas, D.; Luna-Álvarez, A.; de Jesús Rubio, J.; Carvajal-Gámez, B. Noise gradient strategy for an enhanced hybrid convolutional-recurrent deep network to control a self-driving vehicle. *Appl. Soft Comput.* **2020**, *92*, 106258. [CrossRef]
43. Subedi, S. AI in Robotics: Implementing Donkey Car. 2020. Available online: https://red.library.usd.edu/idea/81 (accessed on on 28 November 2022).

MDPI
St. Alban-Anlage 66
4052 Basel
Switzerland
Tel. +41 61 683 77 34
Fax +41 61 302 89 18
www.mdpi.com

Electronics Editorial Office
E-mail: electronics@mdpi.com
www.mdpi.com/journal/electronics

www.ingramcontent.com/pod-product-compliance
Lightning Source LLC
Chambersburg PA
CBHW040828250926
51850CB00007BA/1287
* 9 7 8 3 0 3 6 5 7 0 3 6 5 *